HUMAN BODY

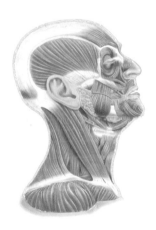

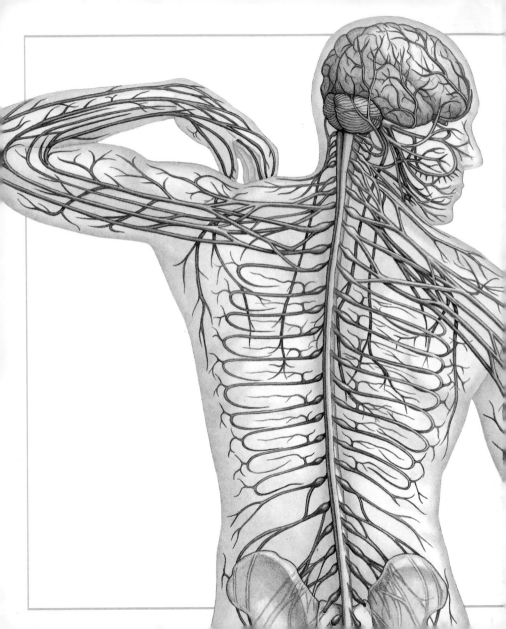

HUMAN BODY

LONDON, NEW YORK, MUNICH, MELBOURNE and DELHI

EDITOR Ann Baggaley
DESIGNER Sara Freeman
US EDITORS Jill Hamilton, Jane Perlmutter
DTP DESIGNERS Julian Dams, Jason Little
DESIGN ASSISTANCE Anna Plucinska
EDITORIAL ASSISTANCE Teresa Pritlove, Hazel Richardson

MEDICAL ADVISERS Daniel Carter, Dr Sue Davidson

SENIOR MANAGING EDITOR Martyn Page
MANAGING ART EDITOR Louise Dick

PRODUCTION MANAGER Michelle Thomas
PICTURE RESEARCHERS Anna Grapes, Lee Thompson
INDEXER Kay Wright

FIRST US EDITION, 2001
6 8 10 9 7

Published in the United States by Dorling Kindersley Publishing, Inc.
375 Hudson St.
New York, New York 10014

A Penguin Company

A Cataloging-in-Publication record is available from the Library of Congress

ISBN 0-7894-7988-5

Color reproduction by Colourscan, Singapore
Printed and bound by Star Standard Industries Pte. Ltd

see our complete catalog at
www.dk.com

Contents

About this book **6**

Anatomy of the human body **8**

Regional atlas of the body **10**
Body systems **36**
Cells, skin, and specialized tissue **44**
Skeletal system **58**
Muscular system **80**
Nervous system **96**
Endocrine system **156**
Cardiovascular system **168**
Lymphatic and immune systems **186**
Respiratory system **194**
Digestive system **206**
Urinary system **236**
Reproductive system **248**
Human life cycle **256**

Diseases and disorders **306**

Skin disorders **308**
Musculoskeletal disorders **314**
Nervous system disorders **330**
Cardiovascular disorders **344**
Infections and immune disorders **362**
Respiratory disorders **376**
Digestive disorders **392**
Urinary disorders **406**
Reproductive disorders **410**
Cancer **424**

Glossary and index **430**

Glossary **432**
Index **438**
Acknowledgments **448**

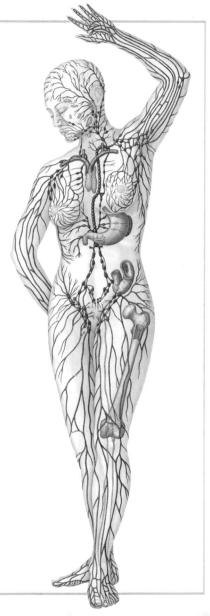

About this book

This book is a guide to the structure, inner workings, and common disorders of the human body. It is divided into two main parts, the first of which covers human anatomy, beginning with a regional atlas of the body. Following the atlas, each body system is examined individually. To conclude this part of the book, there is a section on the human life cycle. The second part of the book describes some of the most common disorders of the body. There is also a glossary of medical terms.

Anatomy of the human body

Regional atlas

These highly detailed anatomical maps cover the entire human body. They illustrate how the inner structures of the body, such as muscles, blood vessels, and major organs, fit together area by area.

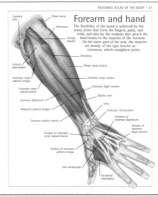
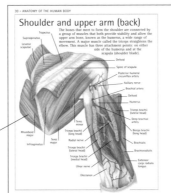

Anatomical maps
Fully annotated illustrations show how particular areas of the body are constructed

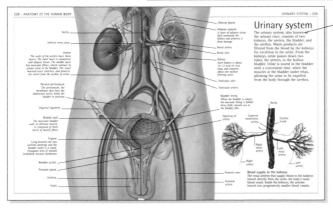

Overview of system
Each body system is introduced with a text outline and an illustration showing how the relevant organs are positioned

Body systems

This section describes and illustrates the major body systems. An initial overview of each system in its entirety is followed by a detailed examination of the main components and their specific roles.

Individual organs

The structure and function of individual body parts and organs in each system are examined in close-up detail.

Cutaway illustrations
Inner structures are revealed in cutaway illustrations

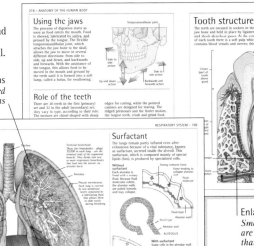

Using the jaws

The process of digestion starts as soon as food enters the mouth. Food is chewed, lubricated by saliva, and pressed by the tongue. The flexible temporomandibular joint, which attaches the jaw bone to the skull, allows the jaw to move in several different directions: from side to side, up and down, and backwards and forwards. With the assistance of the tongue, this allows food to be moved in the mouth and ground by the teeth until it is formed into a soft lump, called a bolus, for swallowing.

Role of the teeth

There are 20 teeth in the first (primary) set and 32 in the adult (secondary) set; they vary in type, according to their role. The incisors are chisel-shaped with sharp edges for cutting, while the pointed canines are designed for tearing. The ridged premolars and the flatter molars, the largest teeth, crush and grind food.

Tooth structure

The teeth are encased in sockets in the jaw bone and held in place by ligaments and thick alveolar gums. At the centre of each tooth there is a soft pulp which contains blood vessels and nerves; this is surrounded by a layer of sensitive tissue called dentine. Above the gum, the tooth has an outer covering of hard enamel. Below the gum, a bone-like tissue called cementum forms the tooth's outer layer.

Lung structure

Each lung is a cone-shaped organ of spongy tissue that contains millions of tiny air sacs called alveoli. Within the lungs is an intricate network of air passages that starts at the trachea below the larynx. The trachea bifurcates to form two primary (main) bronchi, which enter each lung and subdivide into increasingly smaller branches called bronchioles. The bronchioles in turn lead to the alveoli, which they supply with air.

Surfactant

The lungs remain partly inflated even after exhalation because of a vital substance, known as surfactant, secreted inside the alveoli. This surfactant, which is composed mainly of special lipids (fats), is produced by specialized cells.

Enlarged details
Small parts of the body are shown enlarged, so that key details can be seen clearly

Diagrams
Diagrams are used to clarify complex body processes

Diseases and disorders

Disorders of the human body

This part of the book illustrates and describes, system by system, some of the most common disorders of the human body, including injuries, infections, and cancer.

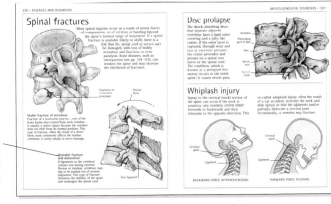

Spinal fractures

Most spinal injuries occur as a result of severe forces of compression, or of rotation or bending beyond the spine's normal range of movement. If a spinal fracture is unstable (likely to shift), there is a risk that the spinal cord or nerves may be damaged, with loss of bodily sensation and function, or even paralysis. Bone diseases, such as osteoporosis (see pp. 318–319), can weaken the spine and may increase the likelihood of fractures.

Disc prolapse

The shock-absorbing discs that separate adjacent vertebrae have a hard outer covering and a jelly-like centre. If the outer layer is ruptured, through wear and tear or excessive pressure, the centre protrudes and presses on a spinal nerve root or nerve or the spinal cord. The condition, which is known as a prolapsed disc, mostly occurs in the lower spine; it causes severe pain.

Whiplash injury

Injury to the cervical (neck) section of the spine can occur if the neck is suddenly and violently forced either forwards or backwards and then rebounds in the opposite direction. This so-called whiplash injury, often the result of a car accident, stretches the neck and may sprain or tear the ligaments and/or partially dislocate a cervical joint. Occasionally, a vertebra may fracture.

Explanatory text
All illustrations are accompanied by straightforward explanatory text

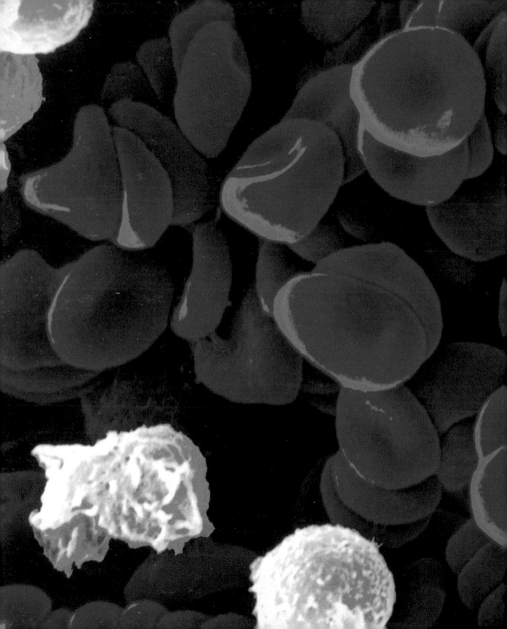

Anatomy of the
HUMAN BODY

Regional atlas of the body 10
Body systems 36
Cells, skin, and specialized tissue 44
Skeletal system 58
Muscular system 80
Nervous system 96
Endocrine system 156
Cardiovascular system 168
Lymphatic and immune systems 186
Respiratory system 194
Digestive system 206
Urinary system 236
Reproductive system 248
Human life cycle 256

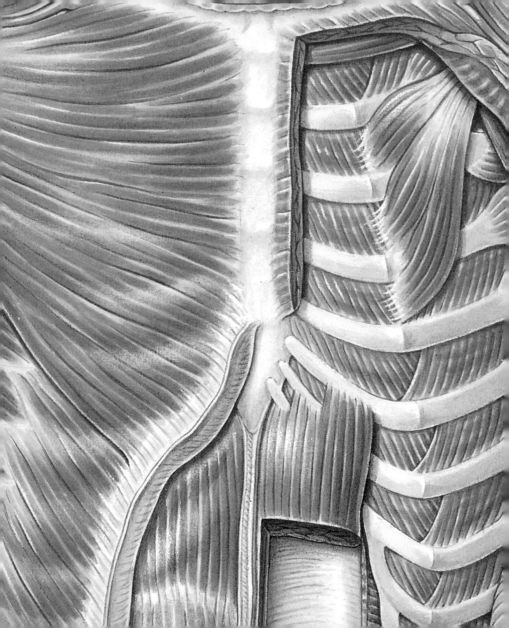

Regional atlas of the body

The individual internal organs and structures of the human body fit together to make a complex, living whole. By dividing the body into regions, such as the head, thorax, abdomen, and pelvis, it is possible to examine in detail how the parts of a particular area are positioned in relation to one another. This atlas illustrates major areas of the entire body, from head to foot.

Trunk
The trunk is the body's central part, consisting of the thorax and the abdomen.

Head and neck 1

The head contains the brain, which is protected by the skull. Overlying the skull bones, layers of muscle controlled by branches of the facial nerve produce a variety of facial expressions. Blood is supplied to, and drained from, the head through a network of vessels branching from the carotid arteries and jugular veins. The neck supports the weight of the head and serves as a connecting channel to and from the rest of the body.

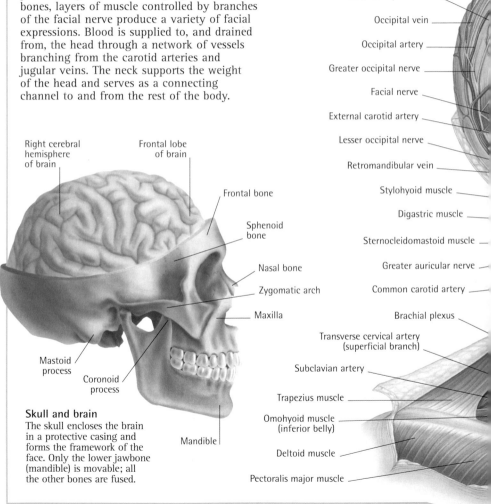

Right cerebral hemisphere of brain

Frontal lobe of brain

Frontal bone

Sphenoid bone

Nasal bone

Zygomatic arch

Maxilla

Mastoid process

Coronoid process

Mandible

Skull and brain
The skull encloses the brain in a protective casing and forms the framework of the face. Only the lower jawbone (mandible) is movable; all the other bones are fused.

Superficial temporal vein (parietal branch)

Superficial temporal artery (parietal branch)

Auriculotemporal nerve

Occipital vein

Occipital artery

Greater occipital nerve

Facial nerve

External carotid artery

Lesser occipital nerve

Retromandibular vein

Stylohyoid muscle

Digastric muscle

Sternocleidomastoid muscle

Greater auricular nerve

Common carotid artery

Brachial plexus

Transverse cervical artery (superficial branch)

Subclavian artery

Trapezius muscle

Omohyoid muscle (inferior belly)

Deltoid muscle

Pectoralis major muscle

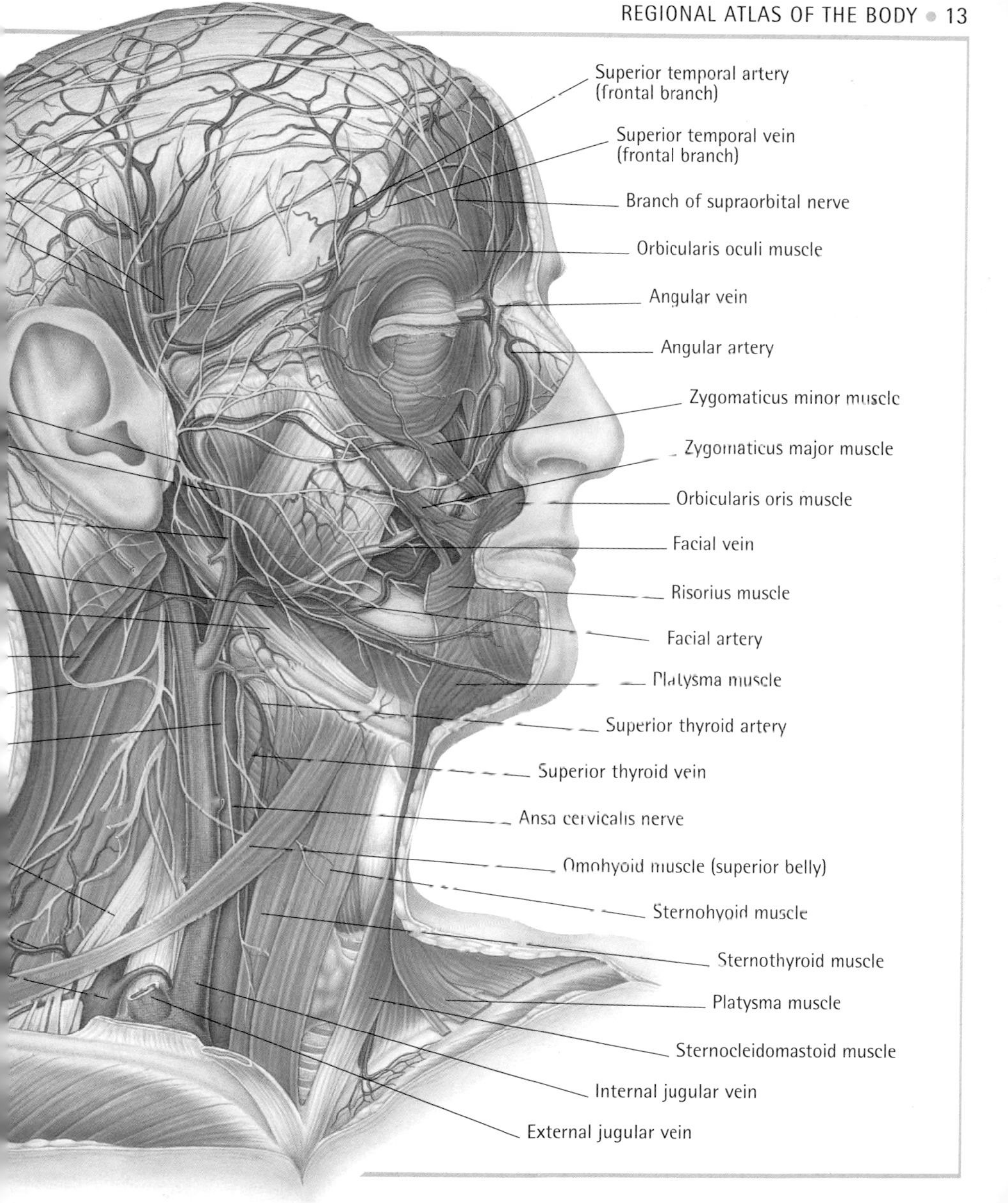

Superior temporal artery
(frontal branch)

Superior temporal vein
(frontal branch)

Branch of supraorbital nerve

Orbicularis oculi muscle

Angular vein

Angular artery

Zygomaticus minor muscle

Zygomaticus major muscle

Orbicularis oris muscle

Facial vein

Risorius muscle

Facial artery

Platysma muscle

Superior thyroid artery

Superior thyroid vein

Ansa cervicalis nerve

Omohyoid muscle (superior belly)

Sternohyoid muscle

Sternothyroid muscle

Platysma muscle

Sternocleidomastoid muscle

Internal jugular vein

External jugular vein

Head and neck 2

The bones of the face help to define the features and make a surrounding framework for the major sense organs: the eyes, nose, and tongue. The upper and lower jaw bones hold the teeth. Powerful neck muscles allow the head to turn from side to side; they also exert a backwards pull on the head, maintaining it in an upright position and preventing it from toppling forwards.

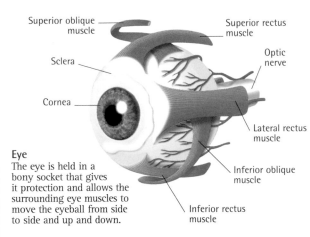

Superior oblique muscle

Superior rectus muscle

Sclera

Optic nerve

Cornea

Eye

The eye is held in a bony socket that gives it protection and allows the surrounding eye muscles to move the eyeball from side to side and up and down.

Lateral rectus muscle

Inferior oblique muscle

Inferior rectus muscle

Superior rectus muscle

Lateral rectus muscle

Inferior oblique muscle

Zygomatic bone

Nasal cartilage

Nasal cavity

Maxilla

Soft palate

Superior longitudinal muscle of tongue

Orbicularis orbis muscle

Genioglossus muscle

Lingual tonsil

Mentalis muscle

Mandible

Geniohyoid muscle

Mylohyoid muscle

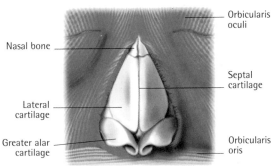

Orbicularis oculi

Nasal bone

Septal cartilage

Lateral cartilage

Greater alar cartilage

Orbicularis oris

Nose

The external structure of the nose is composed largely of cartilage. A small projection of nasal bone forms the bridge of the nose.

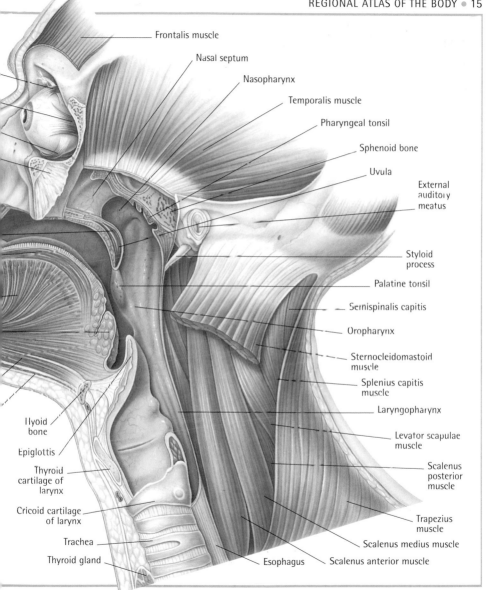

Frontalis muscle

Nasal septum

Nasopharynx

Temporalis muscle

Pharyngeal tonsil

Sphenoid bone

Uvula

External auditory meatus

Styloid process

Palatine tonsil

Semispinalis capitis

Oropharynx

Sternocleidomastoid muscle

Splenius capitis muscle

Laryngopharynx

Levator scapulae muscle

Scalenus posterior muscle

Trapezius muscle

Scalenus medius muscle

Scalenus anterior muscle

Esophagus

Thyroid gland

Trachea

Cricoid cartilage of larynx

Thyroid cartilage of larynx

Epiglottis

Hyoid bone

Neck (front)

The neck contains muscles and various other structures concerned with respiration and swallowing. The trachea, or windpipe, is the body's airway to and from the lungs. Connecting the pharynx (throat) with the trachea, the organ known as the larynx plays a key role in voice production. Salivary glands produce saliva, which aids swallowing and digestion by softening food before it passes down the throat.

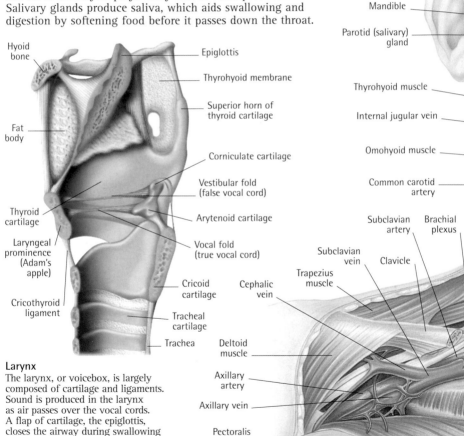

Hyoid bone

Epiglottis

Thyrohyoid membrane

Superior horn of thyroid cartilage

Fat body

Corniculate cartilage

Vestibular fold (false vocal cord)

Thyroid cartilage

Arytenoid cartilage

Laryngeal prominence (Adam's apple)

Vocal fold (true vocal cord)

Cricothyroid ligament

Cricoid cartilage

Tracheal cartilage

Trachea

Mylohyoid muscle

Submandibular (salivary) gland

Mandible

Parotid (salivary) gland

Thyrohyoid muscle

Internal jugular vein

Omohyoid muscle

Common carotid artery

Subclavian artery

Brachial plexus

Subclavian vein

Clavicle

Cephalic vein

Trapezius muscle

Deltoid muscle

Axillary artery

Axillary vein

Pectoralis major muscle

Larynx

The larynx, or voicebox, is largely composed of cartilage and ligaments. Sound is produced in the larynx as air passes over the vocal cords. A flap of cartilage, the epiglottis, closes the airway during swallowing to prevent food entering the lungs.

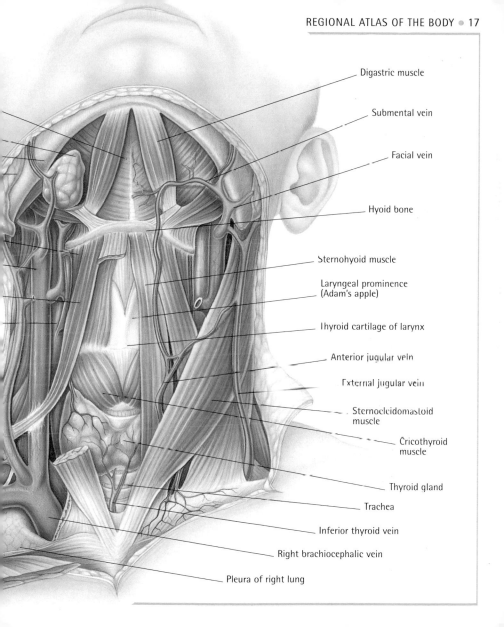

Digastric muscle

Submental vein

Facial vein

Hyoid bone

Sternohyoid muscle

Laryngeal prominence
(Adam's apple)

Thyroid cartilage of larynx

Anterior jugular vein

External jugular vein

Sternocleidomastoid
muscle

Cricothyroid
muscle

Thyroid gland

Trachea

Inferior thyroid vein

Right brachiocephalic vein

Pleura of right lung

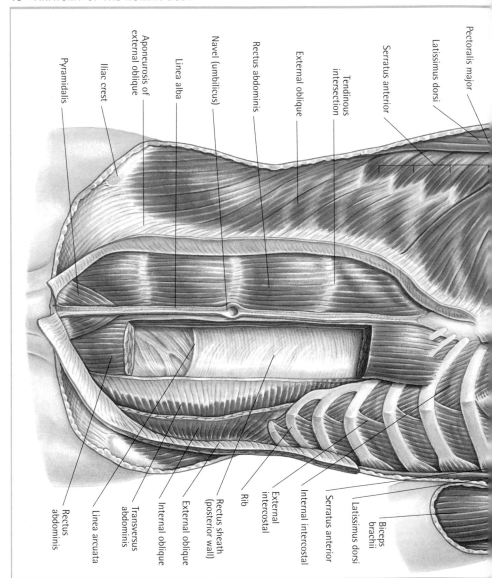

Pectoralis major

Latissimus dorsi

Serratus anterior

External oblique

Tendinous intersection

Rectus abdominis

Navel (umbilicus)

Linea alba

Aponeurosis of external oblique

Iliac crest

Pyramidalis

Biceps brachii

Latissimus dorsi

Serratus anterior

Internal intercostal

External intercostal

Rib

Rectus sheath (posterior wall)

External oblique

Internal oblique

Transversus abdominis

Linea arcuata

Rectus abdominis

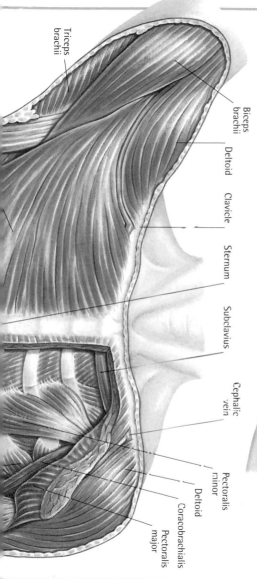

Triceps brachii

Biceps brachii

Deltoid

Clavicle

Sternum

Subclavius

Cephalic vein

Pectoralis minor

Deltoid

Coracobrachialis

Pectoralis major

Trunk (front)

The central region of the body, known as the trunk, consists of the thorax, or chest, and the abdomen. In the upper part of the trunk, the ribcage supports the body and encloses the heart and lungs. Superficial muscles at the front of the body enable the arms to be moved forwards and inwards. Deeper muscles move the ribs during respiration. In the abdominal area below the ribcage, the internal organs are held in entirely by layers of strong muscle.

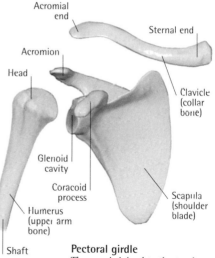

Acromial end

Sternal end

Acromion

Head

Clavicle (collar bone)

Glenoid cavity

Coracoid process

Scapula (shoulder blade)

Humerus (upper arm bone)

Shaft

Pectoral girdle

The arm is joined to the trunk by bones known as the pectoral (shoulder) girdle. These bones are the clavicle, or collar bone, and the scapula, which is a triangular bone forming the shoulder blade.

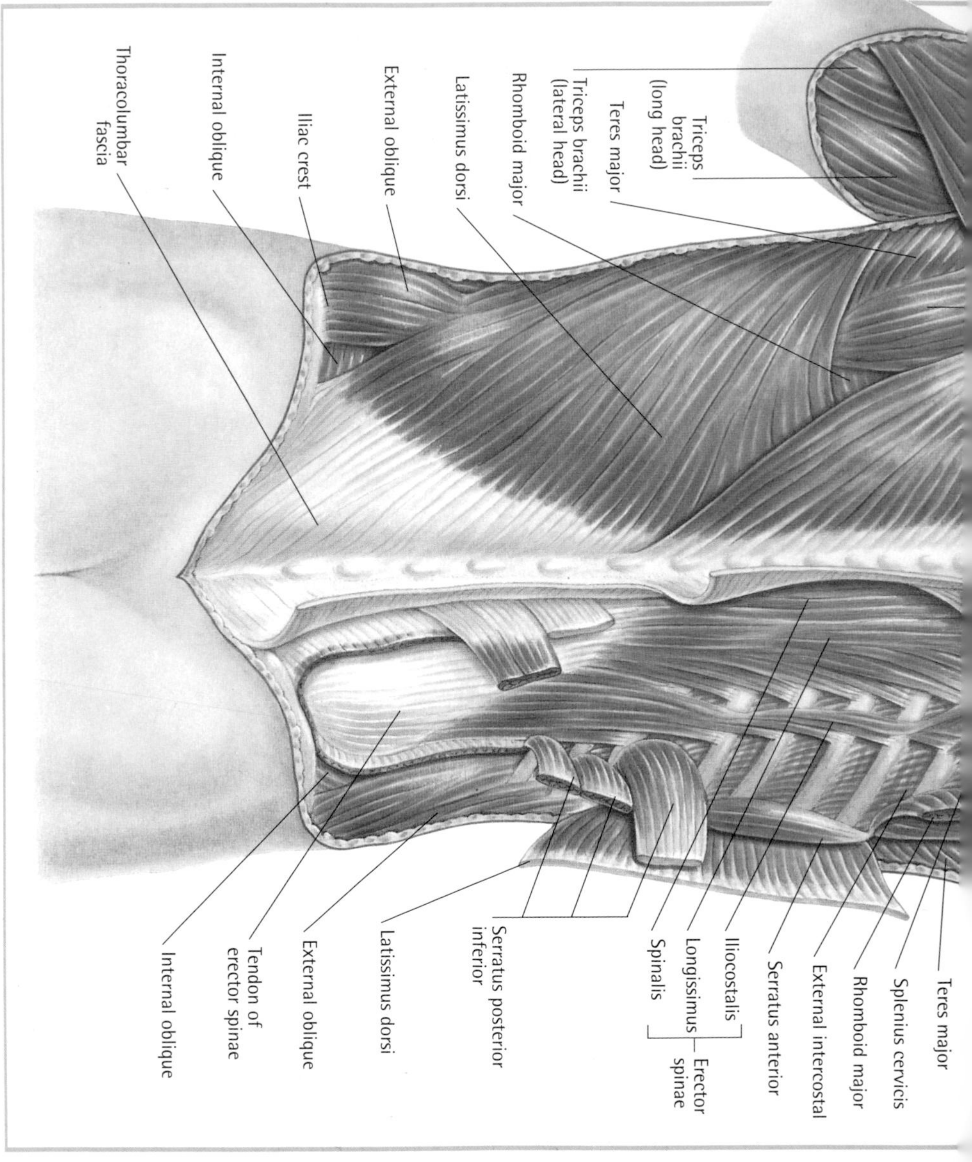

Thoracolumbar
fascia

Internal oblique

External oblique

Iliac crest

Latissimus dorsi

Rhomboid major

Teres major

Triceps brachii
(lateral head)

Triceps
brachii
(long head)

Internal oblique

Tendon of
erector spinae

External oblique

Latissimus dorsi

Serratus posterior
inferior

Spinalis

Longissimus

Iliocostalis

Erector
spinae

Serratus anterior

External intercostal

Rhomboid major

Splenius cervicis

Teres major

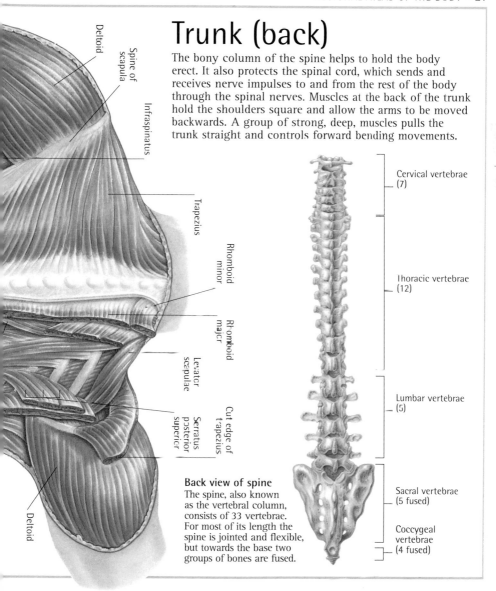

Trunk (back)

The bony column of the spine helps to hold the body erect. It also protects the spinal cord, which sends and receives nerve impulses to and from the rest of the body through the spinal nerves. Muscles at the back of the trunk hold the shoulders square and allow the arms to be moved backwards. A group of strong, deep, muscles pulls the trunk straight and controls forward bending movements.

Deltoid

Spine of scapula

Infraspinatus

Trapezius

Rhomboid minor

Rhomboid major

Levator scapulae

Serratus posterior superior

Cut edge of trapezius

Deltoid

Cervical vertebrae (7)

Thoracic vertebrae (12)

Lumbar vertebrae (5)

Sacral vertebrae (5 fused)

Coccygeal vertebrae (4 fused)

Back view of spine
The spine, also known as the vertebral column, consists of 33 vertebrae. For most of its length the spine is jointed and flexible, but towards the base two groups of bones are fused.

Thoracic cavity

The upper area of the trunk, the thorax, is separated from the abdomen by a muscular sheet called the diaphragm. Within the thoracic cavity, walled by the ribcage and its associated muscles, are the heart, lungs (shown here pulled back), and the body's major blood vessels and airways. Thin membranes (pleurae) surround the lungs to prevent friction during respiration; a membranous sac, the pericardium, protects the heart.

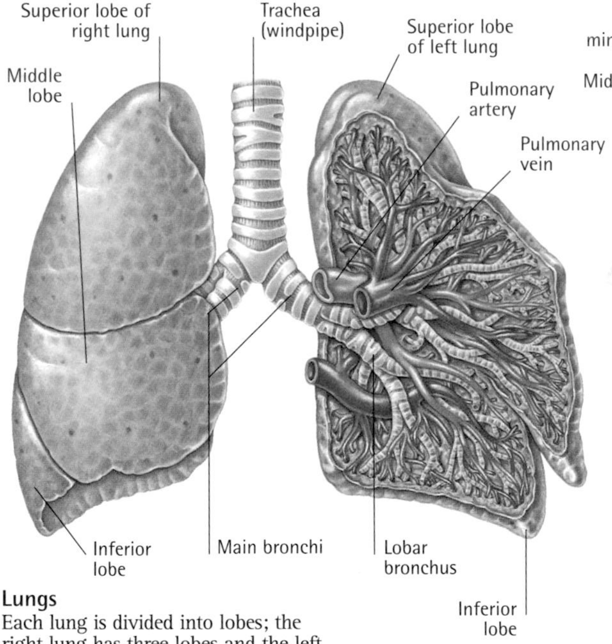

Right common carotid artery

Right internal jugular vein

Right brachiocephalic vein

Superior vena cava

Superior lobe of right lung

Pectoralis major muscle

Superior lobe of right lung

Middle lobe

Trachea (windpipe)

Superior lobe of left lung

Pulmonary artery

Pulmonary vein

Pectoralis minor muscle

Middle lobe of right lung

Inferior lobe

Main bronchi

Lobar bronchus

Inferior lobe

Digitations of serratus anterior muscle

Internal intercostal muscle

External intercostal muscle

Inferior lobe of right lung

External oblique muscle

Pleura

Lungs
Each lung is divided into lobes; the right lung has three lobes and the left has two. Air enters the lungs via the main bronchi, which branch from the trachea and then divide inside the lungs into progressively smaller airways.

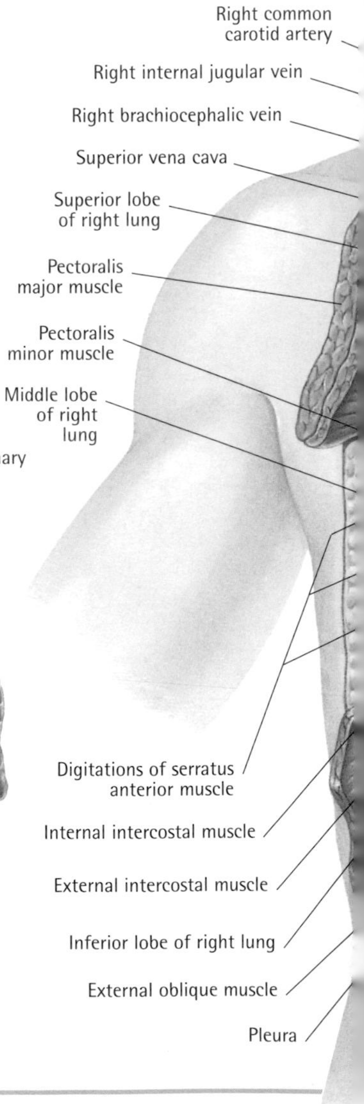

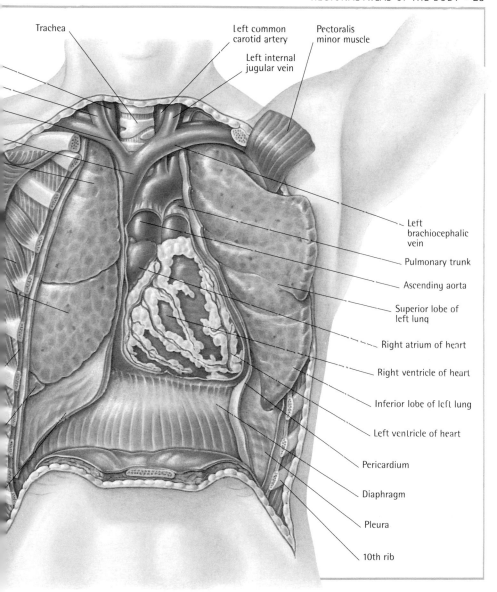

Trachea

Left common
carotid artery

Pectoralis
minor muscle

Left internal
jugular vein

Left
brachiocephalic
vein

Pulmonary trunk

Ascending aorta

Superior lobe of
left lung

Right atrium of heart

Right ventricle of heart

Inferior lobe of left lung

Left ventricle of heart

Pericardium

Diaphragm

Pleura

10th rib

Abdomen 1

The abdomen is the part of the trunk between the chest and the pelvis. Muscles form the abdominal wall, holding in the digestive organs (stomach, intestines, liver, and pancreas) and the spleen, an organ that forms part of the immune system. Behind these organs lie the kidneys. A fatty, folded membrane, known as the greater omentum, hangs in front of the intestines like an apron.

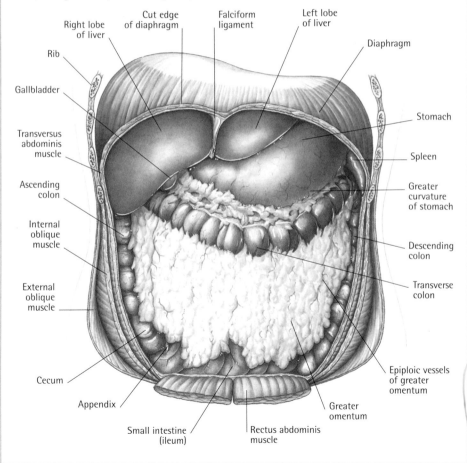

Right lobe of liver

Cut edge of diaphragm

Falciform ligament

Left lobe of liver

Rib

Diaphragm

Gallbladder

Stomach

Transversus abdominis muscle

Spleen

Ascending colon

Greater curvature of stomach

Internal oblique muscle

Descending colon

External oblique muscle

Transverse colon

Cecum

Epiploic vessels of greater omentum

Appendix

Greater omentum

Small intestine (ileum)

Rectus abdominis muscle

Abdomen 2

The abdominal cavity is shown here with the liver removed to show the esophagus. This tube conveys food from the throat to the stomach, passing through the diaphragm that separates the abdomen from the thorax. The aorta, the body's main blood vessel, terminates in the lower part of the abdomen, where it divides into two arteries that supply blood to the pelvic region and legs.

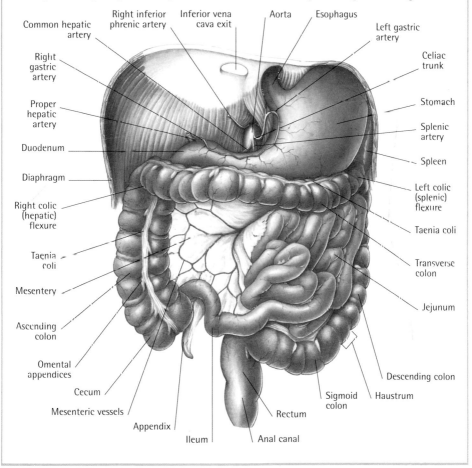

Male pelvic cavity

The pelvic cavity lies below the abdomen. This cavity is framed by the bones of the hip girdle and the fused bones of the lower spine known as the sacrum. In both sexes, the pelvic area contains the lower organs of the digestive and urinary systems, including the rectum and bladder. In males, parts of the reproductive system are located within the pelvic cavity but the major organs are suspended outside the body.

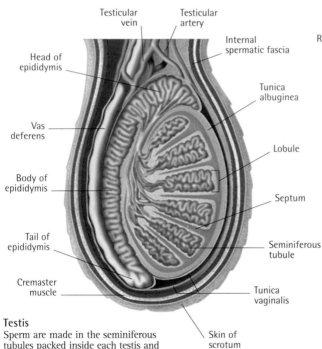

Gluteus medius muscle

Iliacus muscle

External oblique muscle

Internal oblique muscle

Transversus abdominis muscle

Psoas major muscle

External iliac vein

External iliac artery

Rectus abdominis muscle

Parietal peritoneum

Vas deferens

Wall of bladder

Opening of ureter

Linea alba

Pubic symphysis

Prostatic urethra

Suspensory ligament of penis

Corpus spongiosum

Corpus cavernosum

Penile (spongy) urethra

Testicular vein

Testicular artery

Internal spermatic fascia

Head of epididymis

Tunica albuginea

Vas deferens

Lobule

Body of epididymis

Septum

Tail of epididymis

Seminiferous tubule

Cremaster muscle

Tunica vaginalis

Skin of scrotum

Testis

Sperm are made in the seminiferous tubules packed inside each testis and are stored inside a long coiled tube, the epididymis, to mature. From here, the sperm pass to the vas deferens to be ejaculated during sexual intercourse.

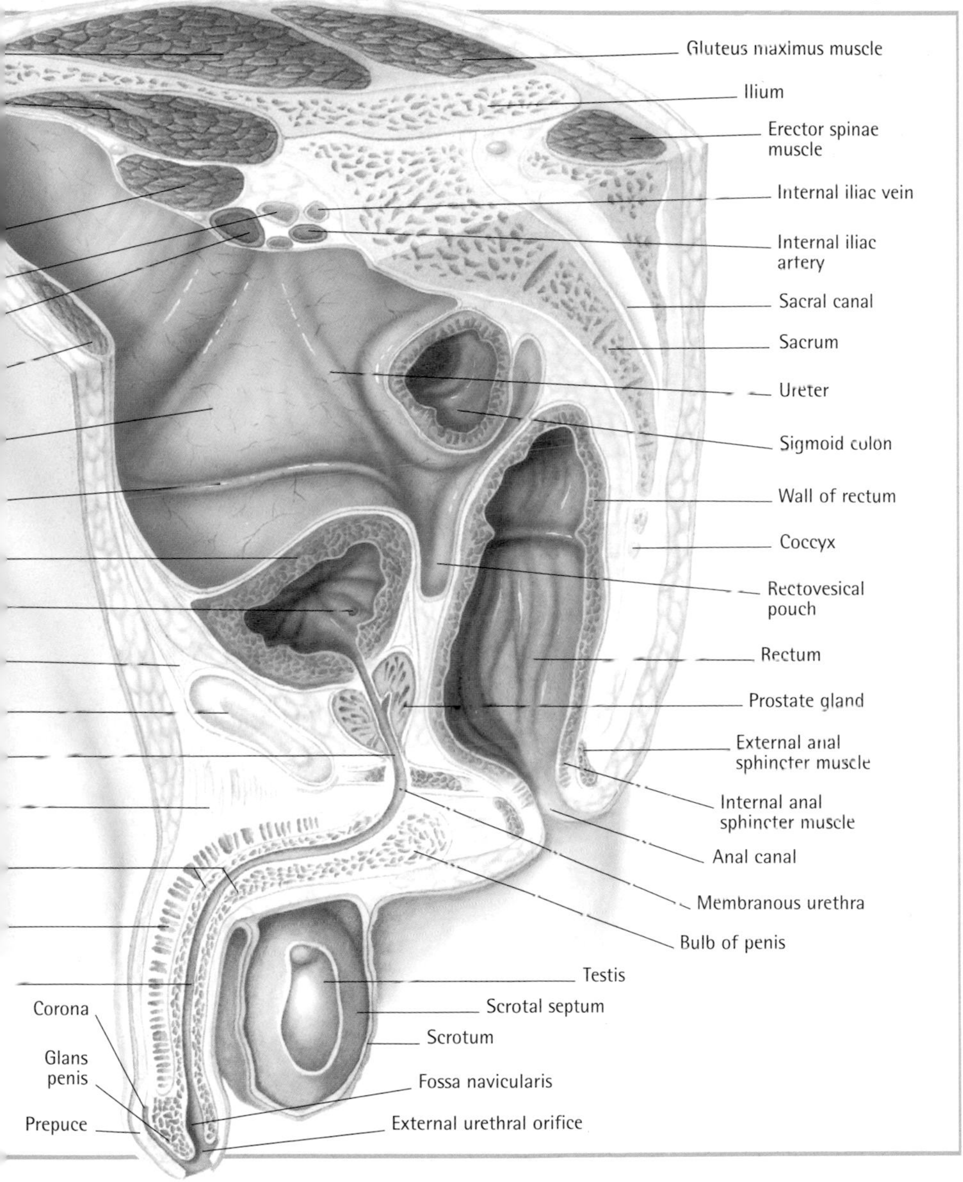

Gluteus maximus muscle

Ilium

Erector spinae muscle

Internal iliac vein

Internal iliac artery

Sacral canal

Sacrum

Ureter

Sigmoid colon

Wall of rectum

Coccyx

Rectovesical pouch

Rectum

Prostate gland

External anal sphincter muscle

Internal anal sphincter muscle

Anal canal

Membranous urethra

Bulb of penis

Testis

Scrotal septum

Scrotum

Fossa navicularis

External urethral orifice

Corona

Glans penis

Prepuce

Female pelvic cavity

In females, the pelvic cavity contains nearly all the reproductive organs. The thick-walled uterus is located between the bladder and the rectum, and usually tilts forwards. It is held in place by ligaments and supported by the pelvic floor muscles. The fallopian tubes extend from either side of the uterus, their free ends terminating just above the ovaries in fingerlike projections called fimbriae. The vagina forms a passage from the uterus to the outside.

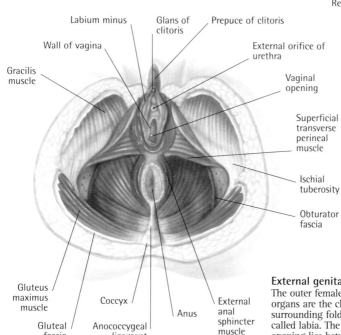

Iliacus muscle

Psoas major muscle

Ovarian artery

External oblique muscle

Ovarian vein

Rectus abdominis muscle

Fallopian tube

Fimbriae

Ovary

Body of uterus

Round ligament of uterus

Wall of bladder

Linea alba

Pubic symphysis

Mons pubis

Labium minus

Glans of clitoris

Prepuce of clitoris

Wall of vagina

External orifice of urethra

Gracilis muscle

Vaginal opening

Superficial transverse perineal muscle

Ischial tuberosity

Obturator fascia

Gluteus maximus muscle

Coccyx

Anus

External anal sphincter muscle

Gluteal fascia

Anococcygeal ligament

External genitalia

The outer female sexual organs are the clitoris and surrounding folds of skin, called labia. The vaginal opening lies between the external orifice of the urethra and the anus.

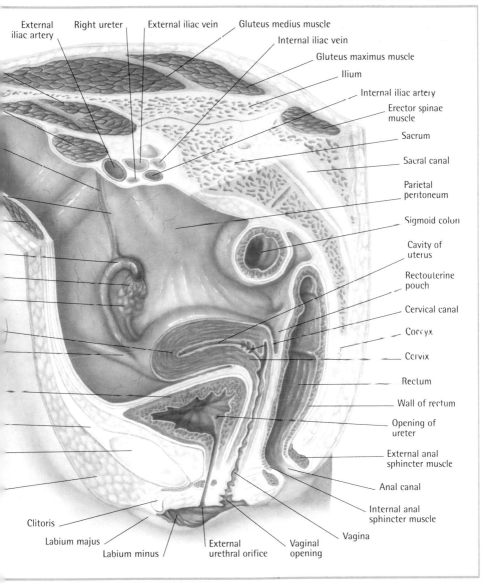

External iliac artery

Right ureter

External iliac vein

Gluteus medius muscle

Internal iliac vein

Gluteus maximus muscle

Ilium

Internal iliac artery

Erector spinae muscle

Sacrum

Sacral canal

Parietal peritoneum

Sigmoid colon

Cavity of uterus

Rectouterine pouch

Cervical canal

Coccyx

Cervix

Rectum

Wall of rectum

Opening of ureter

External anal sphincter muscle

Anal canal

Internal anal sphincter muscle

Clitoris

Labium majus

Labium minus

External urethral orifice

Vaginal opening

Vagina

Shoulder and upper arm (back)

The bones that meet to form the shoulder are connected by a group of muscles that both provide stability and allow the upper arm bone, known as the humerus, a wide range of movement. A major muscle called the triceps straightens the elbow. This muscle has three attachment points: on either side of the humerus and at the scapula (shoulder blade).

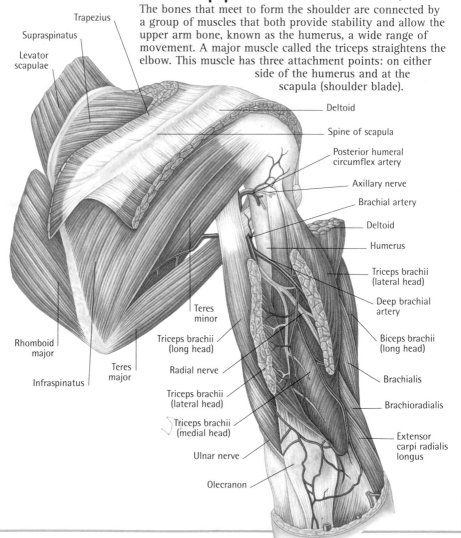

Trapezius

Supraspinatus

Levator scapulae

Deltoid

Spine of scapula

Posterior humeral circumflex artery

Axillary nerve

Brachial artery

Deltoid

Humerus

Triceps brachii (lateral head)

Deep brachial artery

Biceps brachii (long head)

Brachialis

Brachioradialis

Extensor carpi radialis longus

Rhomboid major

Infraspinatus

Teres major

Teres minor

Triceps brachii (long head)

Radial nerve

Triceps brachii (lateral head)

Triceps brachii (medial head)

Ulnar nerve

Olecranon

Forearm and hand

The flexibility of the hand is achieved by the many joints that form the fingers, palm, and wrist, and also by the tendons that attach the hand bones to the muscles of the forearm. On the outer part of the arm, the muscles are mostly of the type known as extensors, which straighten joints.

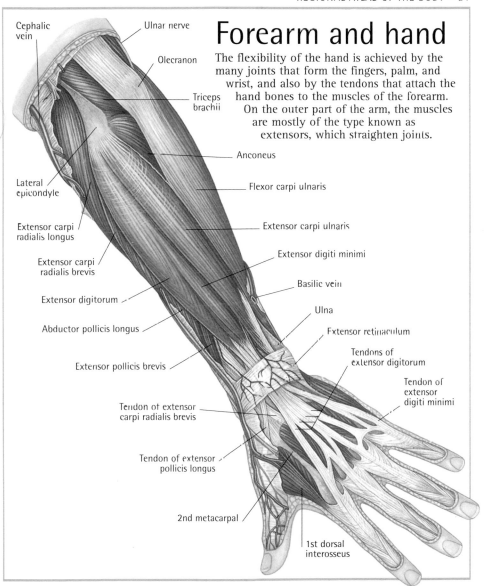

Cephalic vein

Ulnar nerve

Olecranon

Triceps brachii

Anconeus

Flexor carpi ulnaris

Lateral epicondyle

Extensor carpi ulnaris

Extensor carpi radialis longus

Extensor digiti minimi

Extensor carpi radialis brevis

Basilic vein

Extensor digitorum

Ulna

Abductor pollicis longus

Extensor retinaculum

Tendons of extensor digitorum

Extensor pollicis brevis

Tendon of extensor digiti minimi

Tendon of extensor carpi radialis brevis

Tendon of extensor pollicis longus

2nd metacarpal

1st dorsal interosseus

Thigh

Several groups of muscles in the thigh perform various functions, such as straightening and bending the leg during walking, running, and climbing, and holding the trunk steady in a standing position. The largest nerve in the body, the sciatic nerve, runs down the back of the thigh (see far right), branching into two just above the knee.

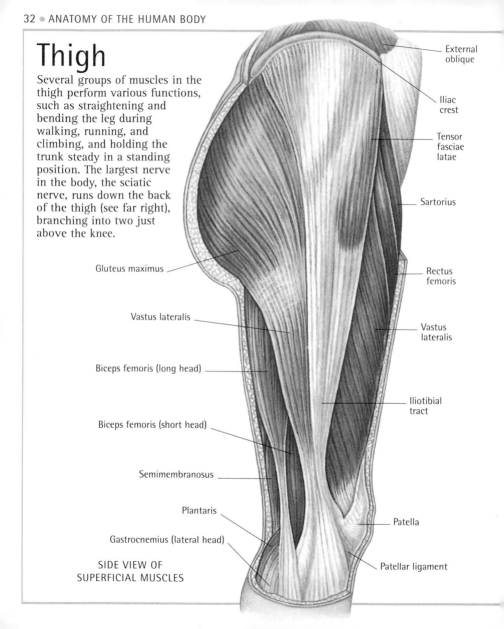

External oblique

Iliac crest

Tensor fasciae latae

Sartorius

Gluteus maximus

Rectus femoris

Vastus lateralis

Vastus lateralis

Biceps femoris (long head)

Iliotibial tract

Biceps femoris (short head)

Semimembranosus

Plantaris

Gastrocnemius (lateral head)

Patella

Patellar ligament

SIDE VIEW OF SUPERFICIAL MUSCLES

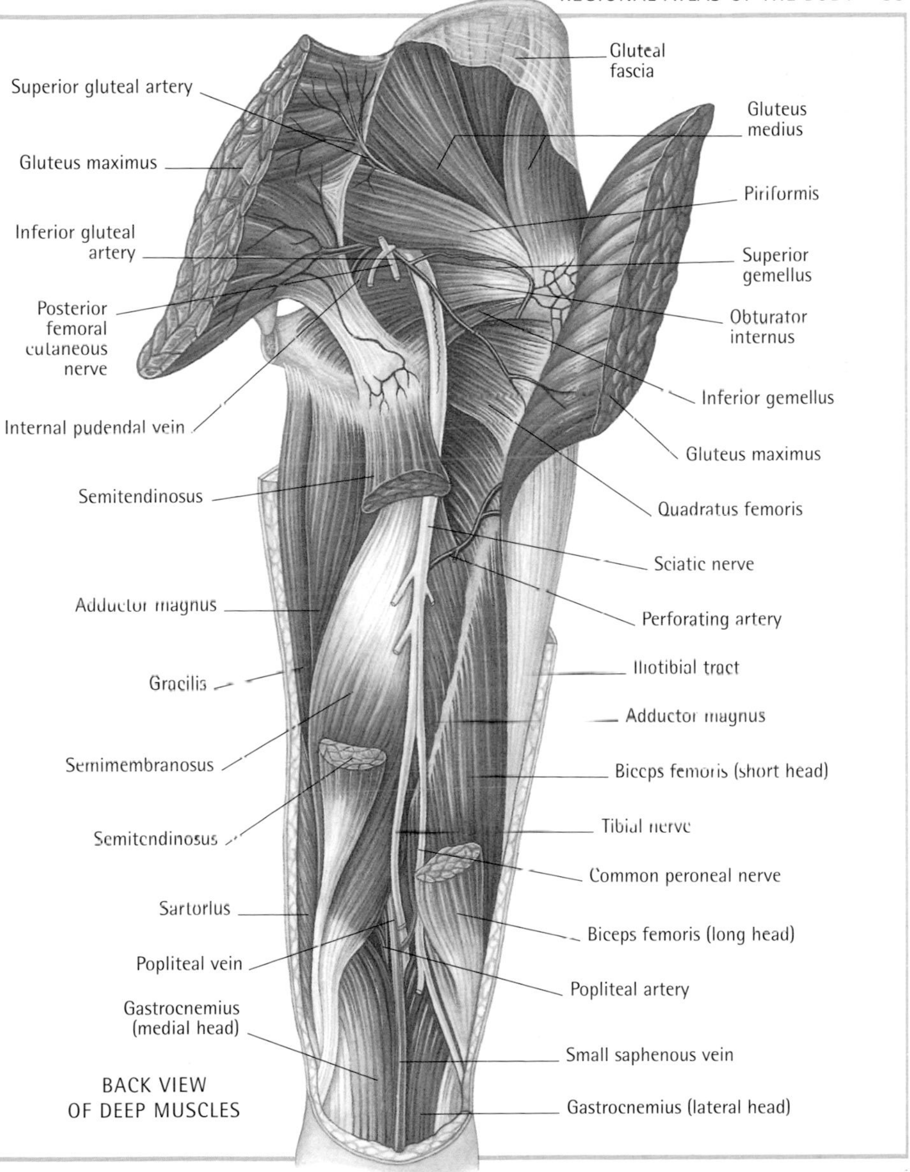

Superior gluteal artery

Gluteus maximus

Inferior gluteal artery

Posterior femoral cutaneous nerve

Internal pudendal vein

Semitendinosus

Adductor magnus

Gracilis

Semimembranosus

Semitendinosus

Sartorius

Popliteal vein

Gastrocnemius (medial head)

Gluteal fascia

Gluteus medius

Piriformis

Superior gemellus

Obturator internus

Inferior gemellus

Gluteus maximus

Quadratus femoris

Sciatic nerve

Perforating artery

Iliotibial tract

Adductor magnus

Biceps femoris (short head)

Tibial nerve

Common peroneal nerve

Biceps femoris (long head)

Popliteal artery

Small saphenous vein

Gastrocnemius (lateral head)

BACK VIEW OF DEEP MUSCLES

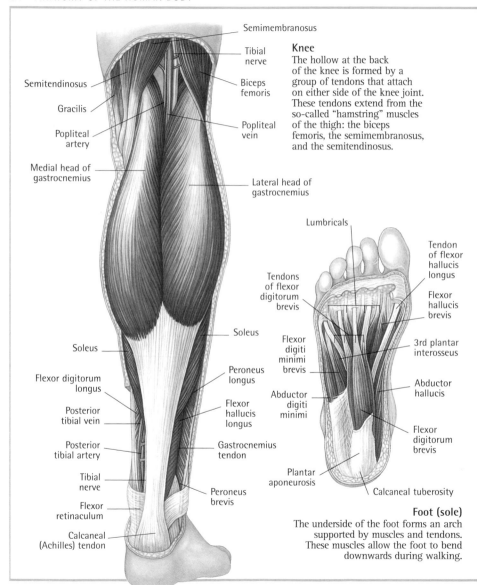

Semimembranosus

Tibial nerve

Semitendinosus

Gracilis

Biceps femoris

Popliteal artery

Popliteal vein

Medial head of gastrocnemius

Lateral head of gastrocnemius

Knee
The hollow at the back of the knee is formed by a group of tendons that attach on either side of the knee joint. These tendons extend from the so-called "hamstring" muscles of the thigh: the biceps femoris, the semimembranosus, and the semitendinosus.

Lumbricals

Tendon of flexor hallucis longus

Tendons of flexor digitorum brevis

Flexor hallucis brevis

Soleus

Flexor digiti minimi brevis

3rd plantar interosseus

Soleus

Peroneus longus

Flexor digitorum longus

Flexor hallucis longus

Abductor digiti minimi

Abductor hallucis

Posterior tibial vein

Posterior tibial artery

Gastrocnemius tendon

Flexor digitorum brevis

Tibial nerve

Plantar aponeurosis

Flexor retinaculum

Peroneus brevis

Calcaneal tuberosity

Calcaneal (Achilles) tendon

Foot (sole)
The underside of the foot forms an arch supported by muscles and tendons. These muscles allow the foot to bend downwards during walking.

Lower leg and foot

The muscles at the back of the leg form the bulge of the calf. Their main function is to exert a pull on the heel bone through the calcaneal (Achilles) tendon, which straightens the ankle and provides forward impetus during walking and running. The blood supply to the lower leg and foot is transported via the tibial arteries.

Foot

The foot is given much of its mobility by muscles that extend from the front of the leg and attach by long tendons to the foot bones. These muscles enable the foot to bend upwards and the toes to straighten. The main blood vessels supplying the foot pass around the ankle joint.

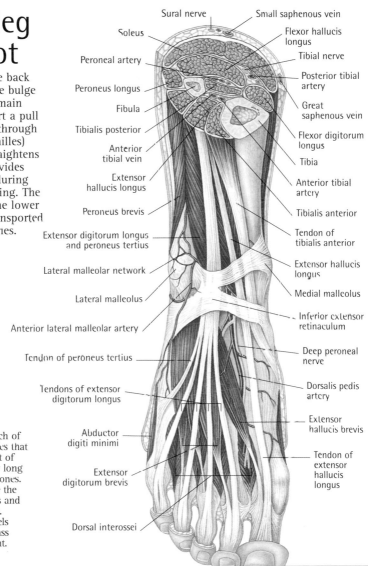

Sural nerve
Small saphenous vein
Soleus
Flexor hallucis longus
Peroneal artery
Tibial nerve
Peroneus longus
Posterior tibial artery
Fibula
Great saphenous vein
Tibialis posterior
Flexor digitorum longus
Anterior tibial vein
Tibia
Extensor hallucis longus
Anterior tibial artery
Peroneus brevis
Tibialis anterior
Extensor digitorum longus and peroneus tertius
Tendon of tibialis anterior
Lateral malleolar network
Extensor hallucis longus
Lateral malleolus
Medial malleolus
Anterior lateral malleolar artery
Inferior extensor retinaculum
Tendon of peroneus tertius
Deep peroneal nerve
Tendons of extensor digitorum longus
Dorsalis pedis artery
Abductor digiti minimi
Extensor hallucis brevis
Extensor digitorum brevis
Tendon of extensor hallucis longus
Dorsal interossei

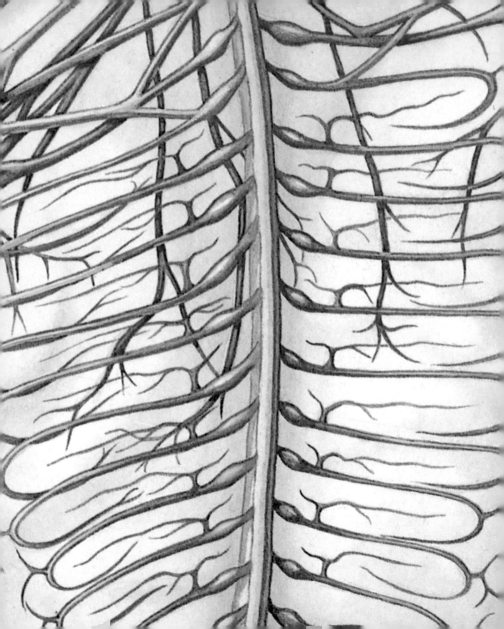

Body systems

A body system is a group of connected parts, including organs and tissues, that together carry out particular functions, such as breathing and digestion. Although the systems can be regarded as separate processes within the body, each is dependent on the others for physical and biochemical support. The human body can maintain health and survive only if all systems work together as an efficiently functioning cooperative.

Nervous system
The body's network of nerves functions as a control and communications system.

Skeletal system

The skeleton is the framework on which the rest of the body is built. Bones also play a role in the other body systems: red and white blood cells grow and develop in fatty inner tissue known as red marrow. Essential minerals, including calcium, are stored in the bones, to be released when the body needs them.

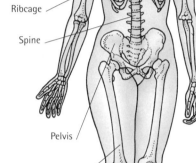

Skull

Ribcage

Spine

Pelvis

Femur

Fibula

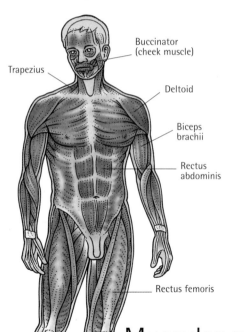

Buccinator
(cheek muscle)

Trapezius

Deltoid

Biceps
brachii

Rectus
abdominis

Rectus femoris

Muscular system

Muscles make up about half the body's bulk. Working with the skeleton, the voluntary muscles enable the body to make precise movements, lift objects, and even speak. Involuntary muscles, which include heart muscle and smooth muscle, provide essential power for the functioning of the respiratory, cardiovascular, and digestive systems.

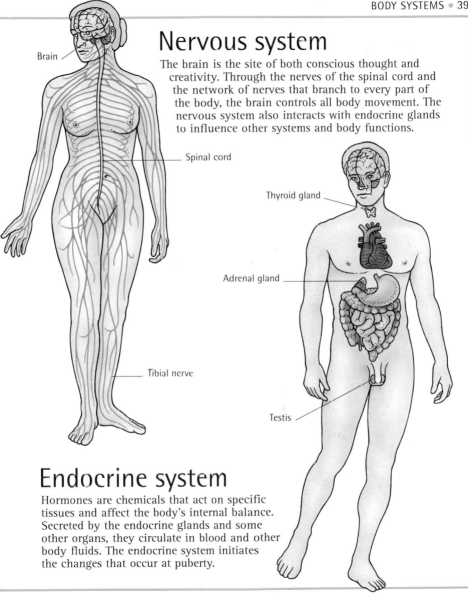

Brain

Spinal cord

Tibial nerve

Nervous system

The brain is the site of both conscious thought and creativity. Through the nerves of the spinal cord and the network of nerves that branch to every part of the body, the brain controls all body movement. The nervous system also interacts with endocrine glands to influence other systems and body functions.

Thyroid gland

Adrenal gland

Testis

Endocrine system

Hormones are chemicals that act on specific tissues and affect the body's internal balance. Secreted by the endocrine glands and some other organs, they circulate in blood and other body fluids. The endocrine system initiates the changes that occur at puberty.

Cardiovascular system

The cardiovascular system's basic function is to pump blood around the body; a pause of more than a few seconds in this life-supporting circulation results in loss of consciousness. The system takes oxygenated blood to all body organs and tissues and can adapt swiftly to changes in demand. Blood circulation also removes waste products from the body.

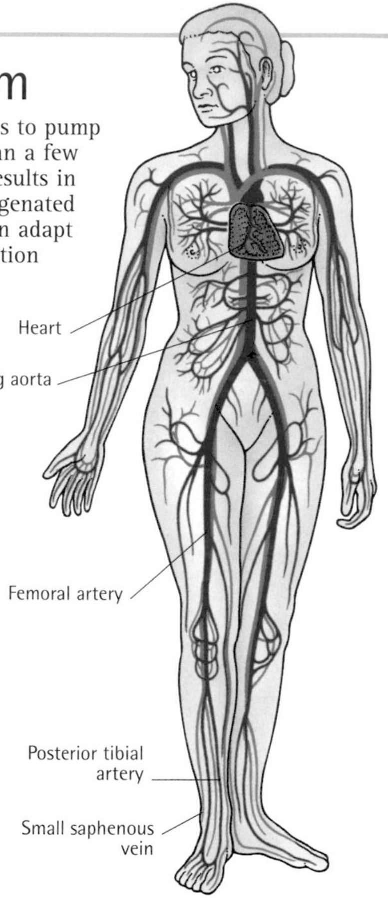

Heart

Descending aorta

Tonsils

Thymus

Femoral artery

Spleen

Posterior tibial artery

Small saphenous vein

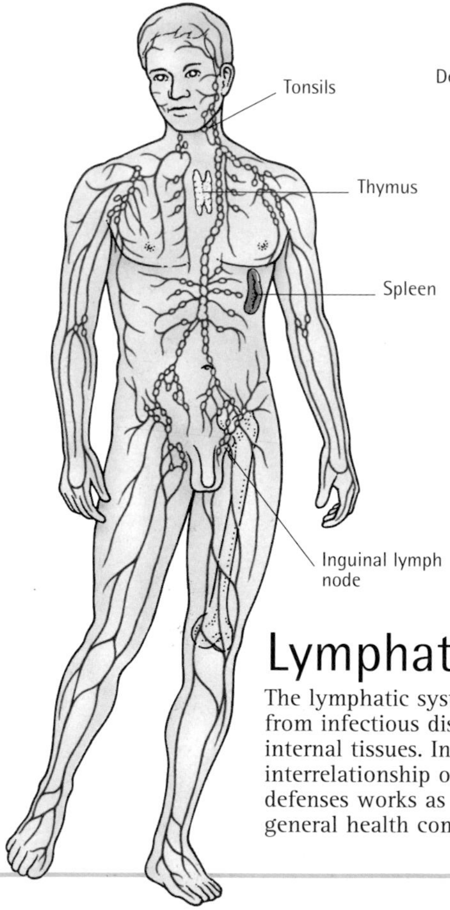

Inguinal lymph node

Lymphatic system

The lymphatic system helps to provide vital protection from infectious disease and to prevent malfunction of internal tissues. In a healthy person, an intricate interrelationship of physical, cellular, and chemical defenses works as a barrier against many threats. Poor general health compromises the system's efficiency.

Respiratory system

The respiratory tract, working together with breathing muscles, carries air into and out of the lungs, where oxygen and carbon dioxide are exchanged. The cardiovascular system transports these gases to and from all body tissues.

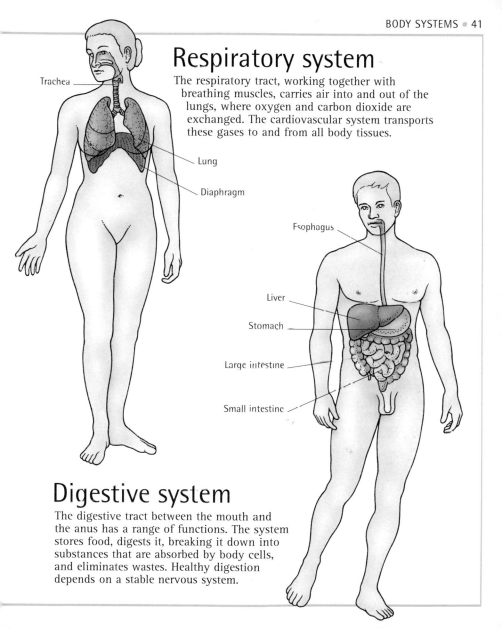

Trachea

Lung

Diaphragm

Esophagus

Liver

Stomach

Large intestine

Small intestine

Digestive system

The digestive tract between the mouth and the anus has a range of functions. The system stores food, digests it, breaking it down into substances that are absorbed by body cells, and eliminates wastes. Healthy digestion depends on a stable nervous system.

Male urinary system

The formation of urine by the kidneys eliminates wastes and excess fluids and helps to maintain the body's chemical balance. Production of urine is influenced by blood flow and blood pressure, hormones, and various body rhythms and cycles. Urine leaves the body through the urethra, which in the male opens at the tip of the penis and also transports semen.

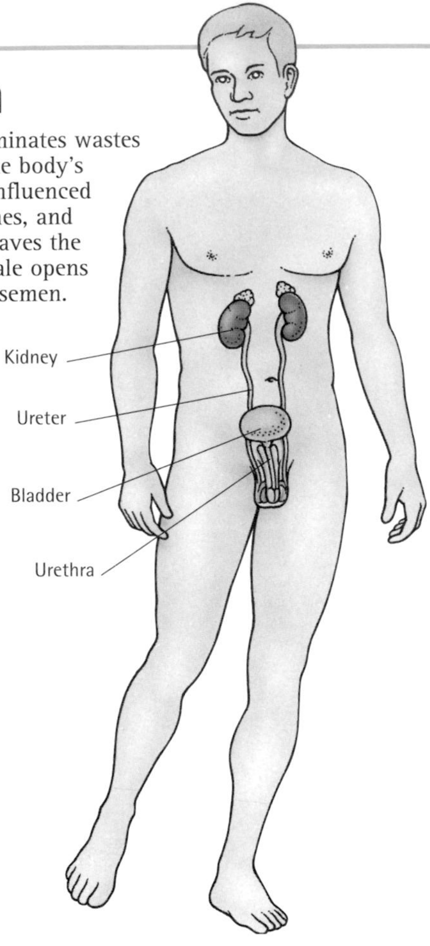

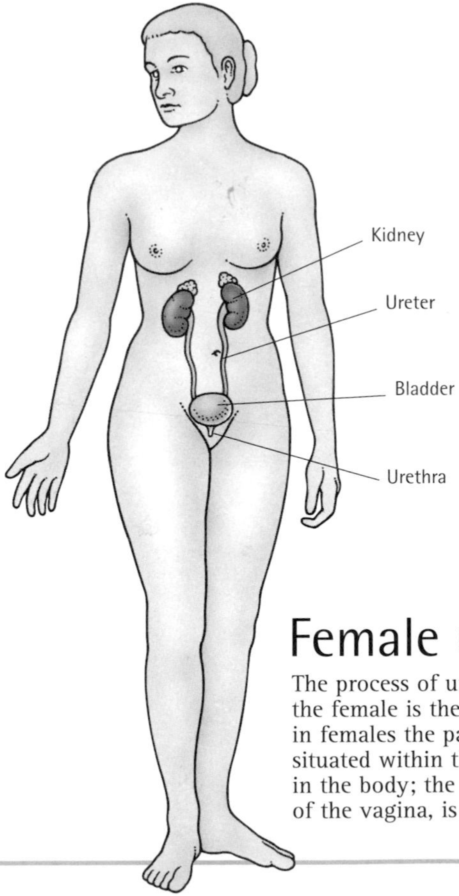

Kidney

Ureter

Bladder

Urethra

Kidney

Ureter

Bladder

Urethra

Female urinary system

The process of urine formation and elimination in the female is the same as in the male. However, in females the parts of the urinary tract that are situated within the pelvic cavity sit slightly lower in the body; the urethra, which opens just in front of the vagina, is much shorter than in the male.

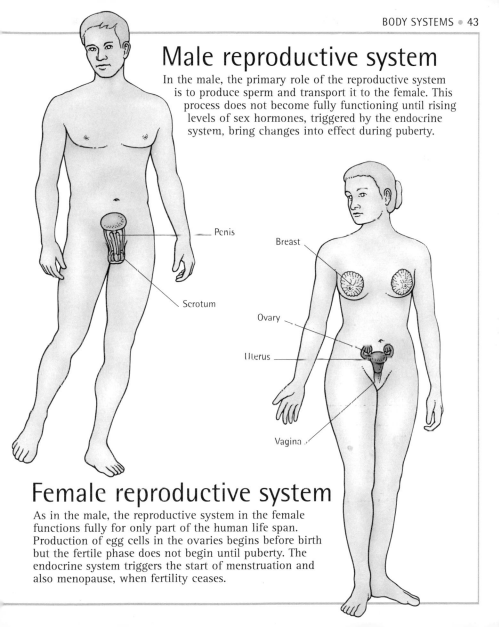

Male reproductive system

In the male, the primary role of the reproductive system is to produce sperm and transport it to the female. This process does not become fully functioning until rising levels of sex hormones, triggered by the endocrine system, bring changes into effect during puberty.

Penis

Scrotum

Breast

Ovary

Uterus

Vagina

Female reproductive system

As in the male, the reproductive system in the female functions fully for only part of the human life span. Production of egg cells in the ovaries begins before birth but the fertile phase does not begin until puberty. The endocrine system triggers the start of menstruation and also menopause, when fertility ceases.

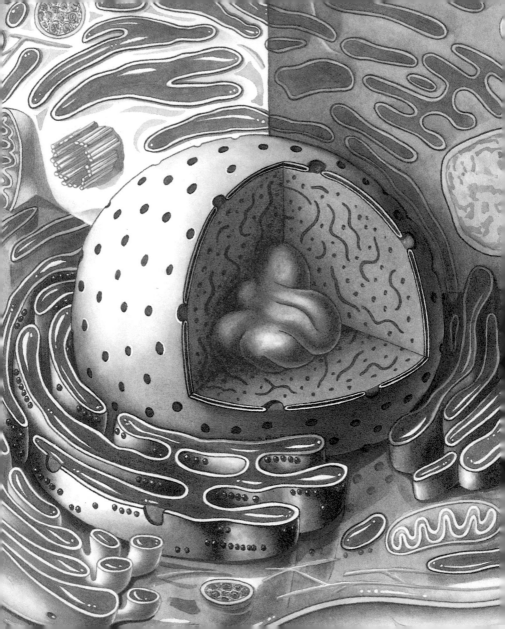

Cells, skin, and specialized tissue

Cells are the basic units of life, capable of performing fundamental processes such as respiration and reproduction; the human body is made up of billions of them. The cells forming the outer surface of the skin and the tissues lining many internal organs are of the type known as epithelial cells. These have diverse structures and are organized in varying numbers of layers.

Cell
The size and shape of human cells vary depending on their particular function. They contain structures such as a nucleus and mitochondria.

Cell structure

Most human cells contain small structures known as organelles ("little organs"), each of which performs a highly specialized task, such as manufacturing protein. Organelles are usually surrounded by a membrane, and they float in a jellylike substance called cytoplasm. Ninety percent of cytoplasm is water; it also contains enzymes, amino acids, and other molecules needed for cell functions.

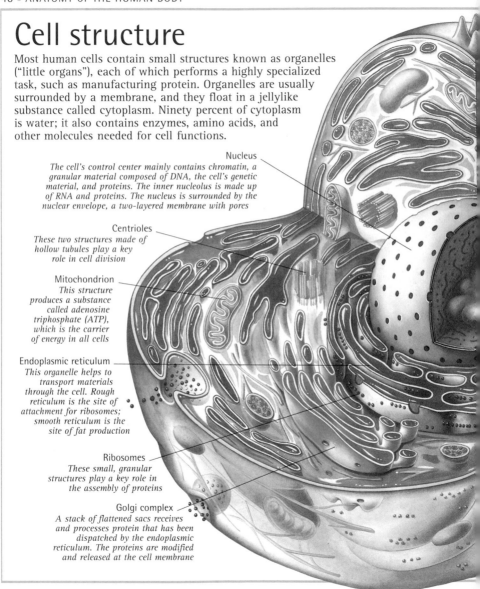

Nucleus
The cell's control center mainly contains chromatin, a granular material composed of DNA, the cell's genetic material, and proteins. The inner nucleolus is made up of RNA and proteins. The nucleus is surrounded by the nuclear envelope, a two-layered membrane with pores

Centrioles
These two structures made of hollow tubules play a key role in cell division

Mitochondrion
This structure produces a substance called adenosine triphosphate (ATP), which is the carrier of energy in all cells

Endoplasmic reticulum
This organelle helps to transport materials through the cell. Rough reticulum is the site of attachment for ribosomes; smooth reticulum is the site of fat production

Ribosomes
These small, granular structures play a key role in the assembly of proteins

Golgi complex
A stack of flattened sacs receives and processes protein that has been dispatched by the endoplasmic reticulum. The proteins are modified and released at the cell membrane

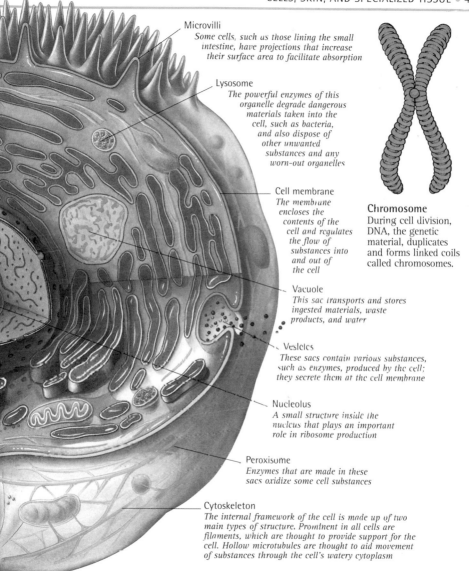

Microvilli
Some cells, such as those lining the small intestine, have projections that increase their surface area to facilitate absorption

Lysosome
The powerful enzymes of this organelle degrade dangerous materials taken into the cell, such as bacteria, and also dispose of other unwanted substances and any worn-out organelles

Cell membrane
The membrane encloses the contents of the cell and regulates the flow of substances into and out of the cell

Chromosome
During cell division, DNA, the genetic material, duplicates and forms linked coils called chromosomes.

Vacuole
This sac transports and stores ingested materials, waste products, and water

Vesicles
These sacs contain various substances, such as enzymes, produced by the cell; they secrete them at the cell membrane

Nucleolus
A small structure inside the nucleus that plays an important role in ribosome production

Peroxisome
Enzymes that are made in these sacs oxidize some cell substances

Cytoskeleton
The internal framework of the cell is made up of two main types of structure. Prominent in all cells are filaments, which are thought to provide support for the cell. Hollow microtubules are thought to aid movement of substances through the cell's watery cytoplasm

Cell types

Each human cell has a characteristic shape, size, and lifespan adapted for its function. Nerve cells have axons along which nerve signals are dispatched. White blood cells have a flexible membrane so that they can squeeze through the tiny spaces between capillaries. Muscle cells can change their length, which varies contractile force. Egg cells develop a protective membrane after fertilization, to prevent further sperm from entering. Sperm have whiplike tails so that they can swim up the female genital tract.

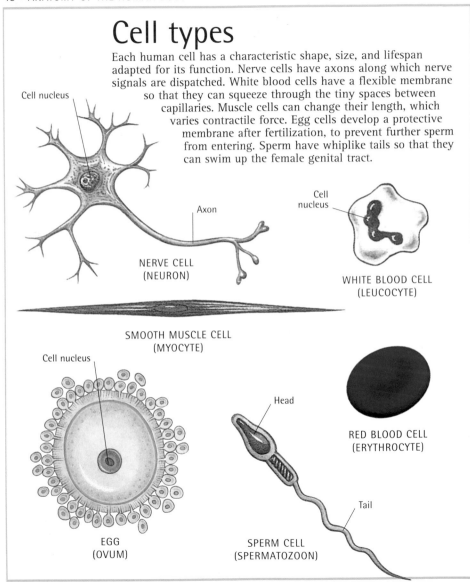

Cell nucleus

Axon

NERVE CELL
(NEURON)

Cell nucleus

WHITE BLOOD CELL
(LEUCOCYTE)

SMOOTH MUSCLE CELL
(MYOCYTE)

Cell nucleus

Head

RED BLOOD CELL
(ERYTHROCYTE)

Tail

EGG
(OVUM)

SPERM CELL
(SPERMATOZOON)

Transport mechanisms

The cell membrane regulates the substances that flow in and out of the cell. Because cell membranes allow only certain substances to pass through, determined in part by the cell's role in the body, they are called selectively permeable. Cell membranes may contain several types of receptor protein, each responding to a specific molecule. Some membrane proteins bind to each other, forming connections between cells.

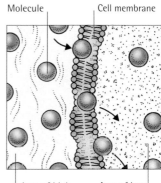

Molecule   Cell membrane

Area of high concentration outside cell

Area of low concentration inside cell

Simple diffusion

Diffusion describes the random movement of molecules from areas of high concentration to areas of low concentration. In simple diffusion, substances such as water and gases pass through spaces in the cell membrane.

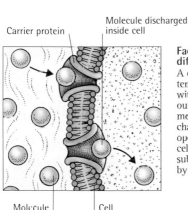

Carrier protein

Molecule discharged inside cell

Molecule outside cell

Cell membrane

Facilitated diffusion

A carrier protein temporarily binds with large molecules outside the cell membrane. It then changes its shape, opening up to the cell interior. Every substance is carried by a specific protein.

Molecule at receptor site

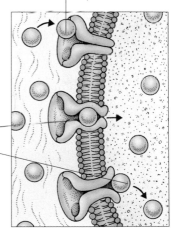

Protein forms channel

Molecule discharged inside cell

Active transport

To move substances from areas of low to high concentration, energy is supplied by ATP. Molecules bind to a receptor site on the cell membrane, triggering a protein to change into a channel through which molecules are squeezed and ejected.

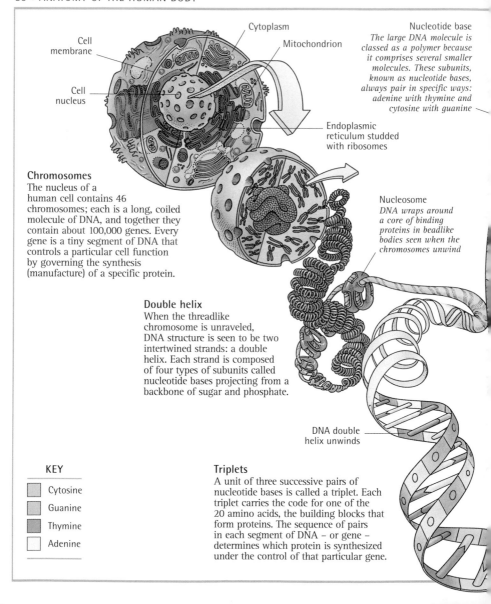

Cell membrane

Cytoplasm

Mitochondrion

Cell nucleus

Nucleotide base
The large DNA molecule is classed as a polymer because it comprises several smaller molecules. These subunits, known as nucleotide bases, always pair in specific ways: adenine with thymine and cytosine with guanine

Endoplasmic reticulum studded with ribosomes

Chromosomes
The nucleus of a human cell contains 46 chromosomes; each is a long, coiled molecule of DNA, and together they contain about 100,000 genes. Every gene is a tiny segment of DNA that controls a particular cell function by governing the synthesis (manufacture) of a specific protein.

Nucleosome
DNA wraps around a core of binding proteins in beadlike bodies seen when the chromosomes unwind

Double helix
When the threadlike chromosome is unraveled, DNA structure is seen to be two intertwined strands: a double helix. Each strand is composed of four types of subunits called nucleotide bases projecting from a backbone of sugar and phosphate.

DNA double helix unwinds

KEY

Cytosine

Guanine

Thymine

Adenine

Triplets
A unit of three successive pairs of nucleotide bases is called a triplet. Each triplet carries the code for one of the 20 amino acids, the building blocks that form proteins. The sequence of pairs in each segment of DNA – or gene – determines which protein is synthesized under the control of that particular gene.

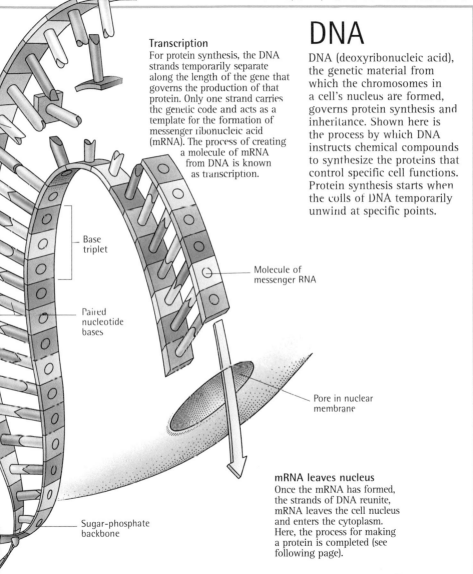

DNA

DNA (deoxyribonucleic acid), the genetic material from which the chromosomes in a cell's nucleus are formed, governs protein synthesis and inheritance. Shown here is the process by which DNA instructs chemical compounds to synthesize the proteins that control specific cell functions. Protein synthesis starts when the coils of DNA temporarily unwind at specific points.

Transcription
For protein synthesis, the DNA strands temporarily separate along the length of the gene that governs the production of that protein. Only one strand carries the genetic code and acts as a template for the formation of messenger ribonucleic acid (mRNA). The process of creating a molecule of mRNA from DNA is known as transcription.

Base triplet

Molecule of messenger RNA

Paired nucleotide bases

Pore in nuclear membrane

mRNA leaves nucleus
Once the mRNA has formed, the strands of DNA reunite, mRNA leaves the cell nucleus and enters the cytoplasm. Here, the process for making a protein is completed (see following page).

Sugar-phosphate backbone

Protein synthesis

After leaving the cell nucleus (see previous page), the new mRNA attaches to structures in the cytoplasm known as ribosomes. As a ribosome moves along the mRNA strand, it produces protein by bringing amino acids into place following the sequence of nucleotide base triplets.

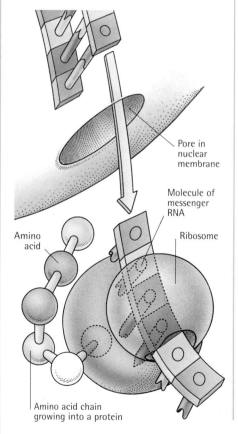

Pore in nuclear membrane

Molecule of messenger RNA

Amino acid

Ribosome

Amino acid chain growing into a protein

Roles of proteins

Proteins are needed for carrying out many vital functions in the body. Some proteins form structures such as hair and muscle; others serve as antibodies, hormones, or enzymes, or, like oxygen-carrying hemoglobin, transport substances in the body.

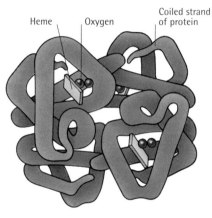

Heme

Oxygen

Coiled strand of protein

HEMOGLOBIN MOLECULE

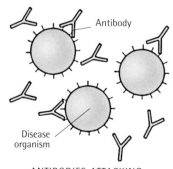

Antibody

Disease organism

ANTIBODIES ATTACKING DISEASE ORGANISMS

Mitosis

Mitosis is a copying process that redistributes DNA when body cells divide; it occurs continuously during all normal growth and tissue replacement. In this process, one cell divides to produce two daughter cells that are identical to each other and to the parent cell. In the stage that precedes the actual division, called interphase, molecules of DNA are replicated and are loosely organized into a network of extended filaments known as chromatin.

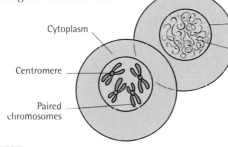

Prophase
DNA strands coil up, forming spiral filaments called chromatids that join at a structure called the centromere. These filaments then condense to form 46 X-shaped pairs of chromosomes.

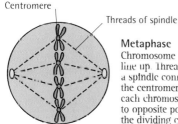

Metaphase
Chromosome pairs line up. Threads create a spindle connecting the centromere of each chromosome pair to opposite poles of the dividing cell.

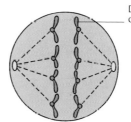

Anaphase
Each centromere splits and the single chromosomes, 46 on each side, move towards opposite sides of the cell.

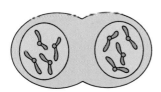

Telophase
Spindle fibers disappear, and a nuclear membrane forms around each group of 46 chromosomes. The cell becomes pinched in the middle and the chromosomes start to uncoil.

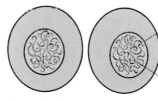

Late telophase
The cell begins to divide; a cell plate forms between the two groups of chromosomes, and the cells splits into two. The 46 chromosomes in each new cell revert to chromatin filaments.

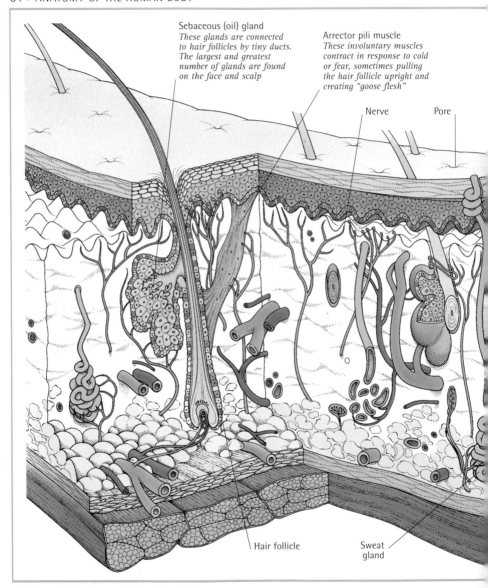

Sebaceous (oil) gland
These glands are connected to hair follicles by tiny ducts. The largest and greatest number of glands are found on the face and scalp

Arrector pili muscle
These involuntary muscles contract in response to cold or fear, sometimes pulling the hair follicle upright and creating "goose flesh"

Nerve

Pore

Hair follicle

Sweat gland

Skin structure

Skin is composed of two main layers. The thin outer layer, or epidermis, is a type of tissue known as stratified squamous epithelium, which consists of sheets of platelike cells. Beneath this lies a deeper layer, the dermis, which comprises fibrous and elastic tissue containing blood vessels, nerve fibers, hair follicles, and sweat glands. The deepest layer of the dermis anchors the skin to underlying tissues.

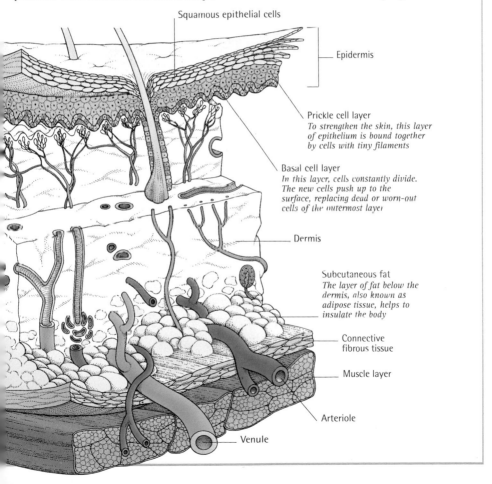

Squamous epithelial cells

Epidermis

Prickle cell layer
To strengthen the skin, this layer of epithelium is bound together by cells with tiny filaments

Basal cell layer
In this layer, cells constantly divide. The new cells push up to the surface, replacing dead or worn-out cells of the outermost layer

Dermis

Subcutaneous fat
The layer of fat below the dermis, also known as adipose tissue, helps to insulate the body

Connective fibrous tissue

Muscle layer

Arteriole

Venule

Nail structure

Nails are made of keratin, a hard, fibrous protein, which is also the main constituent of hair. They lie on a nail bed, an area rich in blood vessels that gives the nail its color. Nails grow from a matrix of active cells beneath a fold of skin called the cuticle and from the lunula, a crescent-shaped area at the base of the nail.

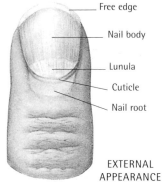

- Free edge
- Nail body
- Lunula
- Cuticle
- Nail root

EXTERNAL APPEARANCE

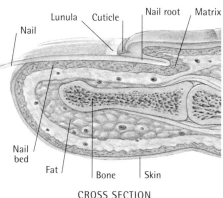

Lunula · Cuticle · Nail root · Matrix · Nail · Nail bed · Fat · Bone · Skin

CROSS SECTION

Hair growth

Hair cells develop in pits in the skin called follicles, which reach down into the dermis. The cells divide and eventually die, forming a hair shaft as the dead cells build up. Each follicle has phases of growth and rest.

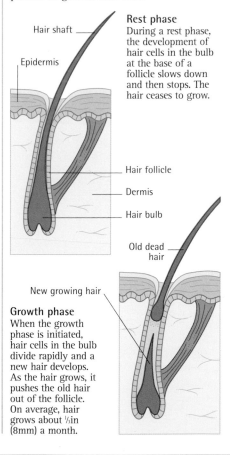

Hair shaft

Epidermis

Rest phase
During a rest phase, the development of hair cells in the bulb at the base of a follicle slows down and then stops. The hair ceases to grow.

- Hair follicle
- Dermis
- Hair bulb

Old dead hair

New growing hair

Growth phase
When the growth phase is initiated, hair cells in the bulb divide rapidly and a new hair develops. As the hair grows, it pushes the old hair out of the follicle. On average, hair grows about ⅓in (8mm) a month.

Specialized tissue

Epithelial tissue, or epithelium, is found in skin and, in diverse forms, has special functions in other body areas. The main types are: simple epithelium, a single layer of squamous (flat), cuboidal (cubelike), or columnar (tall) cells; stratified epithelium, with two or more layers; pseudostratified epithelium, which looks stratified but has only one layer of columnar cells that have surface hairs (cilia) or secrete mucus; and transitional epithelium, which is multi-layered and pliable.

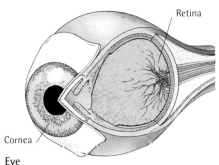

Retina

Cornea

Eye
Stratified squamous epithelium covers the outer cornea of the eye, where it forms a transparent protective coating. Another type of epithelial tissue, simple cuboidal epithelium, is found in the pigmented layer of the retina.

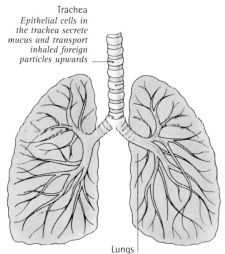

Trachea
Epithelial cells in the trachea secrete mucus and transport inhaled foreign particles upwards

Lungs
The main airways are lined with pseudostratified epithelium

Respiratory system
The epithelial tissue lining the trachea is pseudostratified, consisting of a single layer of cells of different heights. Some of the taller cells either have cilia to trap or move foreign particles or are goblet cells, which secrete mucus.

Ureter
Epithelial cells in the lining of the ureters secrete a mucus that protects the ureters from acidic urine

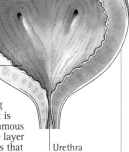

Lining of bladder
Tightly packed, rounded epithelial cells are able to stretch as the bladder fills

Urethra

Urinary system
Transitional epithelium is well-suited as a lining to the urinary system. It is similar to stratified squamous epithelium, but the base layer has rounder surface cells that are extremely pliable and can stretch without tearing.

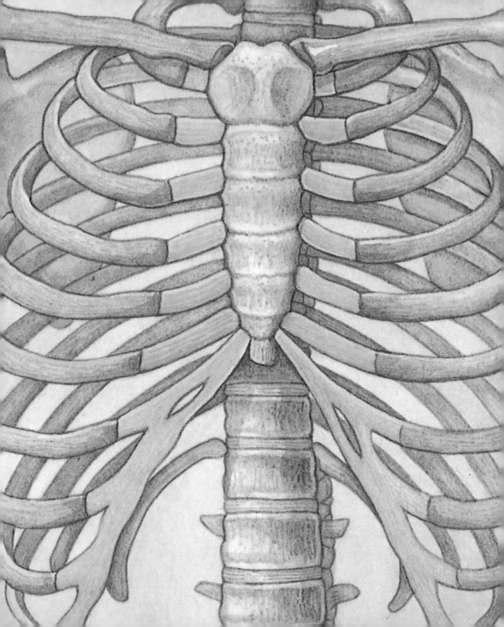

Skeletal system

The living human skeleton is a tough, flexible
structure that supports weight and protects
internal organs. Bone tissue is constantly
renewed and stores minerals essential
to the body, such as calcium and
phosphorus. The joints, where bones
meet, mostly permit a wide range
of movement; the less mobile types
of joint, such as those in the spine,
provide greater stability.

Skeleton
The bones provide a light but
strong framework for the body's
soft tissues. The ribcage (left)
surrounds the heart and lungs.

Skeleton 1

The precise number of bones in the adult human skeleton varies from one person to another, but on average there are 206 bones of varying shapes and sizes. The skeleton is divided into two main parts. The central bones of the skull, ribs, vertebral column (spine), and sternum (breastbone) form the axial skeleton. The bones of the arms and legs, along with the scapula (shoulder blade), clavicle (collar bone), and pelvis, make up the appendicular skeleton.

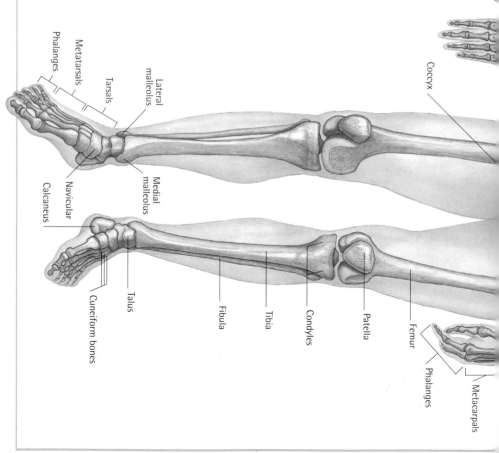

Phalanges

Metatarsals

Tarsals

Lateral malleolus

Coccyx

Medial malleolus

Navicular

Calcaneus

Cuneiform bones

Talus

Fibula

Tibia

Condyles

Patella

Femur

Phalanges

Metacarpals

Bone tissue

Inside, the long bones of the body consist of numerous struts called trabeculae. This structure makes the bones both lightweight and strong.

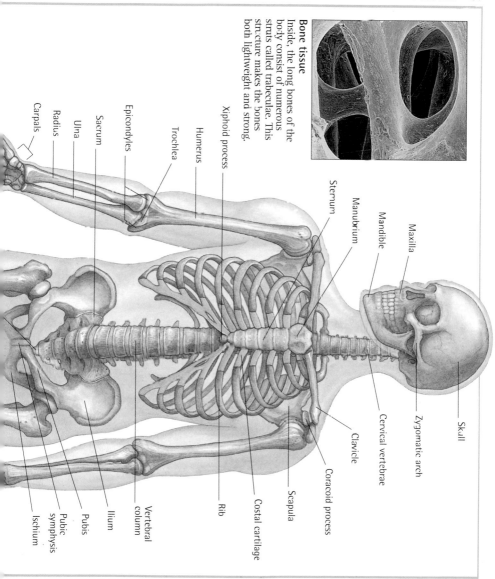

- Carpals
- Radius
- Ulna
- Sacrum
- Epicondyles
- Trochlea
- Humerus
- Xiphoid process
- Sternum
- Manubrium
- Mandible
- Maxilla
- Zygomatic arch
- Skull
- Cervical vertebrae
- Clavicle
- Coracoid process
- Scapula
- Costal cartilage
- Rib
- Vertebral column
- Ilium
- Pubis
- Pubic symphysis
- Ischium

Skeleton 2

The cylindrical, linked vertebrae of the spinal column offer strong, bony protection for the spinal cord. Aided by muscles and ligaments, the vertebrae support the skull and hold the body upright. The spine has a flexible structure that permits twisting and bending of the trunk. The wing-shaped scapulae (shoulder blades) that cover the upper part of the ribcage allow great flexibility in the arms and shoulders.

Parietal bones

Occipital bone

Acromion

Scapula

Rib

Vertebral column

Humerus

Sacrum

Radius

Ilium

Ulna

Foramen magnum

Occipital bone

Carotid canal

Coccyx

Zygomatic arch

Femur

Opening at back of nose

Bony plate

Fibula

Mandible (lower jaw)

Teeth

Tibia

Medial malleolus

Base of skull
The spinal cord reaches the brain through a large opening in the base of the skull called the foramen magnum. Blood vessels pass through openings such as the carotid canals.

Tarsals

Metatarsals

Phalanges

Calcaneus

Bone shapes

The shapes of bones reflect their functions. Long bones act as levers to raise and lower; short bones, such as the talus (ankle bone) are useful as bridges; and flat bones, including those in the skull, form protective shells. Small, rounded, sesamoid bones, such as the patella (kneecap), are embedded within tendons. Irregular bones include vertebrae, the ilium (pelvis), and some skull bones, such as the sphenoid.

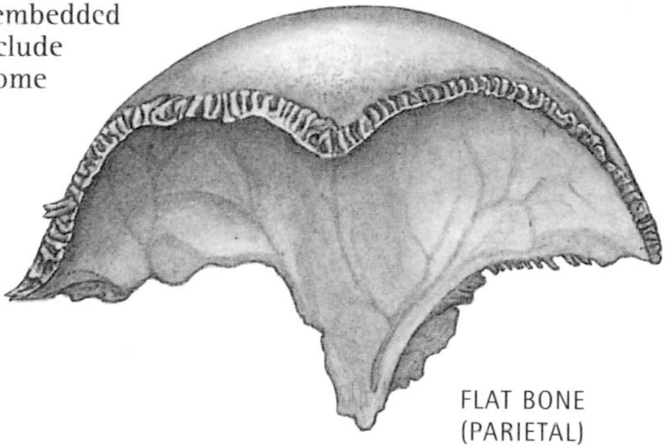

FLAT BONE
(PARIETAL)

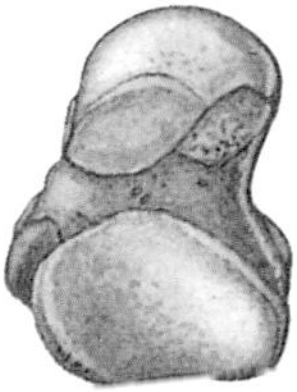

SHORT BONE
(TALUS)

SESAMOID BONE
(PATELLA)

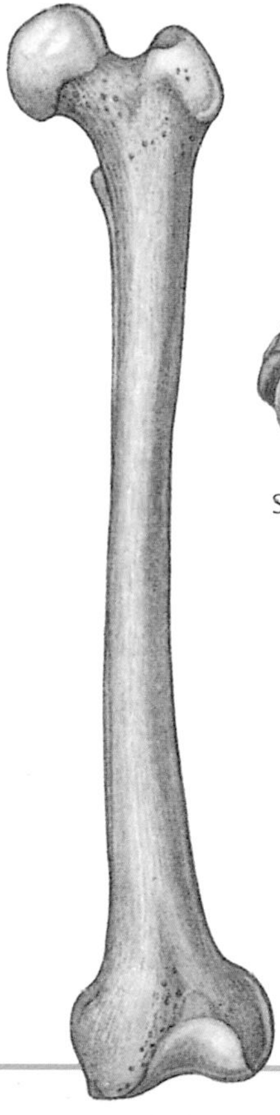

LONG BONE
(FEMUR)

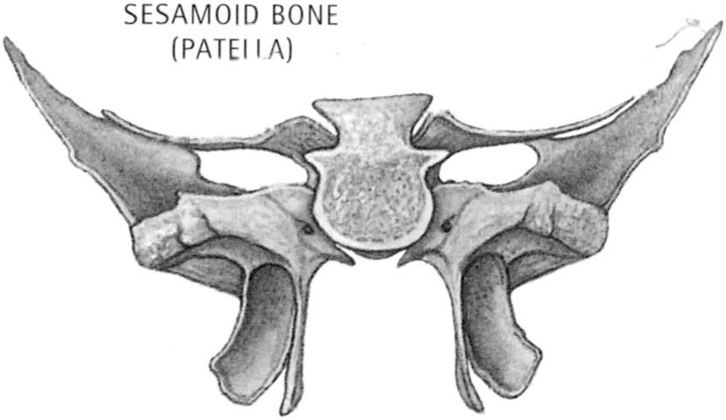

IRREGULAR BONE
(SPHENOID)

Bones of the hand and foot

The skeletal structure of the
hand and foot is similar; in
both cases there is an
interlinking arrangement
of small bones. The hand has
14 phalanges (finger bones),
five metacarpals (palm bones),
and eight carpals (wrist
bones). There are also 14
phalanges (toes) in the foot,
but these bones are usually
shorter than the equivalent
bones in the hand. The rest of
the foot is composed of five
metatarsals (sole bones) and
seven tarsals (ankle bones).

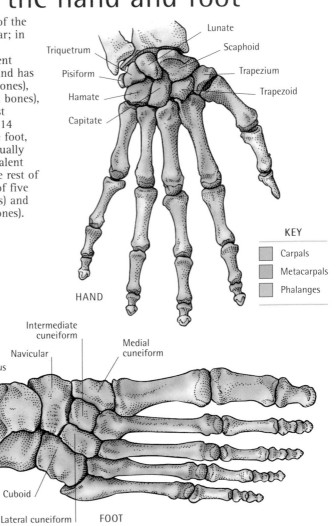

Lunate

Triquetrum

Scaphoid

Pisiform

Trapezium

Hamate

Trapezoid

Capitate

KEY

Carpals

Metacarpals

Phalanges

HAND

KEY

Tarsals

Metatarsals

Phalanges

Intermediate
cuneiform

Navicular

Medial
cuneiform

Talus

Calcaneus

Cuboid

Lateral cuneiform

FOOT

Ribcage

A bony cage within the chest shields the heart, lungs, and other organs. There are 12 pairs of ribs, all attached to the spine. The upper seven pairs of "true ribs" link directly to the sternum by costal cartilage. The next two to three pairs of "false ribs" attach indirectly to the sternum by means of cartilage linked to the ribs above; the remaining "floating ribs" have no links to the sternum. (The lowest ribs are hidden by the liver and stomach in this illustration.)

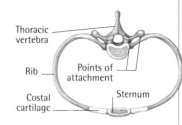

Thoracic vertebra

Rib

Points of attachment

Costal cartilage

Sternum

How the ribs attach
Each rib links to its corresponding thoracic (chest) vertebra at two points. Flexible costal cartilage attaches some ribs to the sternum.

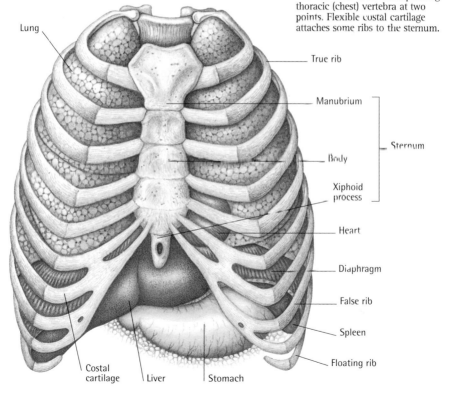

Lung

True rib

Manubrium

Body

Sternum

Xiphoid process

Heart

Diaphragm

False rib

Spleen

Floating rib

Costal cartilage

Liver

Stomach

Pelvic bones

The pelvis varies in shape according to sex. Overall, the structure has a similar appearance in both sexes but takes a shallower and wider form in females to allow for the specialized function of childbearing. Arranged in a ring, the fused pelvic bones provide a strong foundation for the upper body and protection for parts of the reproductive, digestive, and urinary systems.

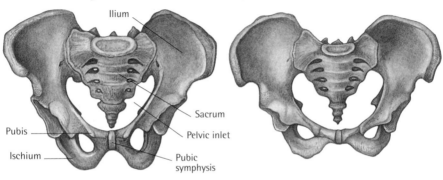

MALE PELVIS

FEMALE PELVIS

Bones of the skull 1

Two separate sets of bones form the intricate structure of the skull. The eight bones enclosing and protecting the brain are called the cranial vault. Another 14 bones make up the skeleton of the face. In adults, all of the skull bones except the mandible (lower jaw) are locked together by joints known as sutures. These seams are visible on the surface of the skull as lines between the bones.

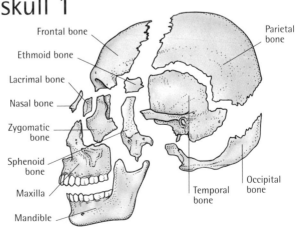

Bones of the skull 2

Viewed from the front, the most prominent skull bones are the frontal bone, which forms the forehead, the zygomatic bones, which give shape to the cheeks, and the upper and lower jaw bones. The back and sides of the cranial vault largely comprise the occipital and parietal bones. In the middle ear there are three tiny bones of the type known as ossicles, which are not technically part of the skull. They conduct sound waves from the eardrum to the inner ear.

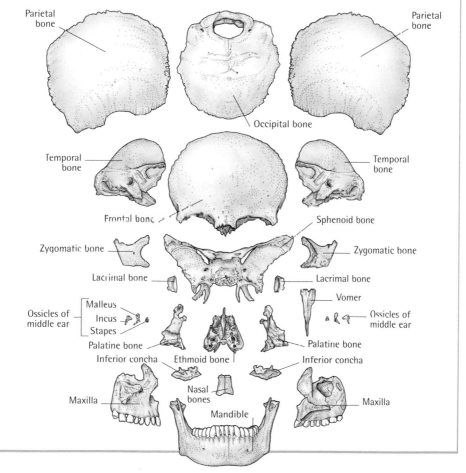

Parietal bone

Parietal bone

Occipital bone

Temporal bone

Temporal bone

Frontal bone

Sphenoid bone

Zygomatic bone

Zygomatic bone

Lacrimal bone

Lacrimal bone

Vomer

Ossicles of middle ear

Malleus
Incus
Stapes

Ossicles of middle ear

Palatine bone

Palatine bone

Inferior concha

Ethmoid bone

Inferior concha

Maxilla

Nasal bones

Maxilla

Mandible

Structure of the spine

The spine is made up of 33 ringlike bones called vertebrae that are linked by a series of mobile joints. Sandwiched between the vertebrae are springy, shock-absorbing discs with a tough outer layer of cartilage. There are three main types of vertebrae: cervical in the neck, thoracic in the upper back, and lumbar in the lower back. Two regions at the base of the spine, the wedge-shaped sacrum and the tail-like coccyx, both consist of several fused vertebrae.

Spinal cord
This vital cable of nerve tissue, which relays messages between the brain and different parts of the body, is protected by the 33 vertebrae of the spinal column

Vertebral processes
These bony knobs extend from the back of each vertebra. Three processes serve as anchor points for muscles; the other four form the linking facet joints between adjacent vertebrae

Spinal nerve
Connected to the spinal cord are 31 pairs of nerves that emerge through gaps between the vertebrae and travel out to body tissues and organs

Sacrum
The five vertebrae of the sacrum are fused

Coccyx
The four tail bones of the spine are much smaller than the other vertebrae

Vertebral body
These bony discs become larger towards the base of the spine to support greater weight.

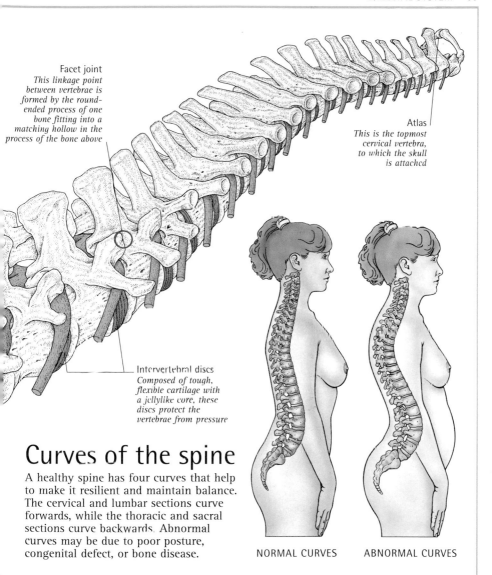

Facet joint
This linkage point between vertebrae is formed by the round-ended process of one bone fitting into a matching hollow in the process of the bone above

Atlas
This is the topmost cervical vertebra, to which the skull is attached

Intervertebral discs
Composed of tough, flexible cartilage with a jellylike core, these discs protect the vertebrae from pressure

Curves of the spine

A healthy spine has four curves that help to make it resilient and maintain balance. The cervical and lumbar sections curve forwards, while the thoracic and sacral sections curve backwards. Abnormal curves may be due to poor posture, congenital defect, or bone disease.

NORMAL CURVES   ABNORMAL CURVES

Regions of the spine

Each section of the spine is adapted to a particular function. The cervical vertebrae support the head and neck, the thoracic vertebrae anchor the ribs, and the strong, weight-bearing regions towards the bottom of the spine provide a stable center of gravity during movement. Although their shapes vary, vertebrae typically comprise a bony disc called the body and projections called processes, to which muscles attach.

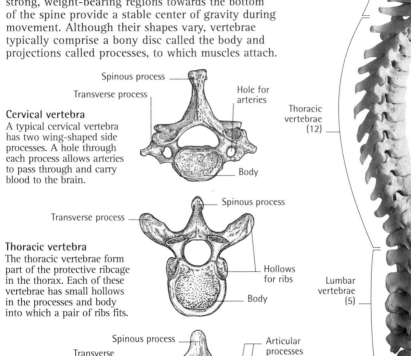

Spinous process

Transverse process

Hole for arteries

Cervical vertebra
A typical cervical vertebra has two wing-shaped side processes. A hole through each process allows arteries to pass through and carry blood to the brain.

Body

Spinous process

Transverse process

Thoracic vertebra
The thoracic vertebrae form part of the protective ribcage in the thorax. Each of these vertebrae has small hollows in the processes and body into which a pair of ribs fits.

Hollows for ribs

Body

Spinous process

Transverse process

Articular processes

Lumbar vertebra
The large body of a lumbar vertebra reflects its role in supporting a major part of the body's weight. The articular processes facilitate movement.

Body

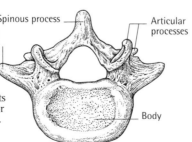

Cervical vertebrae (7)

Thoracic vertebrae (12)

Lumbar vertebrae (5)

Sacrum (5)

Coccyx (4)

Movement of spinal joints

The spine is constructed strongly to hold the head and body upright, but it is also flexible enough to allow the upper body to bend and twist. The cartilage discs between the vertebrae can withstand enormous forces, which can be as much as several hundred pounds per square inch during strenuous movements. Strong ligaments and muscles around the spine stabilize the vertebrae and help to control movement.

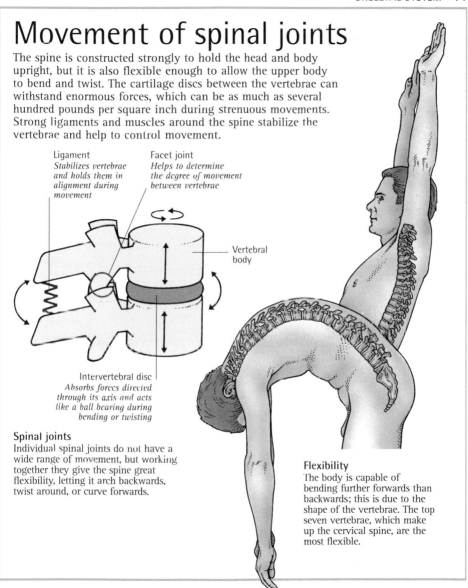

Ligament
Stabilizes vertebrae and holds them in alignment during movement

Facet joint
Helps to determine the degree of movement between vertebrae

Vertebral body

Intervertebral disc
Absorbs forces directed through its axis and acts like a ball bearing during bending or twisting

Spinal joints
Individual spinal joints do not have a wide range of movement, but working together they give the spine great flexibility, letting it arch backwards, twist around, or curve forwards.

Flexibility
The body is capable of bending further forwards than backwards; this is due to the shape of the vertebrae. The top seven vertebrae, which make up the cervical spine, are the most flexible.

Bone structure

Bone is made of specialized cells in a matrix composed mainly of protein fibers, water, and minerals. In the center of a long bone is the medullary canal, which contains bone marrow and blood vessels. Around the marrow are layers of cancellous (spongy) bone, which also contains marrow, and cortical (hard) bone. A membrane, the periosteum, covers the bone surface.

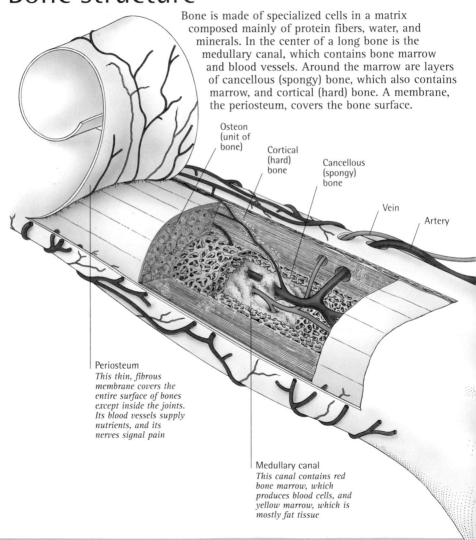

Osteon
(unit of
bone)

Cortical
(hard)
bone

Cancellous
(spongy)
bone

Vein

Artery

Periosteum
*This thin, fibrous
membrane covers the
entire surface of bones
except inside the joints.
Its blood vessels supply
nutrients, and its
nerves signal pain*

Medullary canal
*This canal contains red
bone marrow, which
produces blood cells, and
yellow marrow, which is
mostly fat tissue*

Long bone

At each end of the long bones of the limbs there is an area known as the epiphysis. In childhood, the epiphyses are made mostly of cartilage, which hardens to become cancellous (spongy) bone in a mature adult. The central shaft of a long bone is called the diaphysis.

Osteon

Osteons, which are also called Haversian systems, are the building blocks that make up cortical (hard) bone. They are rod-shaped units with a central canal that is surrounded by concentric layers of bone tissue called lamellae.

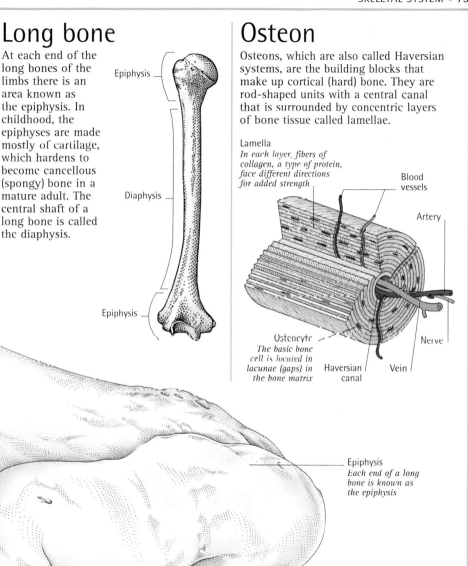

Epiphysis

Diaphysis

Epiphysis

Lamella
In each layer, fibers of collagen, a type of protein, face different directions for added strength

Blood vessels

Artery

Osteocyte
The basic bone cell is located in lacunae (gaps) in the bone matrix

Haversian canal

Vein

Nerve

Epiphysis
Each end of a long bone is known as the epiphysis

Bone growth

Between the shafts (diaphyses) and the ends (epiphyses) of the long bones is a growth area known as the epiphyseal plate. Cartilage cells, known as chondrocytes, proliferate here and form columns that push older cells towards the middle of the bone shaft. The cartilage cells enlarge and eventually die, leaving a space that is filled by new bone cells. Bone growth continues until about the age of 17.

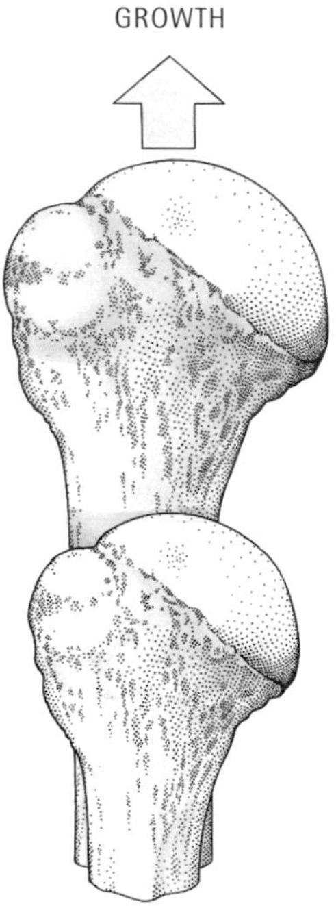

GROWTH

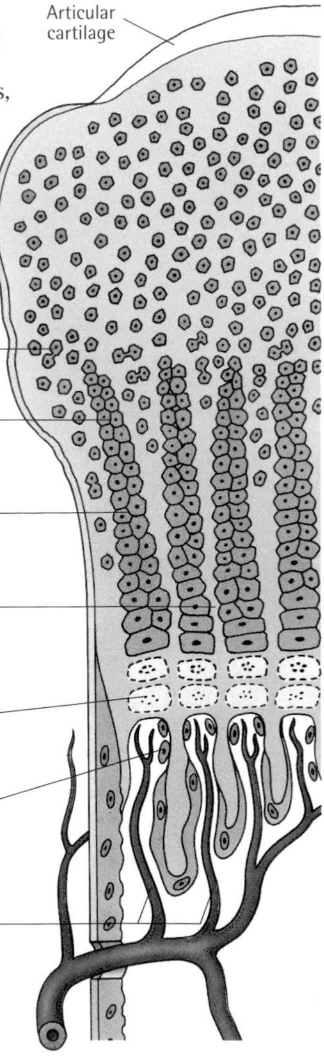

Site of growth plate
The area tinted gray is the epiphyseal growth plate. These areas are found near the ends of the long bones.

Articular cartilage

Cartilage cells divide and increase

↓

Cartilage cells form columns

↓

Cartilage cells enlarge

↓

Calcium is deposited in matrix between cartilage cells

↓

Mature cartilage cells die

↓

New bone cells called osteoblasts attach to calcified matrix

↓

New blood vessels nourish new bone

Bone repair

Bone is a living tissue that is continually broken down and rebuilt throughout a person's life. If it is fractured, bone is able to regrow, and eventually the line of fracture is bridged with new tissue. The repair mechanism is activated rapidly after injury, although the laying down of new bone may take weeks to complete. Once mended, a broken bone may take some months to regain its full strength.

Blood clot develops

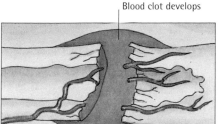

One hour after fracture
When a bone is fractured, the healing process begins almost immediately. Blood leaking from damaged vessels at the site of injury rapidly forms a clot, preventing further bleeding.

Network of fibrous tissue

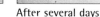

After several days
Cells known as fibroblasts start to build up a network of fibrous tissue called a callus. This gradually bridges the gap between the broken bone ends, replacing the blood clot.

Spongy bone forms

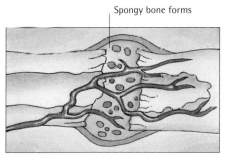

After 1–2 weeks
Soft, spongy bone is deposited on the callus, filling in the fiber network and keeping the bone ends stable. The damaged blood vessels have healed and are growing across the break.

New compact bone

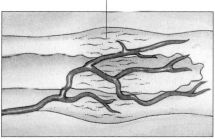

After 2–3 months
New compact bone has replaced the callus, completing the repair. Swelling around the site of the fracture gradually diminishes and the bone continues to regain its full strength.

Joint structure

A joint, or articulation, is where two bones meet. Joints are classified by their structure or by the way they move. Most joints in the body are synovial joints. These are versatile, lubricated joints, such as the knee shown here, in which the surfaces in contact slide over each other easily. Articular cartilage covers the bone ends, ligaments provide stability, and a fibrous capsule encloses the joint. Surrounding muscles produce movement.

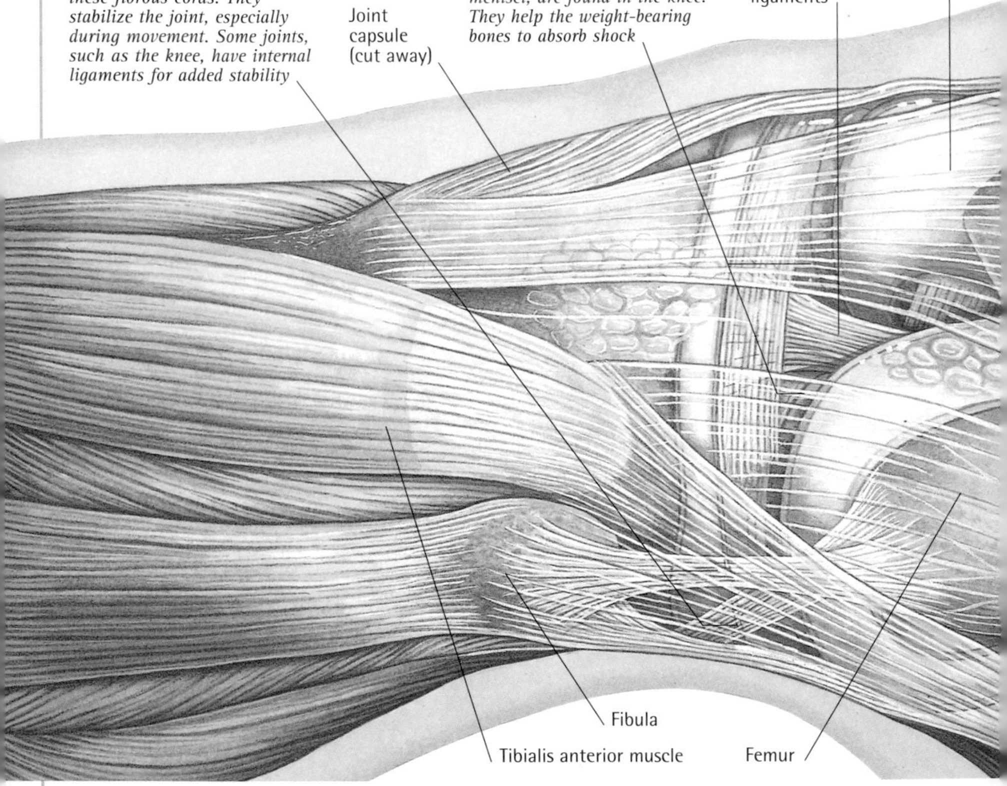

External ligaments
Thickenings of the capsule form these fibrous cords. They stabilize the joint, especially during movement. Some joints, such as the knee, have internal ligaments for added stability

Joint capsule (cut away)

Menisci
Pads of cartilage, called menisci, are found in the knee. They help the weight-bearing bones to absorb shock

Articular cartilage
Where the bone ends are in contact, connective tissue provides a smooth, protective surface for ease of movement

Internal ligaments

Fibula

Tibialis anterior muscle

Femur

Fixed and semi-movable joints

Not all joints are freely movable. After growth is complete, the bones of the skull become fixed together by fibrous tissue, forming immovable suture joints. In the lower leg, the tibia and fibula are stabilized by ligaments that allow only a small amount of movement.

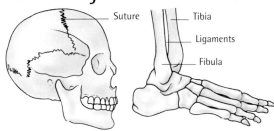

FIXED JOINTS

SEMI-MOVABLE JOINTS

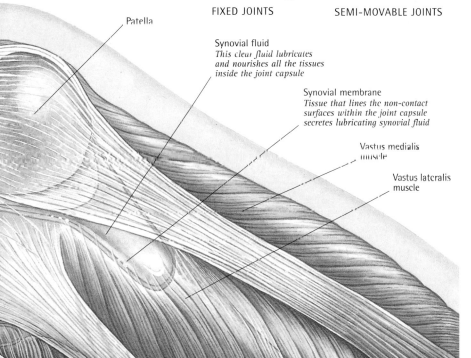

Patella

Synovial fluid
This clear fluid lubricates and nourishes all the tissues inside the joint capsule

Synovial membrane
Tissue that lines the non-contact surfaces within the joint capsule secretes lubricating synovial fluid

Vastus medialis muscle

Vastus lateralis muscle

Types of synovial joint

In a synovial joint, the shape of articular cartilage surfaces and the way they fit together determine the range and direction of the joint's movement. Hinge and pivot joints move in only one plane (for example, from side to side or up and down), while ellipsoidal joints are able to move in two planes at right angles to each other. Most joints in the body can move in more than two planes, which allows a wide range of motion.

Pivot joint
A projection from one bone turns within a ring-shaped socket of another bone, or the ring turns around the bony projection. A pivot joint formed by the top two cervical vertebrae allows the head to turn from side to side, as when shaking the head "no"

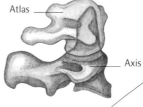

Atlas

Axis

Hinge joint
In this simplest of joints, the convex surface of one bone fits into the concave surface of another bone. This allows for movement like a hinged door in only one plane. Both the elbow and the knee are modified hinge joints: they bend up and down in one plane quite easily but are also capable of very limited rotation

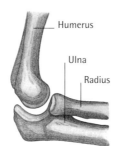

Humerus

Ulna

Radius

Scaphoid   Radius

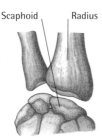

Ellipsoidal joint
An ovoid, or egg-shaped, bone end is held within an elliptical cavity. The radius bone of the forearm and the scaphoid bone of the hand meet in an ellipsoidal joint. This type of joint can be flexed or extended and moved from side to side but rotation is limited

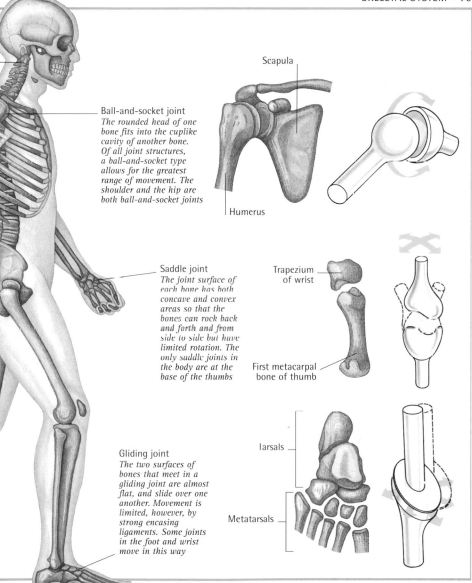

Scapula

Ball-and-socket joint
*The rounded head of one
bone fits into the cuplike
cavity of another bone.
Of all joint structures,
a ball-and-socket type
allows for the greatest
range of movement. The
shoulder and the hip are
both ball-and-socket joints*

Humerus

Saddle joint
*The joint surface of
each bone has both
concave and convex
areas so that the
bones can rock back
and forth and from
side to side but have
limited rotation. The
only saddle joints in
the body are at the
base of the thumbs*

Trapezium
of wrist

First metacarpal
bone of thumb

Gliding joint
*The two surfaces of
bones that meet in a
gliding joint are almost
flat, and slide over one
another. Movement is
limited, however, by
strong encasing
ligaments. Some joints
in the foot and wrist
move in this way*

Tarsals

Metatarsals

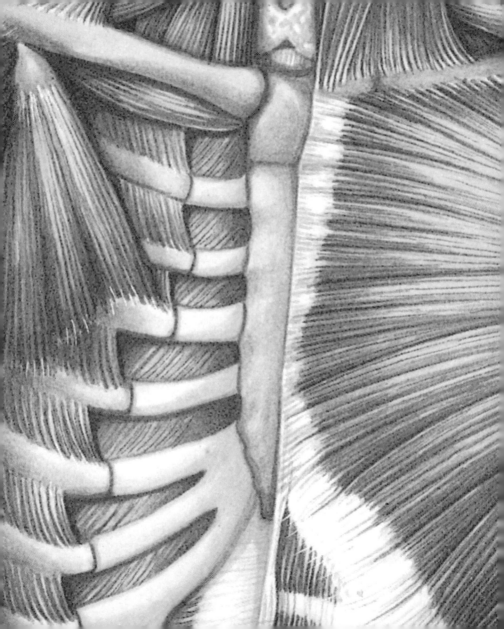

Muscular system

The skeletal muscles make up nearly half the total weight of the human body and provide the forces that enable the body to move and maintain its posture. The majority of these muscles stretch across joints to link one bone with another, and work in groups, in response to nerve impulses; they can usually be controlled voluntarily.

Muscles
Muscles are made of filaments that can stretch and contract, returning to their original shape.

Muscles of the body 1

There are over 600 skeletal muscles in the human body, overlapping each other in intricate layers. A skeletal muscle usually attaches on to one end of a bone and stretches across a joint to attach to another bone. As the muscle contracts, it moves one bone while the other bone remains stable. Muscles just below the skin are described as superficial (shown on the right side of the illustration); beneath these are the deep muscles (left side of the illustration).

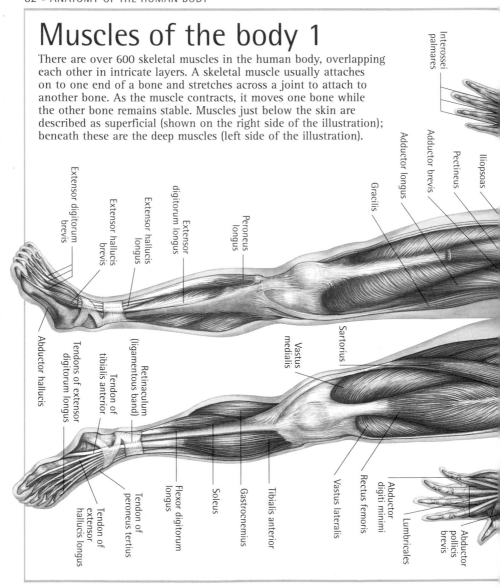

Interossei palmares

Adductor longus

Adductor brevis

Pectineus

Iliopsoas

Gracilis

Extensor digitorum brevis

Extensor hallucis brevis

Extensor digitorum longus

Extensor hallucis longus

Peroneus longus

Abductor hallucis

Tendons of extensor digitorum longus

Tendon of tibialis anterior

Retinaculum (ligamentous band)

Vastus medialis

Sartorius

Tendon of extensor hallucis longus

Tendon of peroneus tertius

Flexor digitorum longus

Soleus

Gastrocnemius

Tibialis anterior

Vastus lateralis

Rectus femoris

Abductor digiti minimi

Lumbricales

Abductor pollicis brevis

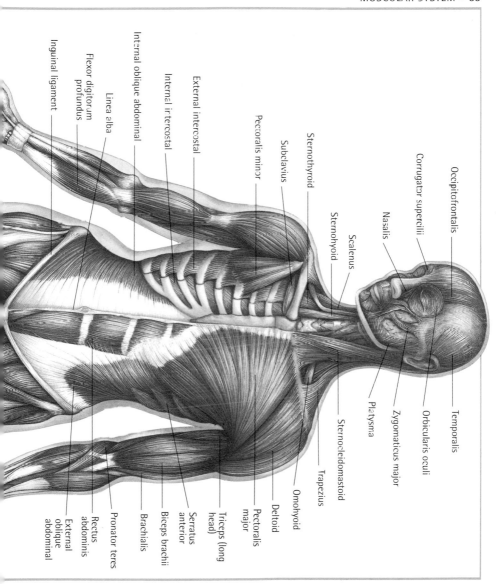

Inguinal ligament

Flexor digitorum profundus

Linea alba

Internal oblique abdominal

Internal intercostal

External intercostal

Pectoralis minor

Subclavius

Sternothyroid

Sternohyoid

Scalenus

Nasalis

Corrugator supercilii

Occipitofrontalis

Temporalis

Orbicularis oculi

Zygomaticus major

Platysma

Sternocleidomastoid

Trapezius

Omohyoid

Deltoid

Pectoralis major

Triceps (long head)

Serratus anterior

Biceps brachii

Brachialis

Pronator teres

Rectus abdominis

External oblique abdominal

Muscles of the body 2

The neck muscles and the massive triangular muscles of the upper back stabilize the head and shoulders and permit a range of complex movements. The most powerful muscles in the human body are those that run along the spine; they maintain posture and provide the strength for lifting and pushing. Deep muscles at the back of the body are shown on the left of the illustration, superficial muscles are shown on the right.

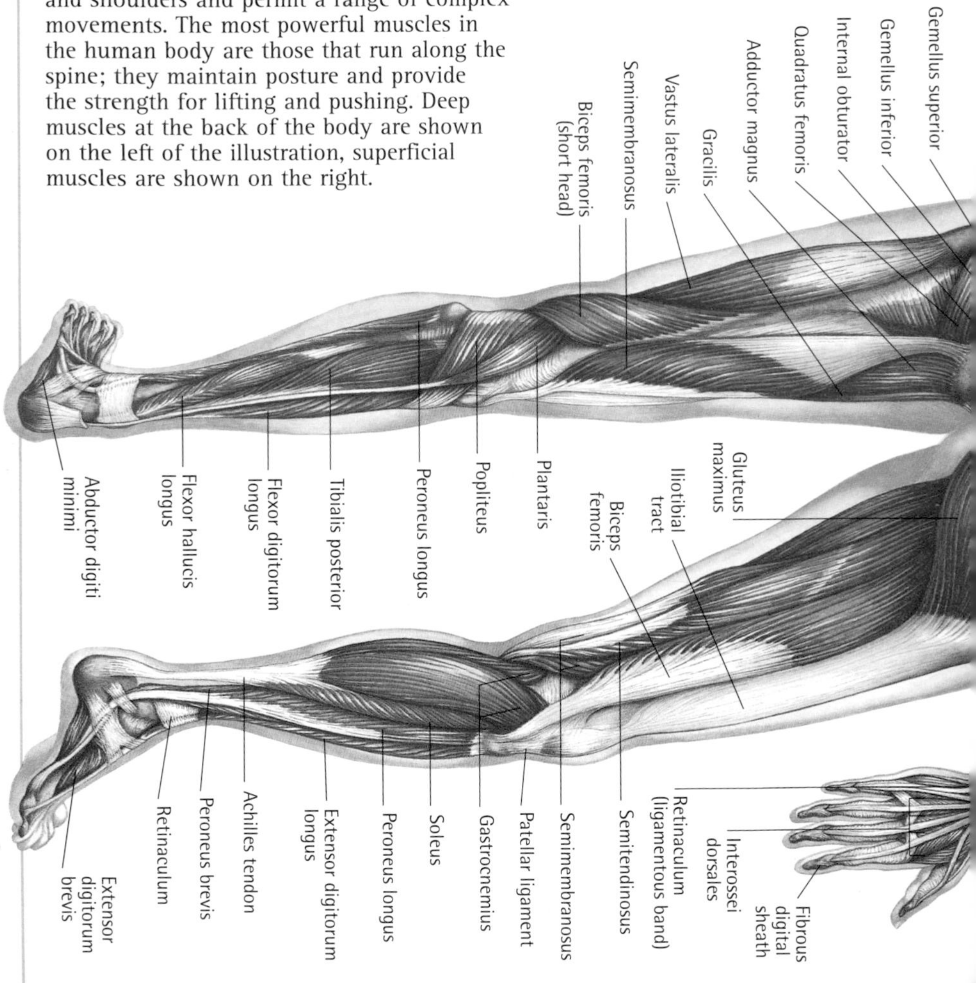

Gemellus superior
Gemellus inferior
Internal obturator
Quadratus femoris
Adductor magnus
Gracilis
Vastus lateralis
Semimembranosus
Biceps femoris (short head)

Abductor digiti minimi
Flexor hallucis longus
Flexor digitorum longus
Tibialis posterior
Peroneus longus
Popliteus
Plantaris
Biceps femoris
Iliotibial tract
Gluteus maximus

Extensor digitorum brevis
Retinaculum
Peroneus brevis
Achilles tendon
Extensor digitorum longus
Peroneus longus
Soleus
Gastrocnemius
Patellar ligament
Semimembranosus
Semitendinosus
Retinaculum (ligamentous band)
Interossei dorsales
Fibrous digital sheath

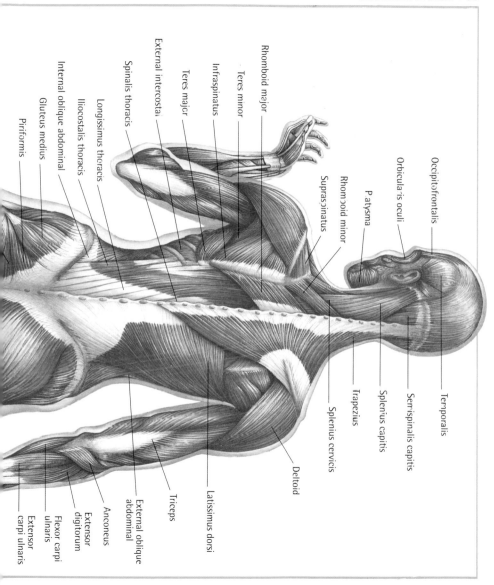

Rhomboid major

Teres minor

Infraspinatus

Teres major

External intercostal

Spinalis thoracis

Longissimus thoracis

Iliocostalis thoracis

Internal oblique abdominal

Gluteus medius

Piriformis

Supraspinatus

Rhomboid minor

Platysma

Orbicularis oculi

Occipitofrontalis

Temporalis

Serispinalis capitis

Splenius capitis

Trapezius

Splenius cervicis

Deltoid

Latissimus dorsi

Triceps

Anconeus

External oblique abdominal

Extensor digitorum

Flexor carpi ulnaris

Extensor carpi ulnaris

Head and facial muscles

The face is supplied with muscles that are attached to the skin and control a wide range of movements. The layers of both superficial and deep muscles are especially complex around the mouth and eyes, where they are involved in producing voluntary actions such as moving the lips in speech or raising the eyebrows. Other muscles in the head and neck control such functions as holding food in the mouth, chewing, swallowing, and moving the tongue (which itself has internal muscles).

Back of head and neck

Large muscles at the back and sides of the neck, for example the sternocleidomastoid, allow the head to bend and rotate. Superficial muscles are shown on the left of the illustration, deep muscles are shown on the right.

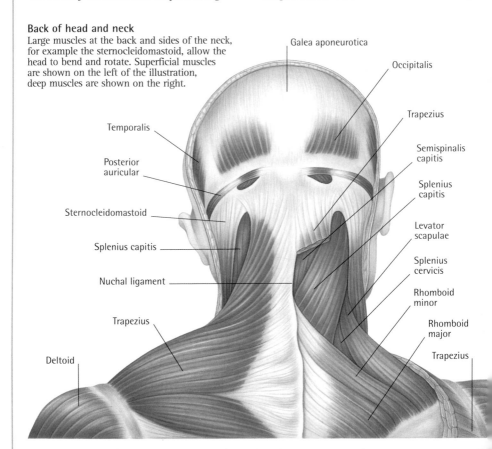

Galea aponeurotica

Occipitalis

Trapezius

Semispinalis capitis

Splenius capitis

Levator scapulae

Splenius cervicis

Rhomboid minor

Rhomboid major

Trapezius

Temporalis

Posterior auricular

Sternocleidomastoid

Splenius capitis

Nuchal ligament

Trapezius

Deltoid

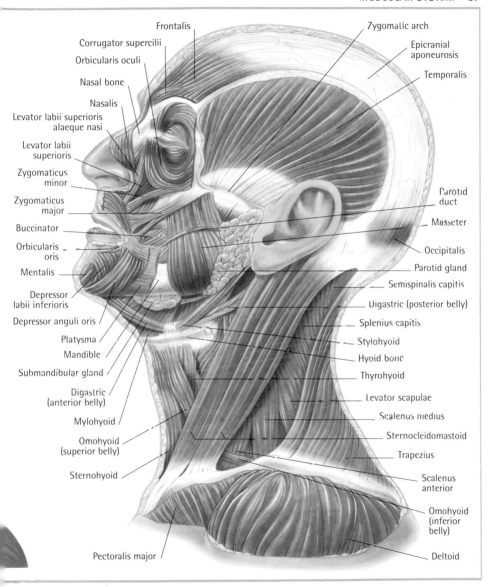

Frontalis
Corrugator supercilii
Orbicularis oculi
Nasal bone
Nasalis
Levator labii superioris alaeque nasi
Levator labii superioris
Zygomaticus minor
Zygomaticus major
Buccinator
Orbicularis oris
Mentalis
Depressor labii inferioris
Depressor anguli oris
Platysma
Mandible
Submandibular gland
Digastric (anterior belly)
Mylohyoid
Omohyoid (superior belly)
Sternohyoid
Pectoralis major

Zygomatic arch
Epicranial aponeurosis
Temporalis
Parotid duct
Masseter
Occipitalis
Parotid gland
Semispinalis capitis
Digastric (posterior belly)
Splenius capitis
Stylohyoid
Hyoid bone
Thyrohyoid
Levator scapulae
Scalenus medius
Sternocleidomastoid
Trapezius
Scalenus anterior
Omohyoid (inferior belly)
Deltoid

Facial expressions

Varying facial expressions are a significant means of communication, conveying a person's mood and emotions. The musculature involved is highly complex, allowing for many subtle nuances of expression. Facial muscles have their insertions (attachments to moving parts) within the skin; this means that even a slight degree of muscle contraction can produce movement of facial skin.

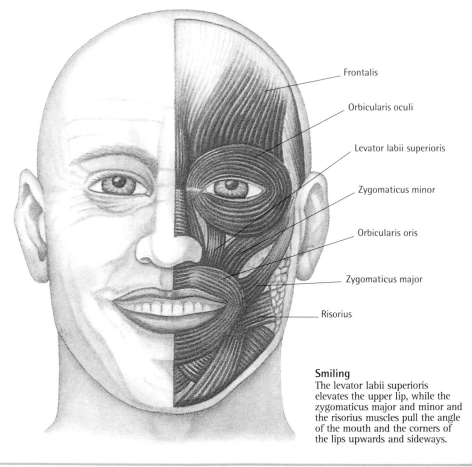

Frontalis

Orbicularis oculi

Levator labii superioris

Zygomaticus minor

Orbicularis oris

Zygomaticus major

Risorius

Smiling
The levator labii superioris elevates the upper lip, while the zygomaticus major and minor and the risorius muscles pull the angle of the mouth and the corners of the lips upwards and sideways.

Frowning

The frontalis and corrugator supercilii furrow the brow, the nasalis widens the nostrils, while the orbicularis oculi narrows the eye. The platysma and depressors pull the mouth and the corners of the lips downwards and sideways and the mentalis puckers the chin.

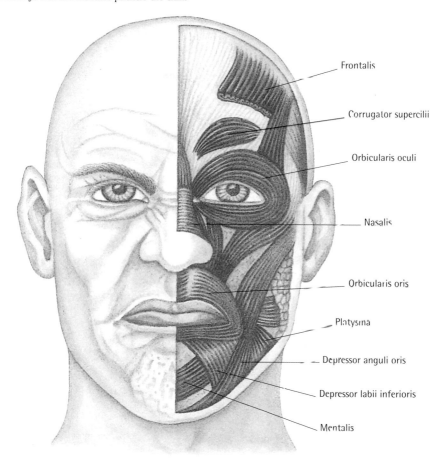

Frontalis

Corrugator supercilii

Orbicularis oculi

Nasalis

Orbicularis oris

Platysma

Depressor anguli oris

Depressor labii inferioris

Mentalis

Muscle-tendon links

Tendons are fibrous cords of connective tissue that link skeletal muscles to bones. Some tendons, especially those located in the hands and feet, are enclosed in self-lubricating sheaths that protect against friction as they move against bone. The tendons attached to the bones of the hand extend up the arm to their controlling muscles near the elbow.

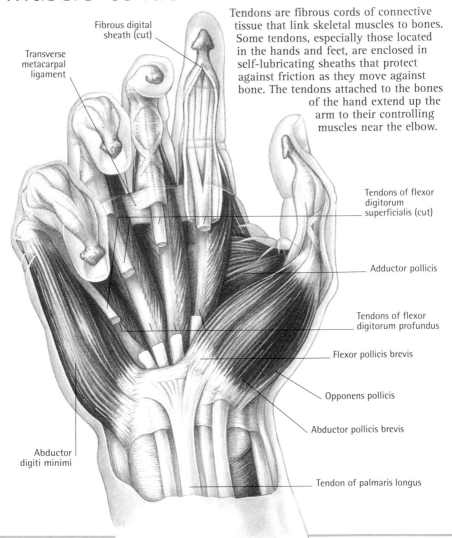

Fibrous digital sheath (cut)

Transverse metacarpal ligament

Tendons of flexor digitorum superficialis (cut)

Adductor pollicis

Tendons of flexor digitorum profundus

Flexor pollicis brevis

Opponens pollicis

Abductor pollicis brevis

Abductor digiti minimi

Tendon of palmaris longus

Tendons of the foot

In the foot, movement of the bones is controlled in a similar way to that in the hand. Long tendons reach down the foot from the muscles at the front and back of the leg, allowing the ankle and the hinge joints of the toes to be flexed (bent) and and extended (straightened).

Retinaculum (ligamentous band)

Peroneus tertius tendon

Tendon of extensor hallucis longus

Tendons of extensor digitorum longus

Tendon–bone links

Tendons are linked strongly to bone by Sharpey's fibers, which are also referred to as perforating fibers. These linking tissues are extensions of the collagen (protein) fibers that largely make up a tendon. They pass through the bone's surface membrane, known as the periosteum, and are embedded in the outer parts of the bone. The strong anchorage provided by Sharpey's fibers keeps the tendons very firmly attached to the bone, even during movement.

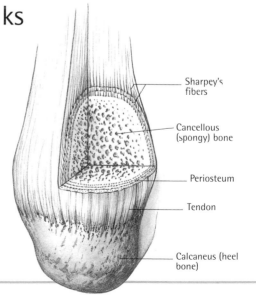

Sharpey's fibers

Cancellous (spongy) bone

Periosteum

Tendon

Calcaneus (heel bone)

Muscle structure

Skeletal muscles consist of densely packed groups of elongated cells, known as muscle fibers, held together by fibrous connective tissue. Numerous capillaries penetrate this tissue to keep muscles supplied with the abundant quantities of oxygen and glucose needed to fuel muscle contraction.

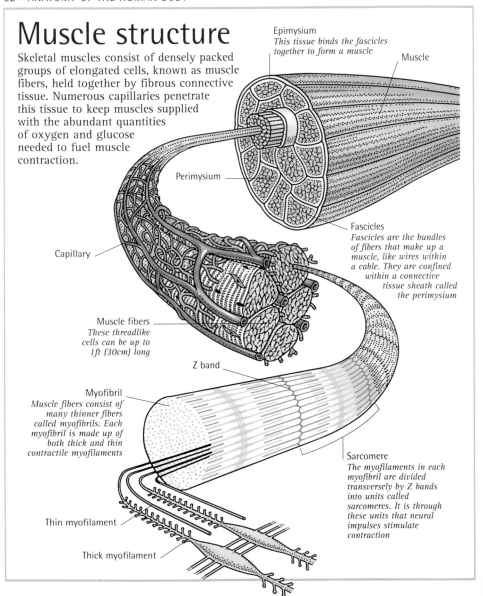

Epimysium
This tissue binds the fascicles together to form a muscle

Muscle

Perimysium

Fascicles
Fascicles are the bundles of fibers that make up a muscle, like wires within a cable. They are confined within a connective tissue sheath called the perimysium

Capillary

Muscle fibers
These threadlike cells can be up to 1ft (30cm) long

Z band

Myofibril
Muscle fibers consist of many thinner fibers called myofibrils. Each myofibril is made up of both thick and thin contractile myofilaments

Sarcomere
The myofilaments in each myofibril are divided transversely by Z bands into units called sarcomeres. It is through these units that neural impulses stimulate contraction

Thin myofilament

Thick myofilament

Muscle types

There are three types of muscle: skeletal, smooth, and cardiac muscle, also known as myocardium. Skeletal muscle is composed of bundles of long, striated fibers (cells). Smooth muscle, which is found in the walls of internal organs such as the intestines, is made of short spindle-shaped fibers packed together in layers. Cardiac muscle, found only in the heart, has short, interconnecting fibers.

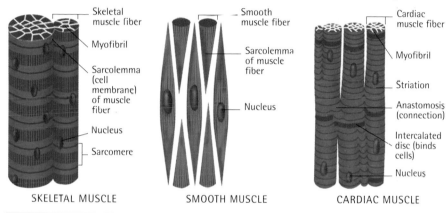

SKELETAL MUSCLE

SMOOTH MUSCLE

CARDIAC MUSCLE

How muscles contract

In a relaxed muscle, the thick and thin myofilaments (fine threads within a muscle fiber) overlap a little. When a muscle contracts, the thick filaments slide further in between the thin filaments like interlacing fingers. This action shortens the entire muscle fiber. The more shortened muscle fibers there are, the greater the contraction in the muscle as a whole.

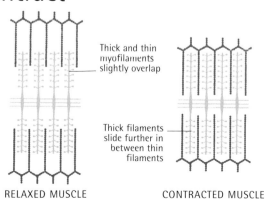

Thick and thin myofilaments slightly overlap

Thick filaments slide further in between thin filaments

RELAXED MUSCLE

CONTRACTED MUSCLE

Lever systems

Most bodily movements employ the mechanical principles by which a force applied to one part of a rigid lever arm is transferred via a pivot point, or fulcrum, to a weight elsewhere on the lever. In the body, muscles apply force, bones serve as levers, and joints function as fulcrums in order to move a body part.

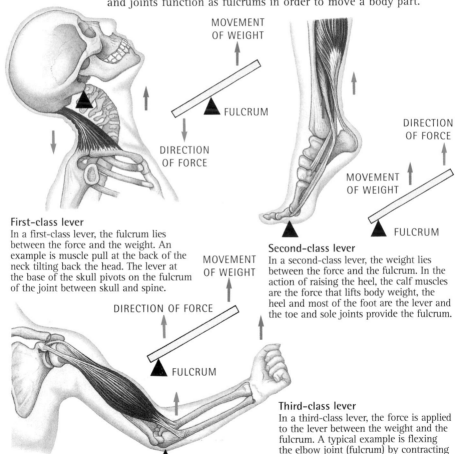

MOVEMENT OF WEIGHT

FULCRUM

DIRECTION OF FORCE

First-class lever

In a first-class lever, the fulcrum lies between the force and the weight. An example is muscle pull at the back of the neck tilting back the head. The lever at the base of the skull pivots on the fulcrum of the joint between skull and spine.

DIRECTION OF FORCE

MOVEMENT OF WEIGHT

FULCRUM

Second-class lever

In a second-class lever, the weight lies between the force and the fulcrum. In the action of raising the heel, the calf muscles are the force that lifts body weight, the heel and most of the foot are the lever and the toe and sole joints provide the fulcrum.

MOVEMENT OF WEIGHT

DIRECTION OF FORCE

FULCRUM

Third-class lever

In a third-class lever, the force is applied to the lever between the weight and the fulcrum. A typical example is flexing the elbow joint (fulcrum) by contracting the biceps to lift the forearm and hand.

Stabilizing muscles

Muscles in the neck and upper back provide strength and support. Those in the neck help to bear the weight of the head and hold it upright. Upper back muscles that attach to the triangular scapula (shoulder blade) help to stabilize the shoulder, which has a greater range of movement than any other joint in the body. In the illustration, superficial muscles are shown on the right, deep muscles are shown on the left.

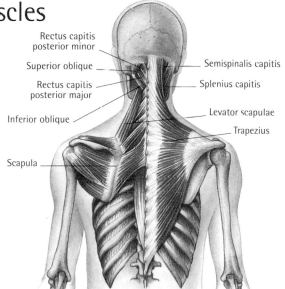

Rectus capitis posterior minor
Superior oblique
Rectus capitis posterior major
Inferior oblique
Scapula
Semispinalis capitis
Splenius capitis
Levator scapulae
Trapezius

Muscles work together

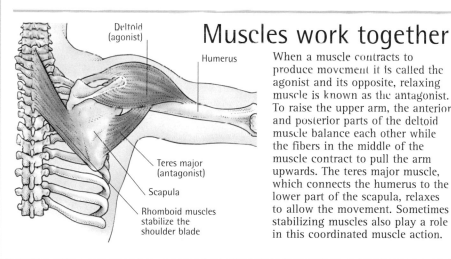

Deltoid (agonist)
Humerus
Teres major (antagonist)
Scapula
Rhomboid muscles stabilize the shoulder blade

When a muscle contracts to produce movement it is called the agonist and its opposite, relaxing muscle is known as the antagonist. To raise the upper arm, the anterior and posterior parts of the deltoid muscle balance each other while the fibers in the middle of the muscle contract to pull the arm upwards. The teres major muscle, which connects the humerus to the lower part of the scapula, relaxes to allow the movement. Sometimes stabilizing muscles also play a role in this coordinated muscle action.

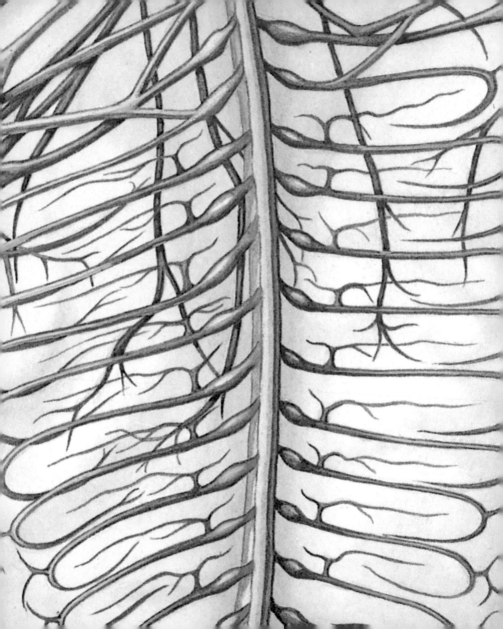

Nervous system

The brain and body are constantly alive with billions of electrical and chemical signals. This incessant activity emanates from neurons, or nerve cells, and their far-reaching fibers. The neurons in the brain and spinal cord make up the central nervous system, which is connected to the rest of the body by the peripheral nerves.

Nerves
A nerve network spreads around the entire body. The nerves that emerge from the spinal cord (left) supply the trunk and limbs.

Nervous system

The nervous system has two main divisions. The central nervous system (CNS) is made up of the brain and spinal cord. Nerve fibers branching out into the body from the CNS form the peripheral nervous system (PNS). These peripheral nerves constantly send information to the CNS, which processes it and sends signals back to the PNS. Some nerve fibers in the PNS form into groups to keep important areas under fine control.

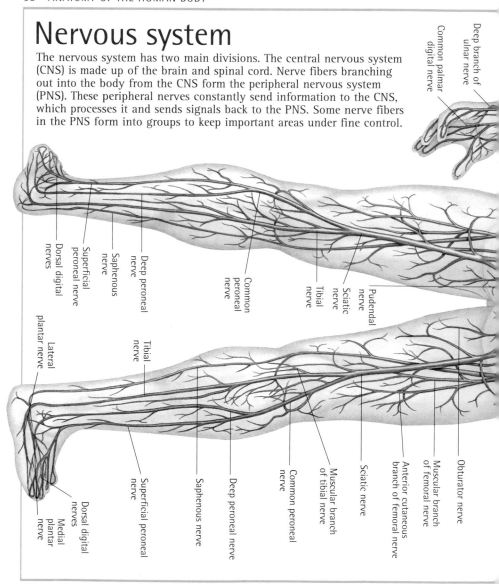

Deep branch of ulnar nerve

Common palmar digital nerve

Dorsal digital nerves

Superficial peroneal nerve

Saphenous nerve

Deep peroneal nerve

Common peroneal nerve

Tibial nerve

Sciatic nerve

Pudendal nerve

Lateral plantar nerve

Tibial nerve

Medial plantar nerve

Dorsal digital nerves

Superficial peroneal nerve

Saphenous nerve

Deep peroneal nerve

Common peroneal nerve

Muscular branch of tibial nerve

Sciatic nerve

Anterior cutaneous branch of femoral nerve

Muscular branch of femoral nerve

Obturator nerve

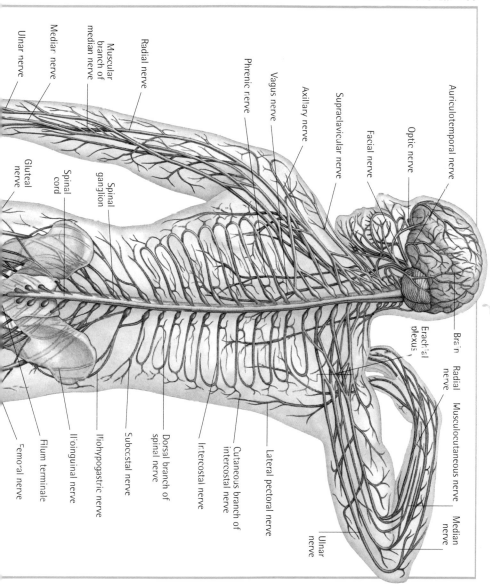

Ulnar nerve

Median nerve

Muscular branch of median nerve

Radial nerve

Phrenic nerve

Vagus nerve

Axillary nerve

Supraclavicular nerve

Facial nerve

Optic nerve

Auriculotemporal nerve

Gluteal nerve

Spinal cord

Spinal ganglion

Brain

Brachial plexus

Radial nerve

Musculocutaneous nerve

Median nerve

Ulnar nerve

Femoral nerve

Filum terminale

Ilioinguinal nerve

Iliohypogastric nerve

Subcostal nerve

Dorsal branch of spinal nerve

Intercostal nerve

Cutaneous branch of intercostal nerve

Lateral pectoral nerve

Peripheral nervous system

The peripheral nervous system has three divisions: autonomic, sensory, and motor. The autonomic nervous system (blue), controls involuntary actions; this system has two divisions, sympathetic and parasympathetic. Sensory nerves (red) transmit information from around the body to the central nervous system. Motor nerves (purple) carry signals from the brain to voluntary skeletal muscles.

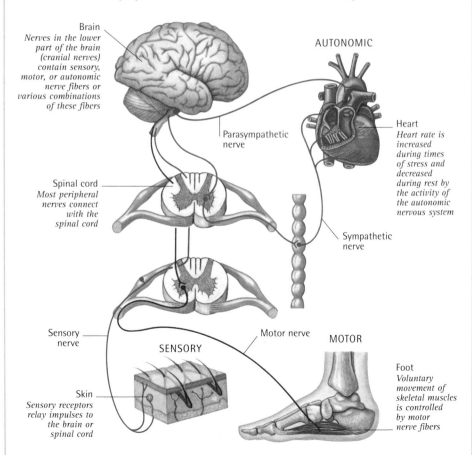

Brain
Nerves in the lower part of the brain (cranial nerves) contain sensory, motor, or autonomic nerve fibers or various combinations of these fibers

AUTONOMIC

Parasympathetic nerve

Heart
Heart rate is increased during times of stress and decreased during rest by the activity of the autonomic nervous system

Spinal cord
Most peripheral nerves connect with the spinal cord

Sympathetic nerve

Sensory nerve

Motor nerve

MOTOR

SENSORY

Foot
Voluntary movement of skeletal muscles is controlled by motor nerve fibers

Skin
Sensory receptors relay impulses to the brain or spinal cord

Nerve structure

Most nerves are formed from bundles of nerve fibers, called fascicles, which are bound together by tissue. The nerves travel to a particular site in the body and carry two types of fiber: sensory (afferent) fibers, which convey impulses from receptors in the skin, sense organs, and internal organs back to the brain and spinal cord; and motor (efferent) fibers, which transmit signals from the brain and spinal cord to a muscle or gland.

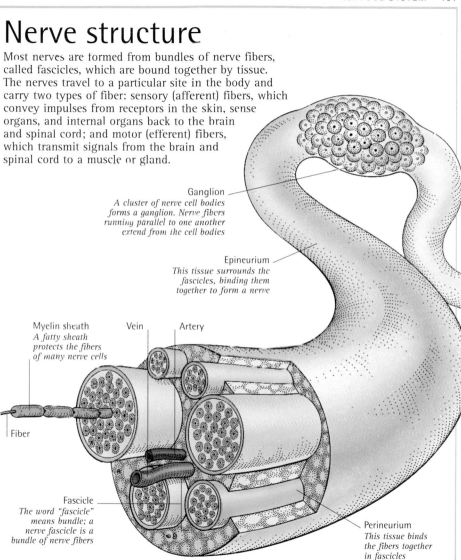

Ganglion
A cluster of nerve cell bodies forms a ganglion. Nerve fibers running parallel to one another extend from the cell bodies

Epineurium
This tissue surrounds the fascicles, binding them together to form a nerve

Myelin sheath
A fatty sheath protects the fibers of many nerve cells

Vein

Artery

Fiber

Fascicle
The word "fascicle" means bundle; a nerve fascicle is a bundle of nerve fibers

Perineurium
This tissue binds the fibers together in fascicles

Neurons

The basic unit of the nervous system is a special cell called a neuron, or nerve cell. This has a cell body with a central nucleus and various other structures that are important for maintaining cell life.

Neurons have long projections, or processes, known as axons (nerve fibers) and dendrites. Axons carry nerve impulses away from the cell; dendrites receive impulses from other neurons.

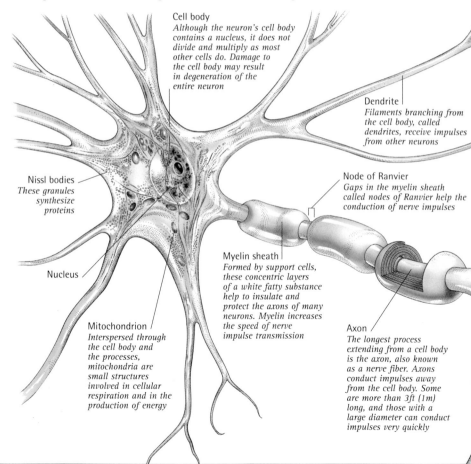

Cell body
Although the neuron's cell body contains a nucleus, it does not divide and multiply as most other cells do. Damage to the cell body may result in degeneration of the entire neuron

Dendrite
Filaments branching from the cell body, called dendrites, receive impulses from other neurons

Nissl bodies
These granules synthesize proteins

Node of Ranvier
Gaps in the myelin sheath called nodes of Ranvier help the conduction of nerve impulses

Nucleus

Myelin sheath
Formed by support cells, these concentric layers of a white fatty substance help to insulate and protect the axons of many neurons. Myelin increases the speed of nerve impulse transmission

Mitochondrion
Interspersed through the cell body and the processes, mitochondria are small structures involved in cellular respiration and in the production of energy

Axon
The longest process extending from a cell body is the axon, also known as a nerve fiber. Axons conduct impulses away from the cell body. Some are more than 3ft (1m) long, and those with a large diameter can conduct impulses very quickly

Neuron types

According to their function and location in the body, the shape and size of neuron cell bodies vary greatly, as do the type, number, and length of their processes. Illustrated right are three main types of neuron: unipolar, bipolar, and multipolar.

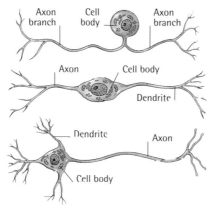

Axon branch | Cell body | Axon branch

Axon | Cell body | Dendrite

Dendrite | Axon | Cell body

Unipolar
These have a projection that divides into one axon; they are usually sensory neurons.

Bipolar
This type, which is found in the retina and the inner ear, has one axon and one dendrite.

Multipolar
Most neurons in the brain and spinal cord are multipolar. They have an axon and several dendrites.

Synaptic knob
These terminal knobs contain vesicles, or sacs, of chemicals called neurotransmitters that transmit impulses from one cell to another

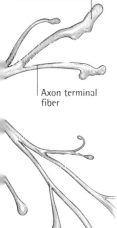

Axon terminal fiber

Support cells

Several types of support cells, or glial cells, act to protect and give structural support to the neurons. The smallest (microglia) destroy microorganisms; others help to insulate axons or regulate the flow of cerebrospinal fluid.

Astrocytes
Delicate processes of cytoplasm extend from these starlike cells. Some cell processes connect with capillaries and help to regulate the flow of substances between neurons and the blood.

Oligodendrocytes
These cells wrap their plasma membranes around neurons of the brain and spinal cord to form myelin sheaths.

Neuron behavior

Neurons must be triggered by a stimulus to produce nerve impulses, which are waves of electrical charge moving along the nerve fibers. When the neuron receives a stimulus, the electrical charge on the inside of the cell membrane changes from negative to positive. A nerve impulse travels down the fiber to a synaptic knob at its end, triggering the release of chemicals (neurotransmitters) that cross the gap between the neuron and the target cell, stimulating a response in the target.

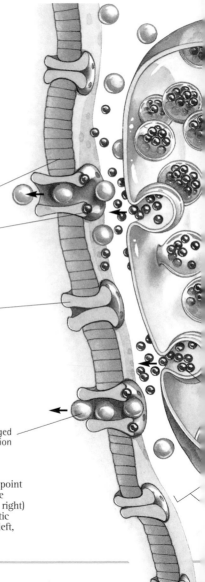

Target cell membrane

Receptor site
The neurotransmitter combines with protein receptors on the target cell membrane, which then becomes permeable to specific ions, such as positively charged sodium ions

Membrane channel
Channels in the target cell membrane become permeable to the sodium ions, which then pass through. This makes the charge inside the membrane change from negative to positive

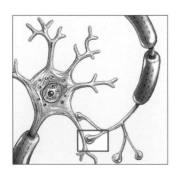

Positively charged sodium ion

Synapse
The communication point between neurons (the synapse, enlarged at right) comprises the synaptic knob, the synaptic cleft, and the target site.

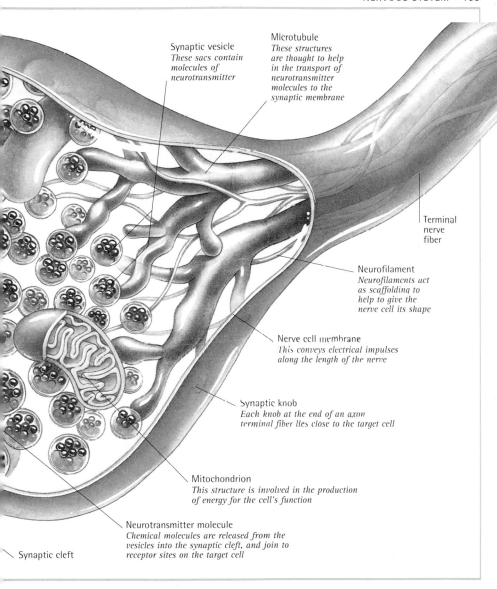

Synaptic vesicle
These sacs contain molecules of neurotransmitter

Microtubule
These structures are thought to help in the transport of neurotransmitter molecules to the synaptic membrane

Terminal nerve fiber

Neurofilament
Neurofilaments act as scaffolding to help to give the nerve cell its shape

Nerve cell membrane
This conveys electrical impulses along the length of the nerve

Synaptic knob
Each knob at the end of an axon terminal fiber lies close to the target cell

Mitochondrion
This structure is involved in the production of energy for the cell's function

Neurotransmitter molecule
Chemical molecules are released from the vesicles into the synaptic cleft, and join to receptor sites on the target cell

Synaptic cleft

Sending a nerve impulse

The level at which a stimulus is strong enough to transmit a nerve impulse is called the threshold. When the threshold is reached, an impulse travels along the entire length of the nerve fiber. The speed of transmission can vary: fibers that are cold (for example, when ice is applied to dull pain), those with small diameters, and those without myelin sheaths conduct impulses more slowly.

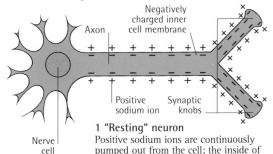

Axon
Negatively charged inner cell membrane
Positive sodium ion
Synaptic knobs
Nerve cell body

1 "Resting" neuron
Positive sodium ions are continuously pumped out from the cell; the inside of the cell membrane is negatively charged.

2 Nerve impulse triggered
Stimulated by an impulse, the cell membrane allows positive sodium ions to rush into the cell. At these local sites, the inside of the membrane becomes positively charged.

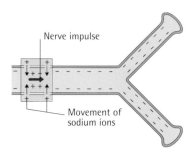

Nerve impulse

Movement of sodium ions

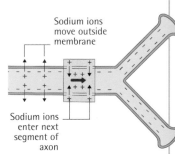

Sodium ions move outside membrane

Sodium ions enter next segment of axon

3 Impulse continues
As the impulse moves along the axon, new segments become positively charged while those that were previously positive return to a negative state.

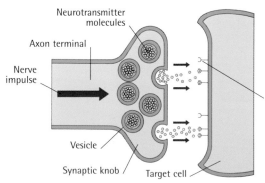

Neurotransmitter molecules
Axon terminal
Nerve impulse
Vesicle
Synaptic knob
Receptor
Target cell

4 Target cell reached
The impulse reaches the synaptic knob, triggering the release of neurotransmitter molecules from the vesicles. The molecules cross the synaptic cleft to activate a response in the target cell.

Inhibition

Neurotransmitters can inhibit nerve signals in some areas of the body, as well as stimulate them. When this happens, channels in a target cell membrane do not open to let sodium ions enter. Instead, the neurotransmitters open channels that either allow positive potassium ions to escape from the cell or let negative chloride ions enter. In both cases, the electrical charge inside the cell membrane stays negative and the neuron cannot be fired.

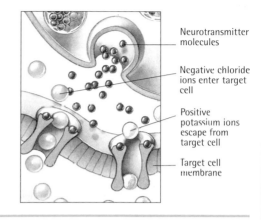

Neurotransmitter molecules

Negative chloride ions enter target cell

Positive potassium ions escape from target cell

Target cell membrane

Regeneration

Peripheral nerve fibers that are crushed or even cut may slowly regenerate if the cell body is undamaged, especially if the connective tissue in which the nerve travels remains intact. Regeneration does not occur in nerves in the brain or spinal cord; these nerve fibers become wrapped in scar tissue.

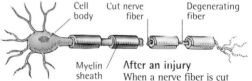

Cell body

Cut nerve fiber

Degenerating fiber

Myelin sheath

After an injury
When a nerve fiber is cut off from the cell body by an injury, the part beyond the cut degenerates. The myelin also degenerates.

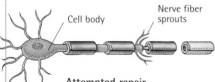

Cell body

Nerve fiber sprouts

Attempted repair
The cell body stimulates the growth of nerve sprouts from the remaining fiber; one of these may join up with its original connection.

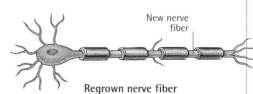

New nerve fiber

Regrown nerve fiber
Growing 0.08–0.16in (2–4mm) a day, the new nerve fiber reaches its previous connection. Function and sensation are slowly restored.

Outer brain

The brain contains over 12 billion neurons and 50 billion supporting glial cells. With the spinal cord, the brain regulates bodily processes and coordinates voluntary movements. Its largest part is the cerebrum, which is divided into four paired lobes: frontal, temporal, occipital, and parietal. The heavily folded surface of the cerebrum has a different pattern of grooves in each person; deep grooves are known as fissures and shallow grooves are called sulci. Ridges on the surface of the brain are called gyri.

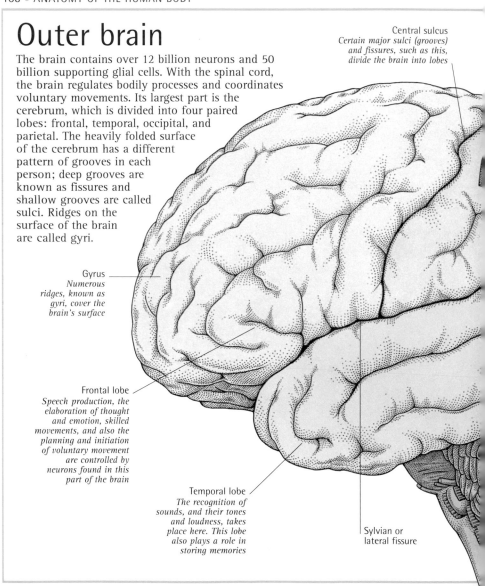

Central sulcus
Certain major sulci (grooves) and fissures, such as this, divide the brain into lobes

Gyrus
Numerous ridges, known as gyri, cover the brain's surface

Frontal lobe
Speech production, the elaboration of thought and emotion, skilled movements, and also the planning and initiation of voluntary movement are controlled by neurons found in this part of the brain

Temporal lobe
The recognition of sounds, and their tones and loudness, takes place here. This lobe also plays a role in storing memories

Sylvian or lateral fissure

Cerebral cortex
Higher intellectual functions, such as interpretation of sensory impulses, are carried out by the complex network of neurons that makes up the brain's outer layer of grey matter

Parietal lobe
Bodily sensations, such as touch, temperature, pressure, and pain, are perceived and interpreted in this area

Halves of the brain

A longitudinal fissure (shown from above) divides the brain into two halves, known as the left and right cerebral hemispheres. These halves are interconnected by a pathway of nerve fibres along which information continuously passes.

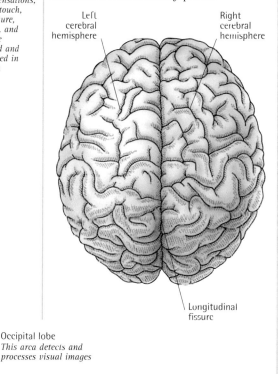

Left cerebral hemisphere

Right cerebral hemisphere

Longitudinal fissure

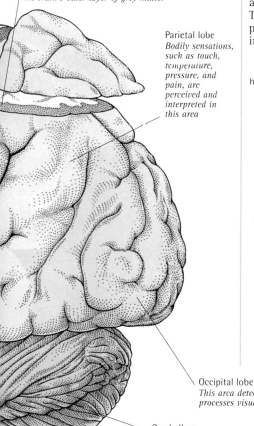

Occipital lobe
This area detects and processes visual images

Cerebellum
This is the second largest part of the brain. Its neurons link up with other regions of the brain and the spinal cord to facilitate smooth, precise movements, and to control balance and posture

Inner brain structures

Beneath the cerebral cortex are several structures. One of these is the thalamus, which lies in the centre of the brain and acts as an information relay station. Surrounding this is a group of other structures known as the limbic system (illustrated in detail on p.136), which plays a role in survival behavior, memory, and emotions such as rage and fright. Closely linked with the limbic system is the hypothalamus, a region that controls automatic body processes.

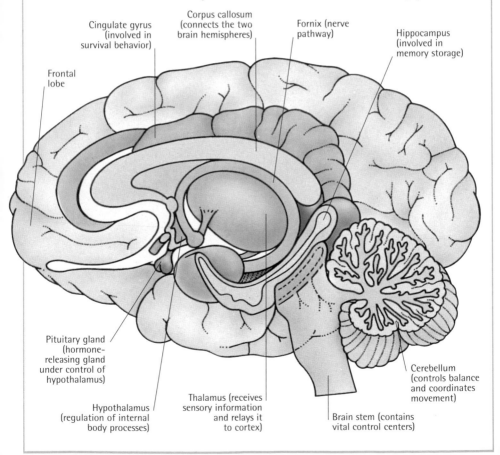

Cingulate gyrus (involved in survival behavior)

Corpus callosum (connects the two brain hemispheres)

Fornix (nerve pathway)

Hippocampus (involved in memory storage)

Frontal lobe

Pituitary gland (hormone-releasing gland under control of hypothalamus)

Hypothalamus (regulation of internal body processes)

Thalamus (receives sensory information and relays it to cortex)

Brain stem (contains vital control centers)

Cerebellum (controls balance and coordinates movement)

Gray and white matter

The entire cerebrum is covered by a layer of gray matter (the cerebral cortex), about 0.08–0.24in (2–6mm) thick, which is made up of groups of nerve cell bodies. Underneath this thin covering is the brain's white matter, with islands of more gray matter. White matter is composed mainly of myelin-covered axons, or nerve fibers, that extend from the nerve cell bodies. The fatty, insulating myelin sheaths act to increase the speed of the transmission of nerve impulses.

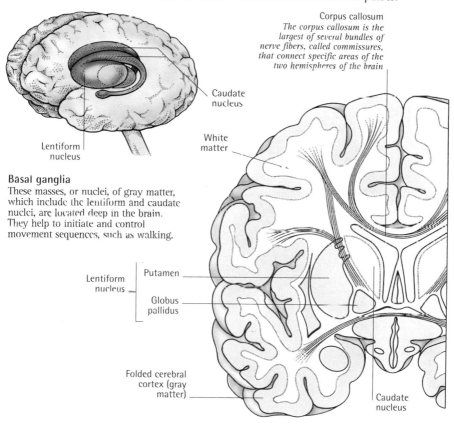

Corpus callosum
The corpus callosum is the largest of several bundles of nerve fibers, called commissures, that connect specific areas of the two hemispheres of the brain

Caudate nucleus

White matter

Lentiform nucleus

Basal ganglia
These masses, or nuclei, of gray matter, which include the lentiform and caudate nuclei, are located deep in the brain. They help to initiate and control movement sequences, such as walking.

Lentiform nucleus {
Putamen

Globus pallidus
}

Folded cerebral cortex (gray matter)

Caudate nucleus

Vertical links

Nerve impulses passing backwards and forwards between the spinal cord, the lower areas of the brain, and the cerebral cortex are transmitted by myelinated nerve fibers organized into tracts. These nerve tracts, known as projection fibers, pass through the upper part of the brain stem and then through a communication link called the internal capsule, which is a compact band of fibers. The projection fibers then intersect the corpus callosum (the band of fibers connecting the hemispheres of the brain), fanning out to form the corona radiata.

Cerebral cortex
This gray matter forms the outer layer of the brain

Corona radiata
The projection fibers spread out in a fan as they travel to the cerebral cortex

Cerebrum

Cranial nerves

Brain stem

Brain stem

Spinal cord

MRI scan of brain
This MRI scan shows the brain stem, a site of vital control centers and also a vertical nerve pathway along which impulses pass between the spinal cord and the brain.

White matter
*The brain's white
matter mainly
comprises myelinated
nerve fibers*

Internal capsule
*Nerve fibers in the
internal capsule
form a dense band*

Information centers

The thalamus is an area of gray matter that acts as
a relay station for sensory nerve signals between the
cerebrum, the brain stem, and the spinal cord. The
brain stem contains centers that regulate several
functions vital for survival, including heartbeat,
respiration, blood pressure, digestion, and certain
reflex actions such as swallowing and vomiting.

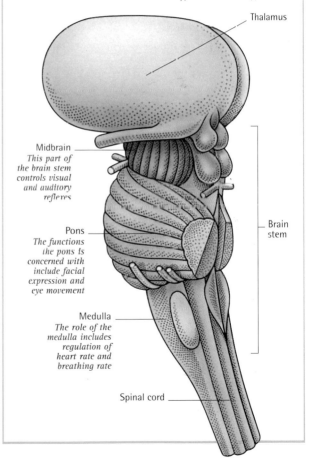

Thalamus

Midbrain
*This part of
the brain stem
controls visual
and auditory
reflexes*

Pons
*The functions
the pons is
concerned with
include facial
expression and
eye movement*

Medulla
*The role of the
medulla includes
regulation of
heart rate and
breathing rate*

Brain
stem

Spinal cord

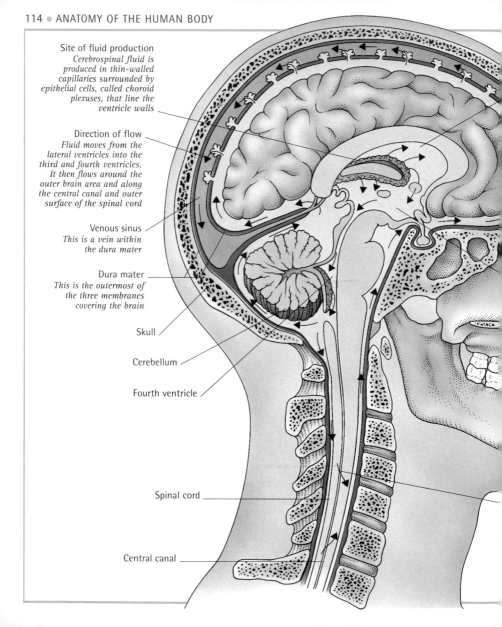

Site of fluid production
Cerebrospinal fluid is produced in thin-walled capillaries surrounded by epithelial cells, called choroid plexuses, that line the ventricle walls

Direction of flow
Fluid moves from the lateral ventricles into the third and fourth ventricles. It then flows around the outer brain area and along the central canal and outer surface of the spinal cord

Venous sinus
This is a vein within the dura mater

Dura mater
This is the outermost of the three membranes covering the brain

Skull

Cerebellum

Fourth ventricle

Spinal cord

Central canal

Lateral ventricle

Cerebrospinal fluid

The soft tissues of the brain and spinal cord are protected by the bony casings of the skull and vertebrae; for additional protection, the tissues are surrounded by a clear, watery fluid called cerebrospinal fluid. This liquid is produced inside the ventricles (chambers) of the brain and is renewed 3–4 times a day. It cushions tissues against excessive blows, and also helps to support the weight of the brain, reducing stress on blood vessels and nerves. The fluid contains glucose to nourish the brain cells.

Site of reabsorption (arachnoid granulations)
After circulating, cerebrospinal fluid is reabsorbed into blood through arachnoid granulations, which are projections from the arachnoid layer of the membranes (meninges) that cover the brain

Third ventricle

Fluid circulation around the spinal cord
Aided by vertebral movement, fluid flows along the central canal and outer surface of the spinal cord

Circulation of fluid

Cerebrospinal fluid produced in the lateral ventricles of the brain drains into the third ventricle via a hole called the interventricular foramen. It then flows through the cerebral aqueduct (canal) and into the fourth ventricle.

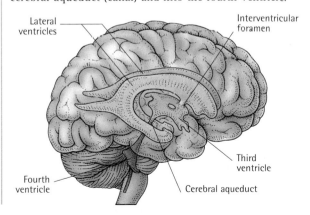

Lateral ventricles

Interventricular foramen

Third ventricle

Cerebral aqueduct

Fourth ventricle

Meninges

Three membranes known as the meninges cover the brain. Lining the inside of the skull is the outermost membrane, the dura mater, which contains veins and arteries that carry the blood supply for the cranial bones. The middle layer is the arachnoid ("spiderlike"). This consists of a membrane from which webbed connective tissue projects into the subarachnoid space, an area containing cerebrospinal fluid as well as blood vessels. Next to the cerebral cortex is the delicate, innermost membrane, known as the pia mater.

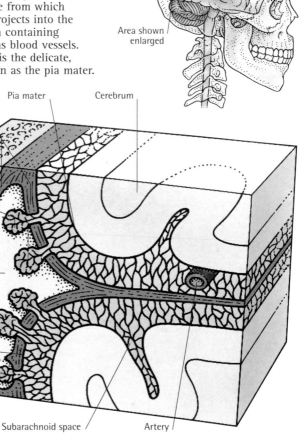

Area shown enlarged

Pia mater

Cerebrum

Dura mater

Arachnoid granulations (projections of arachnoid layer and subarachnoid space through the dura mater)

Venous sinus

Arachnoid

Bone of skull

Subarachnoid space

Artery

Blood supply to the brain

Although the brain accounts for only about 2 percent of the total weight of the body, it requires nearly 20 percent of the body's blood supply. Oxygen and glucose carried by blood are essential to the brain; without a continuous supply, brain function deteriorates rapidly. Oxygenated blood is carried by arteries to all parts of the brain. Deoxygenated blood returning to the heart leaves the brain via a network of veins that drain into channels called sinuses.

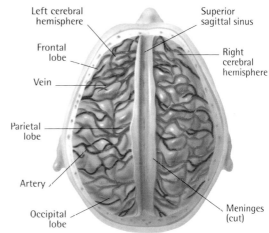

Left cerebral hemisphere

Frontal lobe

Vein

Parietal lobe

Artery

Occipital lobe

Superior sagittal sinus

Right cerebral hemisphere

Meninges (cut)

Blood–brain barrier

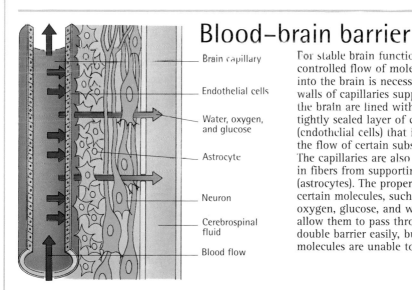

Brain capillary

Endothelial cells

Water, oxygen, and glucose

Astrocyte

Neuron

Cerebrospinal fluid

Blood flow

For stable brain function, a controlled flow of molecules into the brain is necessary. The walls of capillaries supplying the brain are lined with a tightly sealed layer of cells (endothelial cells) that impede the flow of certain substances. The capillaries are also wrapped in fibers from supporting cells (astrocytes). The properties of certain molecules, such as oxygen, glucose, and water, allow them to pass through this double barrier easily, but many molecules are unable to pass.

Brain development

Brain growth, one of the most important parts of embryonic life, occurs very rapidly. From small clusters of tissue, highly specialized areas of brain function emerge. Most of the major development of the brain and nervous system occurs during the early weeks of an embryo's life.

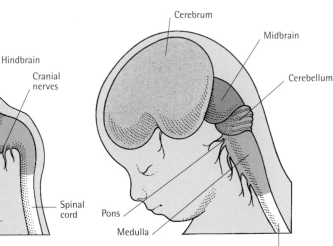

At 4 weeks
A tube of neural tissue has developed along the back of the embryo. Three bulges, called the primary vesicles, that have formed in the top end of this tube will become the main divisions of the brain (forebrain, midbrain, and hindbrain).

At 6 weeks
The neural tube is flexed, and cranial nerves have appeared. Two bulges formed on the forebrain will develop into the two hemispheres of the cerebrum, the largest part of the fully formed brain.

At 11 weeks
The developing cerebellum has begun to grow out from the hindbrain. Meanwhile, the two hemispheres of the cerebrum have been growing rapidly, and are starting to expand over the hindbrain.

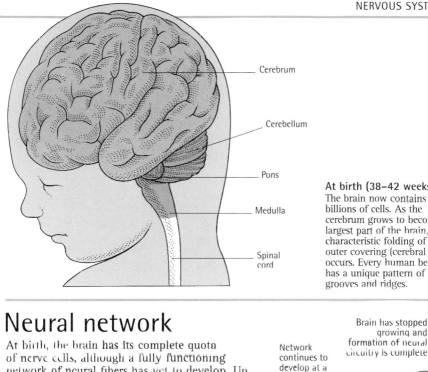

Cerebrum

Cerebellum

Pons

Medulla

Spinal cord

At birth (38–42 weeks)
The brain now contains billions of cells. As the cerebrum grows to become the largest part of the brain, the characteristic folding of the outer covering (cerebral cortex) occurs. Every human being has a unique pattern of grooves and ridges.

Neural network

At birth, the brain has its complete quota of nerve cells, although a fully functioning network of neural fibers has yet to develop. Up to the age of six, brain growth and development of the neural network is rapid; growth then progresses more slowly until the brain is fully mature.

Network continues to develop at a slower rate

Brain has stopped growing and formation of neural circuitry is complete

Nerve network will expand rapidly

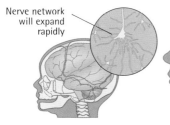

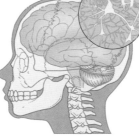

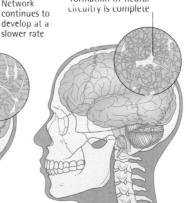

AT BIRTH

AT 6 YEARS

AT 18 YEARS

Spinal cord

The spinal cord is a cable about 17in (43cm) in length running from the brain stem to the lower back. It contains two types of tissue. The inner core is gray matter made up of nerve cell bodies, unmyelinated axons, glial cells, and blood vessels; the outer white matter is mainly composed of tracts of myelinated axons that relay impulses to and from the spinal cord and the brain.

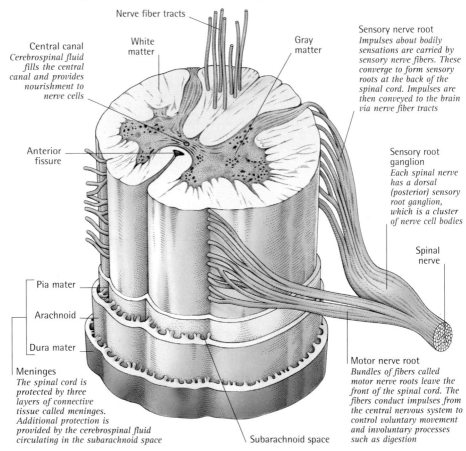

Nerve fiber tracts

Central canal
Cerebrospinal fluid fills the central canal and provides nourishment to nerve cells

White matter

Gray matter

Sensory nerve root
Impulses about bodily sensations are carried by sensory nerve fibers. These converge to form sensory roots at the back of the spinal cord. Impulses are then conveyed to the brain via nerve fiber tracts

Anterior fissure

Sensory root ganglion
Each spinal nerve has a dorsal (posterior) sensory root ganglion, which is a cluster of nerve cell bodies

Spinal nerve

Pia mater

Arachnoid

Dura mater

Meninges
The spinal cord is protected by three layers of connective tissue called meninges. Additional protection is provided by the cerebrospinal fluid circulating in the subarachnoid space

Subarachnoid space

Motor nerve root
Bundles of fibers called motor nerve roots leave the front of the spinal cord. The fibers conduct impulses from the central nervous system to control voluntary movement and involuntary processes such as digestion

Extent of spinal cord

During the body's growth and development, the spinal cord does not continue to lengthen the way that the spinal bones do; it therefore occupies only the first two-thirds of the vertebral column. Cylindrical and slightly flattened, the spinal cord is about as thick as a finger for most of its length; extending from the lower part of the spinal cord is a cluster of individual nerve fibers called the cauda equina.

Spinal nerves

Through 31 pairs of spinal nerves, the spinal cord is connected to the rest of the body; via these nerves, the spinal cord sends and receives information to and from the brain. The spinal nerves run through gaps between adjacent vertebrae and enter the back and front of the spinal cord as spinal nerve roots.

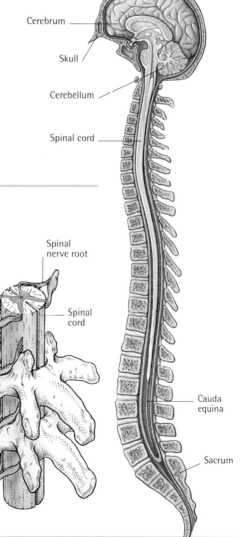

Cerebrum

Skull

Cerebellum

Spinal cord

Spinal nerve root

Spinal cord

Vertebra

Spinal nerve

Cauda equina

Sacrum

Protection of the spinal cord

The spinal cord is protected primarily by the bony segments of the vertebral column and its supporting ligaments. Further protection is provided by cerebrospinal fluid, which acts as a shock absorber, and the epidural space, a cushioning area of fat and connective tissue that lies in between the periosteum (the membrane that covers the vertebral bones) and the dura mater, the outer layer of the meninges (membranes covering the spinal cord).

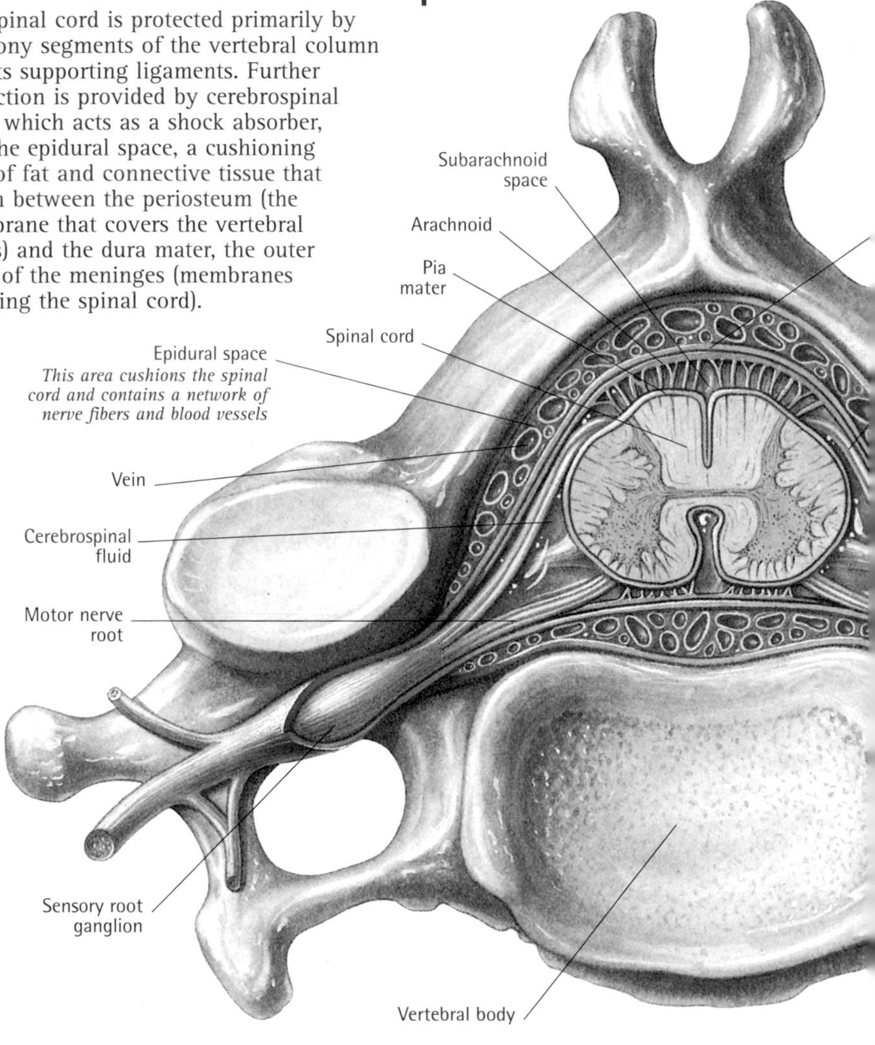

Subarachnoid space

Arachnoid

Pia mater

Spinal cord

Epidural space
This area cushions the spinal cord and contains a network of nerve fibers and blood vessels

Vein

Cerebrospinal fluid

Motor nerve root

Sensory root ganglion

Vertebral body

Fiber tracts of the spinal cord

Myelinated nerve fibers in the spinal cord are grouped together according to the direction and the type of impulse they transmit or receive. Some of these tracts connect and relay impulses between just a few pairs of spinal nerves. The gray matter of the spinal cord is organized into areas called horns.

Dura mater
The dura mater is the outer of the three membranes covering the spinal cord

Sensory nerve root

Periosteum
This thin membrane covers the surface of the vertebra

Dorsal (back) horns
Neuron cell bodies here receive information from sensory nerve fibers all around the body about sensations, including touch, temperature, awareness of muscle activity, and balance

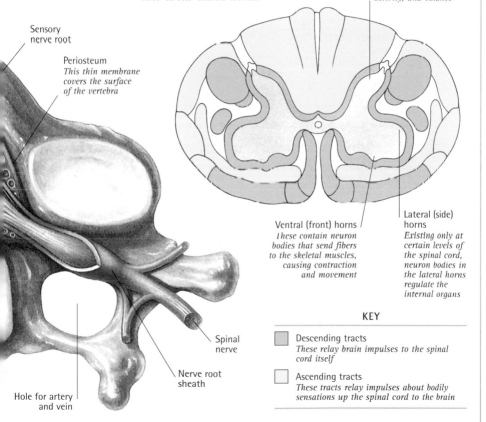

Ventral (front) horns
These contain neuron bodies that send fibers to the skeletal muscles, causing contraction and movement

Lateral (side) horns
Existing only at certain levels of the spinal cord, neuron bodies in the lateral horns regulate the internal organs

Spinal nerve

Nerve root sheath

Hole for artery and vein

KEY

☐ **Descending tracts**
These relay brain impulses to the spinal cord itself

☐ **Ascending tracts**
These tracts relay impulses about bodily sensations up the spinal cord to the brain

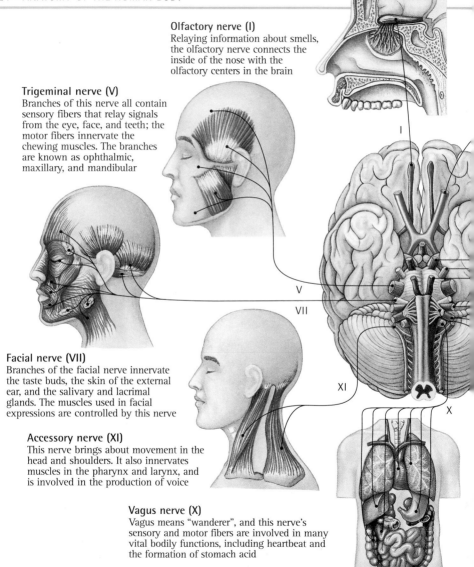

Olfactory nerve (I)
Relaying information about smells, the olfactory nerve connects the inside of the nose with the olfactory centers in the brain

Trigeminal nerve (V)
Branches of this nerve all contain sensory fibers that relay signals from the eye, face, and teeth; the motor fibers innervate the chewing muscles. The branches are known as ophthalmic, maxillary, and mandibular

Facial nerve (VII)
Branches of the facial nerve innervate the taste buds, the skin of the external ear, and the salivary and lacrimal glands. The muscles used in facial expressions are controlled by this nerve

Accessory nerve (XI)
This nerve brings about movement in the head and shoulders. It also innervates muscles in the pharynx and larynx, and is involved in the production of voice

Vagus nerve (X)
Vagus means "wanderer", and this nerve's sensory and motor fibers are involved in many vital bodily functions, including heartbeat and the formation of stomach acid

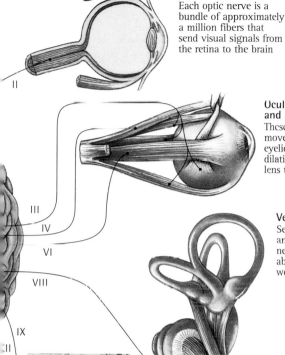

Optic nerve (II)
Each optic nerve is a bundle of approximately a million fibers that send visual signals from the retina to the brain

Oculomotor (III), trochlear (IV), and abducent nerves (VI)
These three nerves regulate voluntary movements of the eye muscles and eyelids. They also control pupil dilation and adjust the shape of the lens to focus on near objects

Vestibulocochlear nerve (VIII)
Sensory fibers in the vestibular and cochlear branches of this nerve transmit information about sound and balance as well as orientation of the head

Glossopharyngeal (IX) and hypoglossal nerves (XII)
Motor fibers of these nerves are involved in swallowing, while the sensory fibers relay information about taste, touch, and heat from both the tongue and the pharynx

Cranial nerves

On the undersurface of the brain, 12 pairs of cranial nerves (I–XII) emerge from the central nervous system (CNS) to form part of the peripheral nervous system (PNS). The cranial nerves perform sensory and/or motor functions, mainly in the head and neck region. The nine nerves with motor fibers also contain sensory fibers; these sensory fibers, which are known as proprioceptors, relay information to the CNS about muscle tension and the balance of the body.

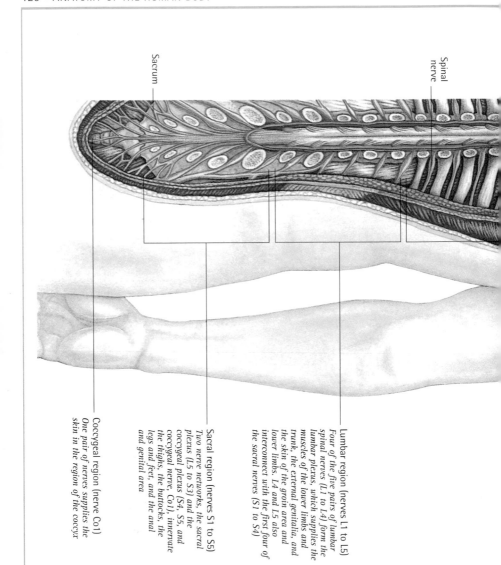

Sacrum

Spinal nerve

Lumbar region (nerves L1 to L5)
Four of the five pairs of lumbar spinal nerves (L1 to L4) form the lumbar plexus, which supplies the muscles of the lower limbs and trunk, the external genitalia, and the skin of the groin area and lower limbs. L4 and L5 also interconnect with the first four of the sacral nerves (S1 to S4)

Sacral region (nerves S1 to S5)
Two nerve networks, the sacral plexus (L5 to S3) and the coccygeal plexus (S4, S5, and coccygeal nerve, Co1), innervate the thighs, the buttocks, the legs and feet, and the anal and genital area

Coccygeal region (nerve Co1)
One pair of nerves supplies the skin in the region of the coccyx

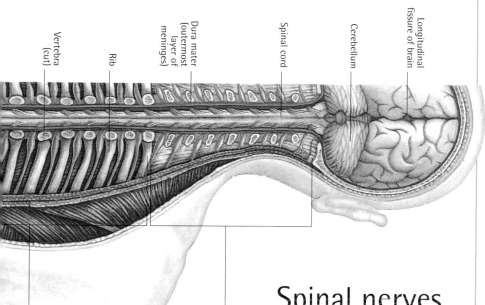

Longitudinal
fissure of brain

Cerebellum

Spinal cord

Dura mater
(outermost
layer of
meninges)

Rib

Vertebra
(cut)

Cervical region (nerves C1 to C8)
The eight pairs of cervical spinal nerves
interconnect, forming two networks, the
cervical plexus (C1 to C4) and the brachial
plexus (C5 to C8 and T1). These innervate
the back of the head, the neck, shoulders,
arms and hands, as well as the diaphragm

Thoracic region (nerves T1 to T12)
Thoracic spinal nerves are directly
connected to the muscles between
the ribs, the deep back muscles,
and regions of the abdomen and
thorax. T1 also contributes to
the brachial plexus

Spinal nerves

The 31 pairs of peripheral spinal
nerves emerge from the spinal
cord and extend through spaces
between the vertebrae. Each
nerve divides and subdivides into
a number of branches; two main
divisions serve the front and the
back of the body in the region
innervated by that particular
nerve. The branches of one
spinal nerve may join up with
other nerves to form groups
called plexuses; these innervate
certain areas of complex
function or movement, such as
the shoulder and neck.

Areas of sensation

The body "map" shown here delineates the skin in zones, called dermatones, that are served by specific spinal nerves (see pp.126-127). The dermatones in the trunk appear roughly horizontal, while those in the limbs are longitudinal. In real life, the distribution of nerve roots, and therefore of sensation, overlaps slightly.

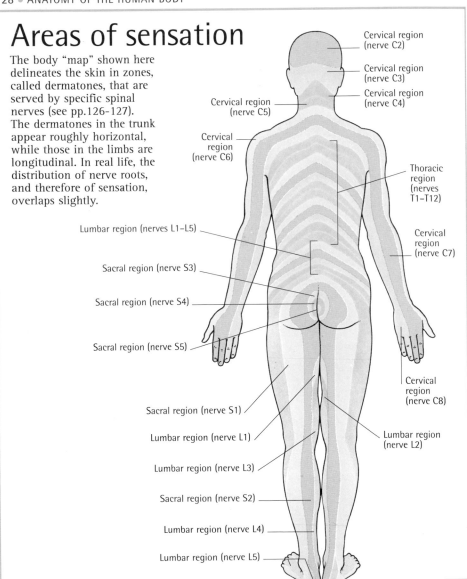

Cervical region (nerve C2)

Cervical region (nerve C3)

Cervical region (nerve C4)

Cervical region (nerve C5)

Cervical region (nerve C6)

Thoracic region (nerves T1–T12)

Cervical region (nerve C7)

Lumbar region (nerves L1–L5)

Sacral region (nerve S3)

Sacral region (nerve S4)

Sacral region (nerve S5)

Cervical region (nerve C8)

Sacral region (nerve S1)

Lumbar region (nerve L1)

Lumbar region (nerve L2)

Lumbar region (nerve L3)

Sacral region (nerve S2)

Lumbar region (nerve L4)

Lumbar region (nerve L5)

Spinal reflex

A reflex is an involuntary and predictable response to a stimulus. The patellar spinal reflex (also known as the "knee jerk" reflex) is commonly tested to assess the response of the spine's neural pathways. Tapping the patellar ligament in the knee stretches the front thigh muscle, stimulating a sensory neuron that transmits a nerve signal to the spinal cord. Motor nerve fibers then relay the signal to the muscle, which contracts and causes a slight forward kick of the leg and foot.

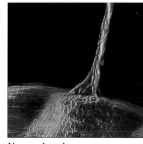

Nerve signal
The microscopic image above shows motor nerve fibers (pink) relaying a signal to skeletal muscle fibers (red).

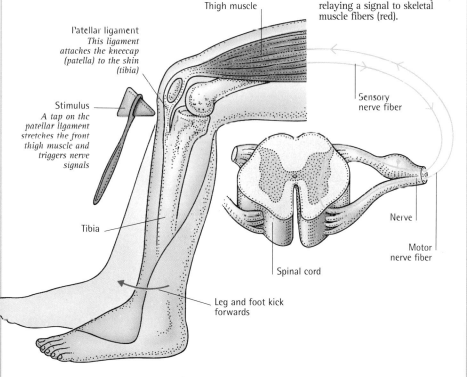

Thigh muscle

Patellar ligament
This ligament attaches the kneecap (patella) to the shin (tibia)

Stimulus
A tap on the patellar ligament stretches the front thigh muscle and triggers nerve signals

Tibia

Sensory nerve fiber

Nerve

Motor nerve fiber

Spinal cord

Leg and foot kick forwards

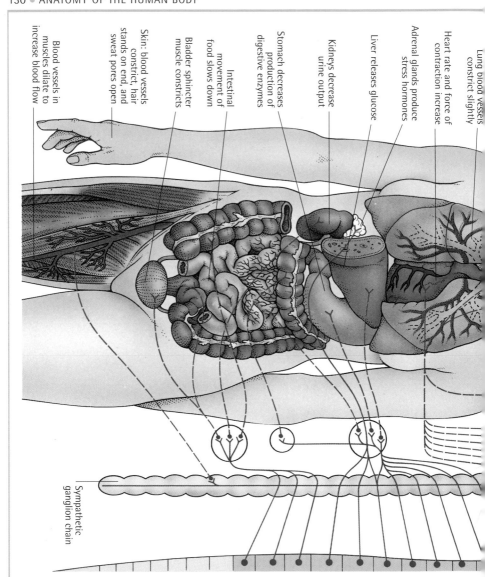

Lung blood vessels constrict slightly

Heart rate and force of contraction increase

Adrenal glands produce stress hormones

Liver releases glucose

Kidneys decrease urine output

Stomach decreases production of digestive enzymes

Intestinal movement of food slows down

Bladder sphincter muscle constricts

Skin: blood vessels constrict, hair stands on end, and sweat pores open

Blood vessels in muscles dilate to increase blood flow

Sympathetic ganglion chain

Sympathetic nerves

The sympathetic nervous system is one of two divisions of the autonomic nervous system (ANS), which controls involuntary body functions. Sympathetic nerves act on organs and blood vessels to prepare the body to react to stressful situations. The nerves arise mainly in the thoracic (chest) segments of the spinal cord and their axons pass through chains of ganglia (nerve clusters) on either side of the spinal column. There, they can branch off to join other axons and stimulate many organs.

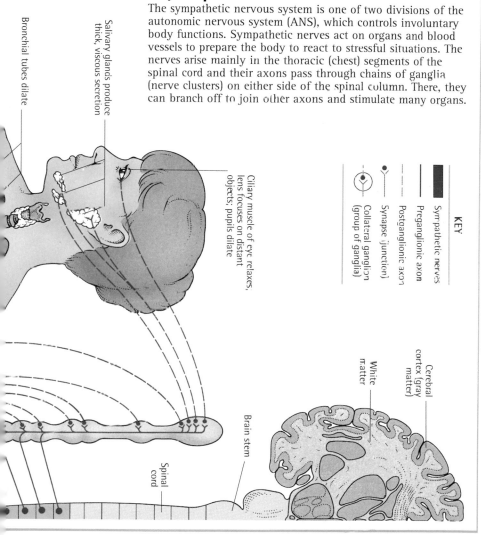

Bronchial tubes dilate

Salivary glands produce thick, viscous secretion

Ciliary muscle of eye relaxes, lens focuses on distant objects; pupils dilate

KEY

Sympathetic nerves

Preganglionic axon

Postganglionic axon

Synapse (junction)

Collateral ganglion (group of ganglia)

Cerebral cortex (gray matter)

White matter

Brain stem

Spinal cord

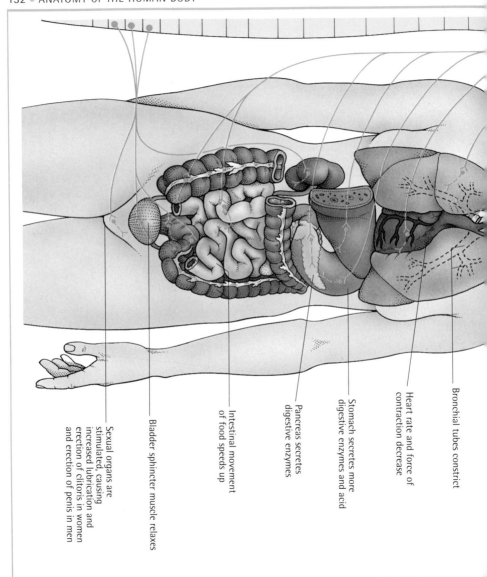

Sexual organs are stimulated, causing increased lubrication and erection of clitoris in women and erection of penis in men

Bladder sphincter muscle relaxes

Intestinal movement of food speeds up

Pancreas secretes digestive enzymes

Stomach secretes more digestive enzymes and acid

Heart rate and force of contraction decrease

Bronchial tubes constrict

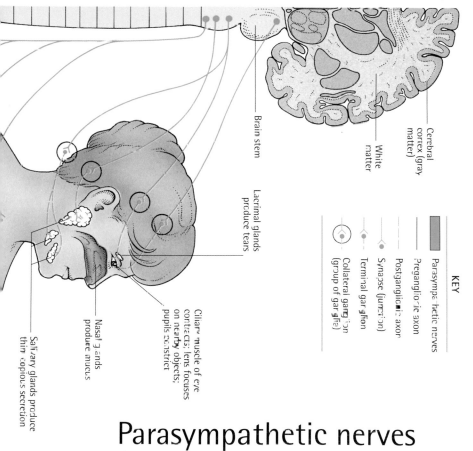

Cerebral
cortex (gray
matter)

White
matter

Brain stem

Lacrimal glands
produce tears

KEY

Parasympathetic nerves

Preganglionic axon

Postganglionic axon

Synapse (junction)

Terminal ganglion

Collateral ganglion
(group of ganglia)

Ciliary muscle of eye
contracts; lens focuses
on nearby objects;
pupils constrict

Nasal glands
produce mucus

Salivary glands produce
thin copious secretion

Parasympathetic nerves

The parasympathetic division of the autonomic nervous system
(ANS) usually has an opposing effect to the sympathetic division.
It operates mainly in quiet, non-stressful conditions and its
activity predominates during sleep. The parasympathetic nerves
arise in the brain stem and the lower spinal cord and their axons
are very long. The ganglia, where synapses are formed, are very
near to the target organs, which means that parasympathetic
nerves usually affect one organ only.

Pathway structures

In the sympathetic and parasympathetic nervous systems, signals pass down pathways made up of two neurons. The first neuron is a preganglionic neuron, which carries signals from the central nervous system to a ganglion (nerve cluster). In the ganglion, the neuron synapses with the second neuron, the postganglionic neuron. Most of the ganglia in the sympathetic system lie in two chains, one on either side of the vertebral column. From these chain ganglia, neurons can send branches to stimulate several target organs or vessels at once. Parasympathetic ganglia are near or embedded into the target organs or vessels, and so only one organ is usually stimulated.

Trachea

Sympathetic ganglion chain

Inferior vena cava

Sympathetic ganglion chains
Most of the ganglia in the sympathetic nervous system are joined by fibers to form two chains. These chains run down the back of the chest and torso on either side of the spinal column.

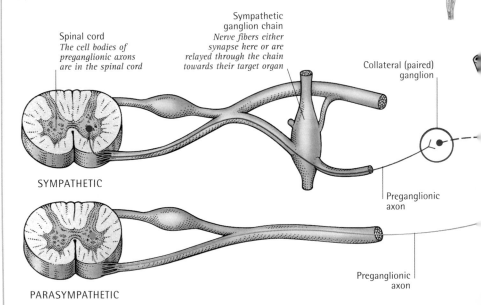

Spinal cord
The cell bodies of preganglionic axons are in the spinal cord

Sympathetic ganglion chain
Nerve fibers either synapse here or are relayed through the chain towards their target organ

Collateral (paired) ganglion

SYMPATHETIC

Preganglionic axon

Preganglionic axon

PARASYMPATHETIC

Coordination of response

The sympathetic and parasympathetic nervous systems usually have opposing effects. Many organs and tissues are stimulated by both systems, allowing them to be controlled very precisely. The amount of light entering the eyes through the pupils is one factor that is constantly controlled in this way.

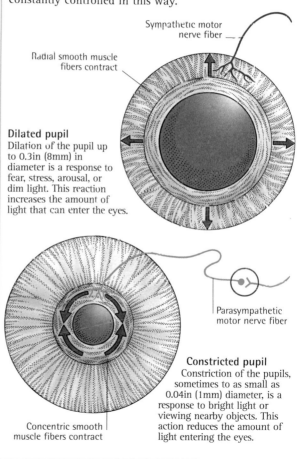

Sympathetic motor nerve fiber

Radial smooth muscle fibers contract

Dilated pupil
Dilation of the pupil up to 0.3in (8mm) in diameter is a response to fear, stress, arousal, or dim light. This reaction increases the amount of light that can enter the eyes.

Postganglionic axon

Visceral organ (urinary bladder)

Smooth muscle cell

Postganglionic axon

Parasympathetic ganglion

Parasympathetic motor nerve fiber

Constricted pupil
Constriction of the pupils, sometimes to as small as 0.04in (1mm) diameter, is a response to bright light or viewing nearby objects. This action reduces the amount of light entering the eyes.

Concentric smooth muscle fibers contract

Limbic system

The limbic system is a primitive part of the brain that controls instinctive behavior, such as "fight-or-flight" stress responses. The components of this ring-shaped system play a complex and important role in the expression of drives and emotions, the effects of moods on external behavior, and the formation of memories.

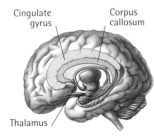

Cingulate gyrus

Corpus callosum

Thalamus

LOCATION OF LIMBIC SYSTEM

Cingulate gyrus
This area, together with the parahippocampal gyrus, comprises the limbic cortex, which modifies behavior and emotions

Fornix
This pathway of nerve fibers transmits information from the hippocampus and other areas of the limbic system to the mamillary body

Midbrain
Limbic areas of the midbrain influence physical activity via the basal ganglia, the large clusters of nerve cell bodies below the cortex

Septum pellucidum
This is a thin sheet of nervous tissue that connects the fornix to the corpus callosum

Column of fornix

Pons

Hippocampus
This curved band of gray matter is involved with learning, the recognition that something is new, and memory

Olfactory bulb
The connection of these sensory organs to the limbic system helps to explain why smells evoke memories and emotional responses

Mamillary body
This acts as a relay station, transmitting information between the fornix and the thalamus

Amygdala
This structure influences behavior and activities, such as sexual interest and feeding, and emotions such as anger

Parahippocampal gyrus
With other structures, this area helps to modify the expression of emotions such as rage and fright

Hypothalamus

The hypothalamus is an area of the brain the size of a lump of sugar. It is composed of numerous tiny clusters of nerve cells called nuclei. Together with the pituitary gland, these nuclei monitor and regulate the body's temperature, food intake, water–salt balance, blood flow, sleep–wake cycle, and the activity of hormones. They are also involved in determining responses to emotions such as anger and fear.

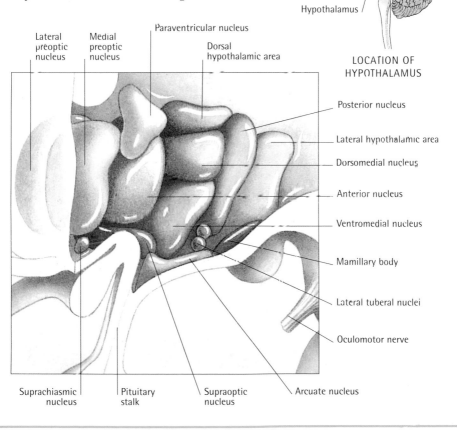

Hypothalamus

LOCATION OF HYPOTHALAMUS

Lateral preoptic nucleus

Medial preoptic nucleus

Paraventricular nucleus

Dorsal hypothalamic area

Posterior nucleus

Lateral hypothalamic area

Dorsomedial nucleus

Anterior nucleus

Ventromedial nucleus

Mamillary body

Lateral tuberal nuclei

Oculomotor nerve

Suprachiasmic nucleus

Pituitary stalk

Supraoptic nucleus

Arcuate nucleus

Brain stem functions

One of the functions of the brain stem is to keep the brain awake and alert. Consciousness is maintained by an arousal system in the brain stem known as the reticular activating system (RAS). This comprises pathways of nerve fibers that detect incoming sensory information and send activating signals through the midbrain to the cerebral cortex. The brain stem also controls sleep, maintains posture, and sustains breathing and heart rate.

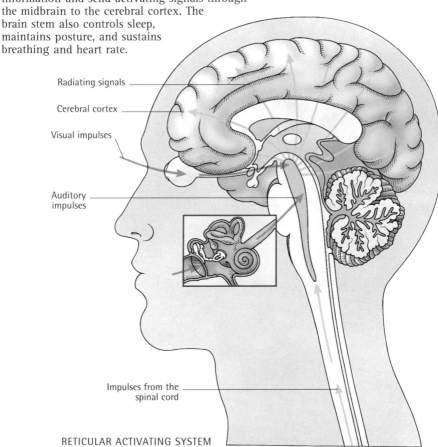

Radiating signals

Cerebral cortex

Visual impulses

Auditory impulses

Impulses from the spinal cord

RETICULAR ACTIVATING SYSTEM

Sleep

Nerve cells in the brain do not rest during sleep; instead, they carry out activities that are different from those of the waking hours. Distinctive patterns of non-rapid eye movement (NREM) as well as rapid eye movement (REM) sleep, when most dreams occur, can be detected by recording the electrical activity of the brain (shown right).

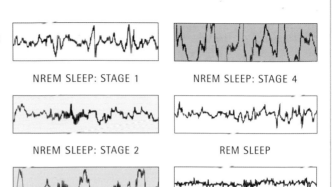

NREM SLEEP: STAGE 1

NREM SLEEP: STAGE 2

NREM SLEEP: STAGE 3

NREM SLEEP: STAGE 4

REM SLEEP

ALERT

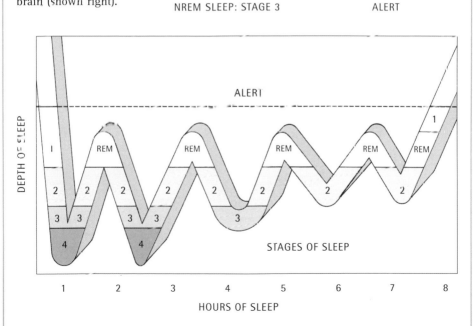

Information processing

Information received from the senses or generated by thought is processed in many parts of the brain. Some areas are responsible for processing sensory data, such as sight and sound. Others, such as the cortex and cerebellum, issue commands that initiate or coordinate voluntary movement. Important data are stored away as memory in other areas.

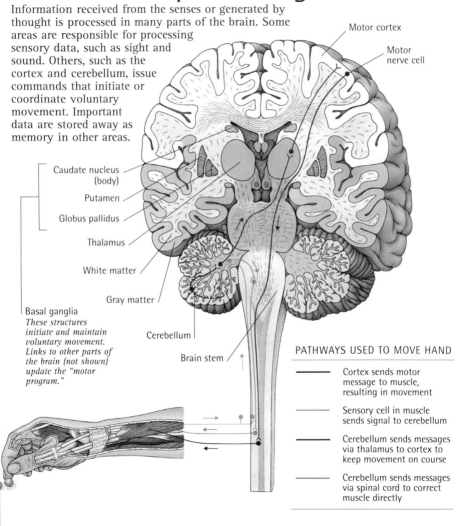

Motor cortex

Motor nerve cell

Caudate nucleus (body)

Putamen

Globus pallidus

Thalamus

White matter

Gray matter

Basal ganglia
These structures initiate and maintain voluntary movement. Links to other parts of the brain (not shown) update the "motor program."

Cerebellum

Brain stem

PATHWAYS USED TO MOVE HAND

— Cortex sends motor message to muscle, resulting in movement

— Sensory cell in muscle sends signal to cerebellum

— Cerebellum sends messages via thalamus to cortex to keep movement on course

— Cerebellum sends messages via spinal cord to correct muscle directly

Movement and touch

The thin, outer, crumpled layer of the cerebrum (the cerebral cortex) is the site of conscious behavior. Particular areas of the cortex on each side of the brain are responsible for movement and touch. The size of these areas varies according to the complexity of movement or degree of sensitivity involved. Scientists have been able to map the motor and sensory regions of the cortex by observing the effects of damage to or removal of certain parts of the brain.

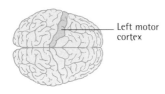

Left motor cortex

TOP VIEW

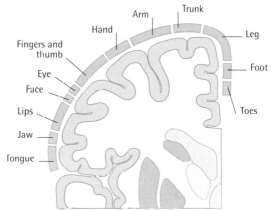

Arm
Trunk
Hand
Leg
Fingers and thumb
Foot
Eye
Face
Lips
Toes
Jaw
Tongue

Motor map of the brain
Each side of the brain controls the motor functions of the opposite side of the body. The most important motor areas, such as the fingers and hands, are controlled by a larger area of cortex.

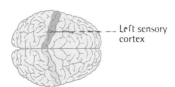

Left sensory cortex

TOP VIEW

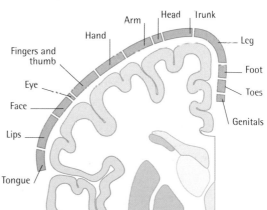

Arm
Head
Trunk
Hand
Leg
Fingers and thumb
Foot
Eye
Toes
Face
Lips
Genitals
Tongue

Touch map of the brain
The areas of the cortex that respond to touch lie just behind the motor areas. Areas of the body where touch sensations are very important, such as the hands, fingers, and feet, have a larger area of the cortex devoted to them.

Association areas

Large parts of the cerebral cortex are taken up by so-called "association areas", which analyze information received from the primary sensory sites. For example, a site called the primary auditory cortex registers basic information about the pitch and volume of sound (such as speech); Wernicke's area, part of the auditory association cortex, analyzes speech so that it can be understood.

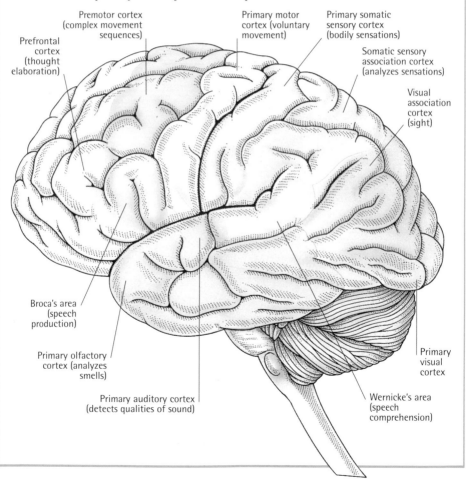

Premotor cortex (complex movement sequences)

Prefrontal cortex (thought elaboration)

Primary motor cortex (voluntary movement)

Primary somatic sensory cortex (bodily sensations)

Somatic sensory association cortex (analyzes sensations)

Visual association cortex (sight)

Broca's area (speech production)

Primary olfactory cortex (analyzes smells)

Primary auditory cortex (detects qualities of sound)

Primary visual cortex

Wernicke's area (speech comprehension)

Memory

To create memories, nerve cells are thought to form new protein molecules and new interconnections. No one region of the brain stores all memories because the storage site depends on the type of memory: how to type or ride a bike, for example, are memories held in motor areas, while those about music are held in the auditory areas. A band of gray matter called the hippocampus helps to transfer facts or aspects of events into long-term memory storage.

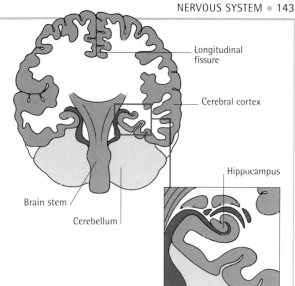

Longitudinal fissure

Cerebral cortex

Hippocampus

Brain stem

Cerebellum

Types of memory

Sensory memory, such as the brief recognition of words on a page, is stored only for milliseconds. If retained and interpreted, this sensory input may become short-term memory for a few minutes. The transfer of short-term to long-term memory is known as consolidation, and requires attention, repetition, and associative ideas. How easily information is recalled depends upon how it was consolidated.

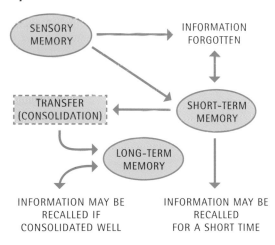

SENSORY MEMORY

INFORMATION FORGOTTEN

TRANSFER (CONSOLIDATION)

SHORT-TERM MEMORY

LONG-TERM MEMORY

INFORMATION MAY BE RECALLED IF CONSOLIDATED WELL

INFORMATION MAY BE RECALLED FOR A SHORT TIME

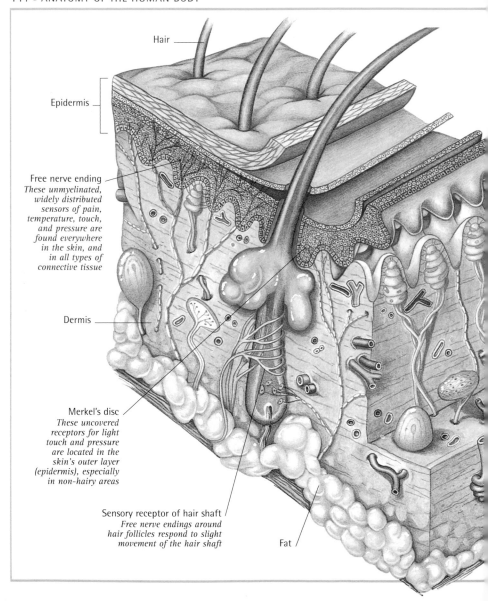

Hair

Epidermis

Free nerve ending
*These unmyelinated,
widely distributed
sensors of pain,
temperature, touch,
and pressure are
found everywhere
in the skin, and
in all types of
connective tissue*

Dermis

Merkel's disc
*These uncovered
receptors for light
touch and pressure
are located in the
skin's outer layer
(epidermis), especially
in non-hairy areas*

Sensory receptor of hair shaft
*Free nerve endings around
hair follicles respond to slight
movement of the hair shaft*

Fat

Touch receptors

The sense of touch operates by means of sensory receptors in the skin or in deeper tissues. These receptors relay signals to the spinal cord and brain stem; from there they travel to the higher areas of the brain. Some receptors are enclosed in a capsule of connective tissue, while others are uncovered.

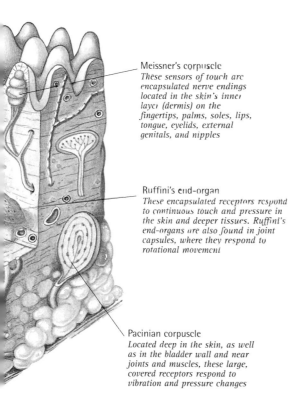

Meissner's corpuscle
These sensors of touch are encapsulated nerve endings located in the skin's inner layer (dermis) on the fingertips, palms, soles, lips, tongue, eyelids, external genitals, and nipples

Ruffini's end-organ
These encapsulated receptors respond to continuous touch and pressure in the skin and deeper tissues. Ruffini's end-organs are also found in joint capsules, where they respond to rotational movement

Pacinian corpuscle
Located deep in the skin, as well as in the bladder wall and near joints and muscles, these large, covered receptors respond to vibration and pressure changes

Pain

Pain receptors are sensory nerve endings that are stimulated by chemicals (such as histamine) released when tissues are damaged. They transmit pain signals to the brain, and may trigger the release of pain-blocking substances called endorphins.

Damaged tissue   Histamine   Nerve ending

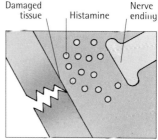

Blocking pain
Analgesic drugs such as aspirin block production of chemicals that stimulate pain receptors.

Brain cell   Narcotic drug   Nerve ending   Pain signal

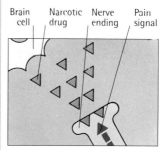

Imitating endorphins
Narcotic drugs such as morphine mimic the effect of endorphins and block pain signals between cells.

Taste

Taste receptor cells, known as taste buds, are located mainly within nodules called papillae on the surface of the tongue. There are four types of taste buds, on different parts of the tongue, which are sensitive to sweet, bitter, sour, and salty flavors. Taste signals are picked up by nerve fibers from one of three cranial nerves. The impulses then travel to taste centers in the brain.

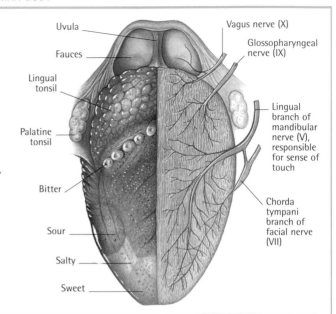

Taste buds

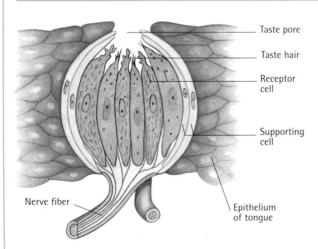

Taste buds consist of a cluster of receptor, or "taste", cells and supporting cells. Projecting from the top of a receptor cell are tiny taste hairs, which are exposed to saliva that enters through taste pores. Substances taken into the mouth and dissolved in saliva interact with the receptor sites on the taste hairs, generating a nerve impulse that is transmitted to the brain.

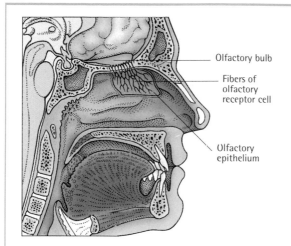

Olfactory bulb

Fibers of
olfactory
receptor cell

Olfactory
epithelium

Smell

The olfactory (smell) receptor
cells are located high in the
nasal cavity within a
specialized area of mucous
membrane known as the
olfactory epithelium. Fibers
from these cells extend into
the olfactory bulbs (swellings
at the end of the olfactory
nerves), which link up with
olfactory regions in the
brain. The human sense of
smell can detect more than
10,000 odors.

Mechanism of smell

Odor molecules entering
the nose dissolve in nasal
mucus and stimulate the
hairlike endings (cilia)
of the olfactory receptor
cells, generating a nerve
impulse. The impulse
travels along the fibers
of the cells, which pass
through holes in the bone
roofing the nasal cavity
and enter the olfactory
bulbs. Within the bulbs,
the receptor cell fibers
connect with olfactory
nerves, sending the
impulse on to the brain.

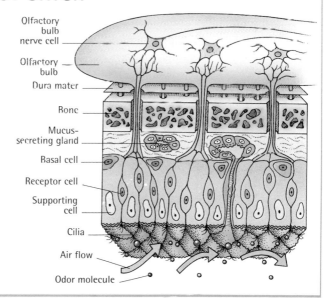

Olfactory
bulb
nerve cell

Olfactory
bulb

Dura mater

Bone

Mucus-
secreting gland

Basal cell

Receptor cell

Supporting
cell

Cilia

Air flow

Odor molecule

Ear structure and hearing

The ear has three parts: outer, middle, and inner. The visible part of the outer ear, the pinna, funnels sound waves down the ear canal to the eardrum. The middle ear, an air-filled cavity between the eardrum and the inner ear, contains three tiny bones (ossicles) that conduct sounds to the inner ear, or labyrinth. Here, nerve fibers in a snail-like organ called the cochlea pick up vibrations and relay them to the brain.

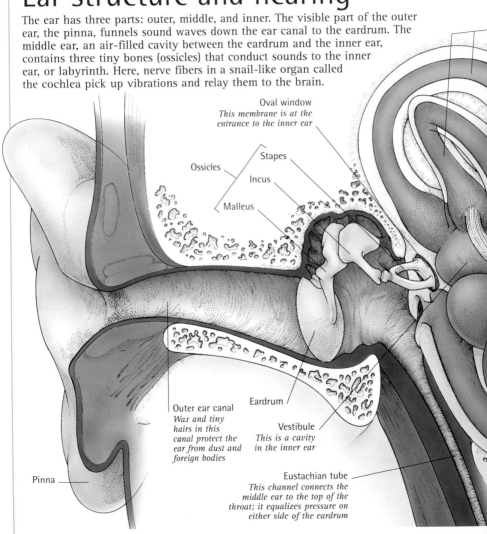

Oval window
This membrane is at the entrance to the inner ear

Ossicles

Stapes

Incus

Malleus

Outer ear canal
Wax and tiny hairs in this canal protect the ear from dust and foreign bodies

Eardrum

Vestibule
This is a cavity in the inner ear

Pinna

Eustachian tube
This channel connects the middle ear to the top of the throat; it equalizes pressure on either side of the eardrum

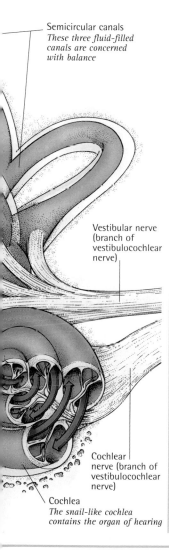

Semicircular canals
These three fluid-filled canals are concerned with balance

Vestibular nerve (branch of vestibulocochlear nerve)

Cochlear nerve (branch of vestibulocochlear nerve)

Cochlea
The snail-like cochlea contains the organ of hearing

Cochlea

The cochlea has three fluid-filled tubes around a bony core. The central tube, the cochlear duct, contains the spiral organ of Corti (the organ of hearing), in which there are sensory hair cells that respond to sound.

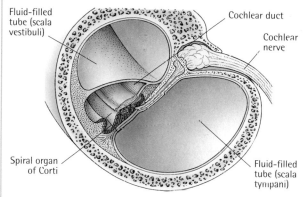

Fluid-filled tube (scala vestibuli)

Cochlear duct

Cochlear nerve

Spiral organ of Corti

Fluid-filled tube (scala tympani)

Spiral organ of Corti

Tiny sensory hairs project from each hair cell of the spiral organ. When sound vibrations set off wavelike motions in the cochlear fluid, the hairs translate the movement into impulses that are conveyed to the brain.

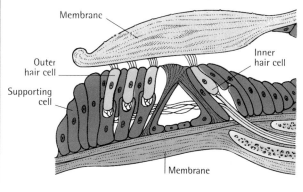

Membrane

Outer hair cell

Supporting cell

Inner hair cell

Membrane

Balance

The sense of balance relies in part on visual information about the body's position, and on sensory receptors in the skin and joints. There are also several structures in the inner ear that help to maintain balance. Within the vestibule, two sacs (the utricle and saccule) each contain a sensory patch (macula) that registers the position of the head; and in the bulge, or ampulla, at the base of each semicircular canal, a receptor structure (crista) detects rotation of the head.

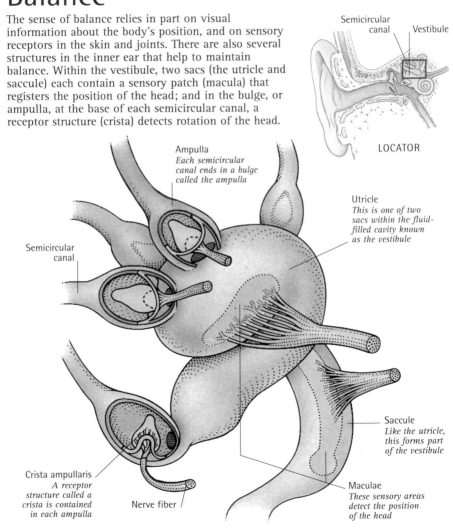

Semicircular canal

Vestibule

LOCATOR

Ampulla
Each semicircular canal ends in a bulge called the ampulla

Utricle
This is one of two sacs within the fluid-filled cavity known as the vestibule

Semicircular canal

Saccule
Like the utricle, this forms part of the vestibule

Crista ampullaris
A receptor structure called a crista is contained in each ampulla

Nerve fiber

Maculae
These sensory areas detect the position of the head

Role of maculae

Maculae (sensory patches) in the inner ear monitor the position of the head. They contain tiny hairs that project from sensory cells and are embedded in a gelatinous mass. If the head is tipped, gravity pulls the mass down, stimulating the hair cells.

UPRIGHT

DISPLACED

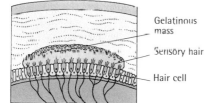

Gelatinous mass

Sensory hair

Hair cell

UPRIGHT MACULA

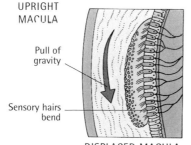

Pull of gravity

Sensory hairs bend

DISPLACED MACULA

Role of cristae

The crista ampullaris, located at the end of each semicircular canal in the inner ear, responds to rotational movements. Each crista contains hair cells embedded in a conical gelatinous mass known as the cupula. When the fluid in the semicircular canals swirls during movement, it displaces the cupula, stimulating the hair cells.

ROTATING

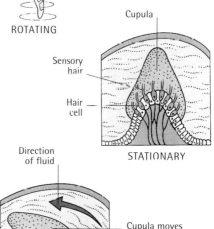

Cupula

Sensory hair

Hair cell

STATIONARY

Direction of fluid

Cupula moves

Sensory hairs bend

ROTATING

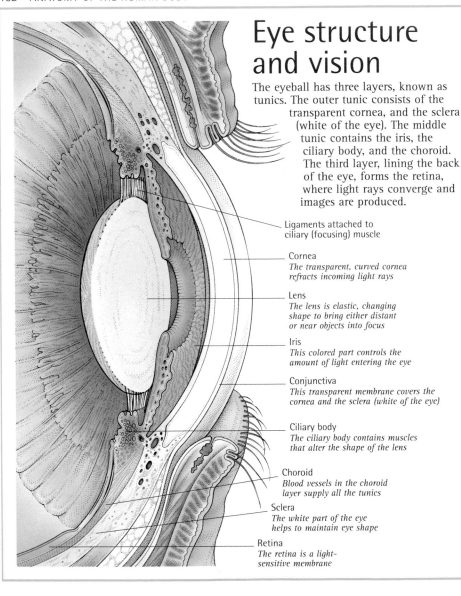

Eye structure and vision

The eyeball has three layers, known as tunics. The outer tunic consists of the transparent cornea, and the sclera (white of the eye). The middle tunic contains the iris, the ciliary body, and the choroid. The third layer, lining the back of the eye, forms the retina, where light rays converge and images are produced.

Ligaments attached to ciliary (focusing) muscle

Cornea
The transparent, curved cornea refracts incoming light rays

Lens
The lens is elastic, changing shape to bring either distant or near objects into focus

Iris
This colored part controls the amount of light entering the eye

Conjunctiva
This transparent membrane covers the cornea and the sclera (white of the eye)

Ciliary body
The ciliary body contains muscles that alter the shape of the lens

Choroid
Blood vessels in the choroid layer supply all the tunics

Sclera
The white part of the eye helps to maintain eye shape

Retina
The retina is a light-sensitive membrane

Cavities of the eye

The front cavity of the eye has two chambers (anterior and posterior) filled with aqueous humor, a fluid providing oxygen, glucose, and proteins. This fluid is produced by the ciliary body, which controls the focusing power of the lens. The back cavity of the eye is filled with a clear gel called vitreous humor. Both humors contribute to the constant internal pressure that maintains the shape of the eye.

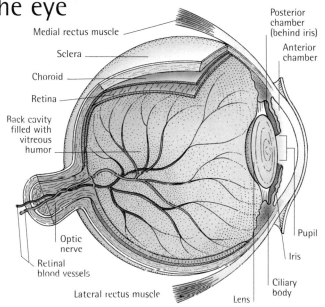

Medial rectus muscle

Sclera

Choroid

Retina

Back cavity filled with vitreous humor

Optic nerve

Retinal blood vessels

Lateral rectus muscle

Posterior chamber (behind iris)

Anterior chamber

Pupil

Iris

Ciliary body

Lens

Accessory eye structures

Several accessory structures support, move, lubricate, and protect the eyes from damage and infection. These include the orbital bones of the eye socket, the muscles of the eyeball, lacrimal (tear) glands and ducts, eyebrows, eyelids, and eyelashes.

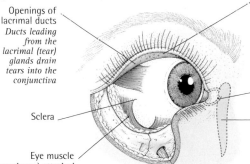

Openings of lacrimal ducts
Ducts leading from the lacrimal (tear) glands drain tears into the conjunctiva

Sclera

Eye muscle

Six muscles (three are shown) attached to the sclera control movement

Conjunctiva
A transparent mucous membrane covers and moistens the sclera, the cornea, and the inside of the eyelids

Lacrimal canal

Nasolacrimal duct
Excess tears evaporate or drain through canals into a duct that is connected to the nasal cavity

Visual pathways

Light rays enter the eye and converge on the retina, where an upside-down image is created. The retina of each eye relays impulses to the brain through the optic nerves, which cross paths at a junction called the optic chiasm; half the nerve fibers from the right eye cross to the left, and vice versa, before passing on to the brain. In the visual cortex the image is turned upright.

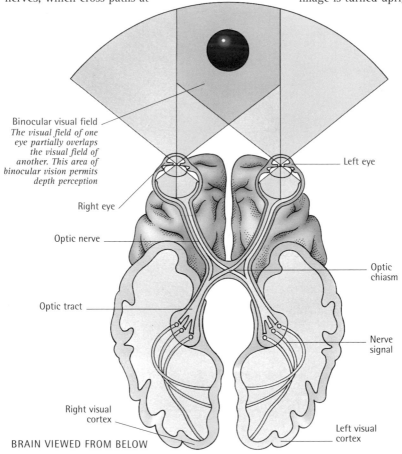

Binocular visual field
The visual field of one eye partially overlaps the visual field of another. This area of binocular vision permits depth perception

Left eye

Right eye

Optic nerve

Optic chiasm

Optic tract

Nerve signal

Right visual cortex

Left visual cortex

BRAIN VIEWED FROM BELOW

Accommodation

The eye's ciliary muscles automatically respond to the proximity or distance of an object by altering the shape of the lens. This adjustment, which is known as accommodation, changes the angle of incoming light rays, so that they create a sharp image on the retina, whether the object is near or far.

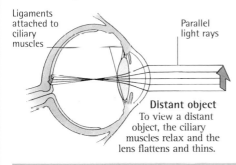

Ligaments attached to ciliary muscles

Parallel light rays

Distant object
To view a distant object, the ciliary muscles relax and the lens flattens and thins.

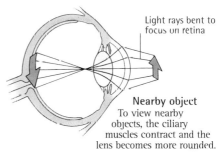

Light rays bent to focus on retina

Nearby object
To view nearby objects, the ciliary muscles contract and the lens becomes more rounded.

Rods and cones

There are two types of nerve cells in the retina: rods and cones. Rods contain only one type of light-sensitive pigment, and cannot discern color. Cones, on which color vision depends, are of three classes that respond to green, red, or blue light. When stimulated by light, rods and cones produce electrical signals, triggering connecting nerve cells to relay further impulses to the brain.

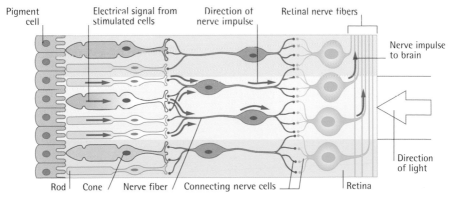

Pigment cell

Electrical signal from stimulated cells

Direction of nerve impulse

Retinal nerve fibers

Nerve impulse to brain

Direction of light

Rod   Cone   Nerve fiber   Connecting nerve cells   Retina

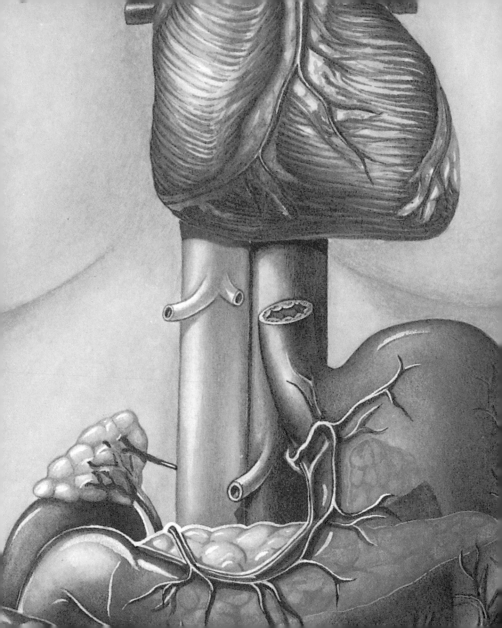

Endocrine system

The endocrine system is a collection of hormone-producing glands and cells located in various parts of the body, such as the pancreas and the ovaries. Hormones are complex chemical substances that are secreted into the bloodstream and regulate body functions such as metabolism, growth, and sexual reproduction.

Hormones
Hormones are produced in many parts of the body, including the heart, stomach, kidneys, and pancreas.

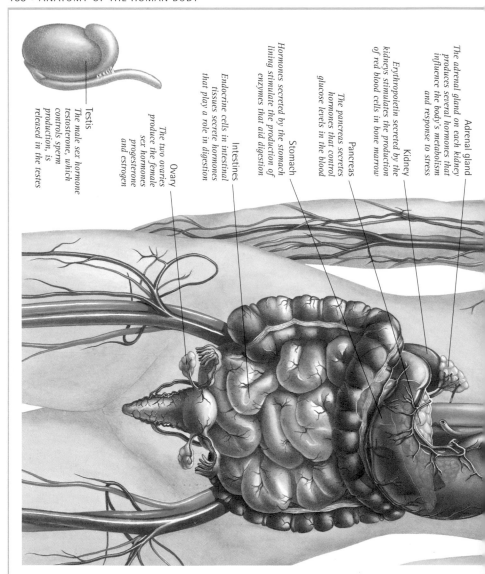

Adrenal gland
The adrenal gland on each kidney produces several hormones that influence the body's metabolism and response to stress

Kidney
Erythropoietin secreted by the kidneys stimulates the production of red blood cells in bone marrow

Pancreas
The pancreas secretes hormones that control glucose levels in the blood

Stomach
Hormones secreted by the stomach lining stimulate the production of enzymes that aid digestion

Intestines
Endocrine cells in intestinal tissues secrete hormones that play a role in digestion

Ovary
The two ovaries produce the female sex hormones progesterone and estrogen

Testis
The male sex hormone testosterone, which controls sperm production, is released in the testes

Hormone producers

Hormones are complex chemical substances produced by specialized glands, called endocrine glands, or by cells in some body organs, for example, the heart or a part of the gastrointestinal tract. Secreted into internal body fluids, many hormones target specific tissues, controlling their function and, in some organs, stimulating the production of further hormones. A special class of hormones, the prostaglandins, produce only local effects at their site of production.

Heart
The heart produces a hormone called atriopeptin, which reduces blood volume and blood pressure

Hypothalamus
Most hormones from this cluster of nerve cells at the base of the brain stimulate other glands to produce their own hormones

Pituitary gland
Called the "master gland", this organ controls many other endocrine glands

Thyroid gland
This gland controls metabolism, including the maintenance of body weight, the rate of energy use, and heart rate. Unlike most other glands, it can store the hormones it produces

Pineal gland
This tiny gland secretes melatonin, a hormone that may influence sexual development

Parathyroid gland
The four parathyroid glands at the back of the thyroid gland produce a hormone that regulates blood-calcium levels

Pituitary gland

The pituitary gland, sometimes called the "master gland", controls the activities of many other endocrine glands and cells. This pea-sized structure hangs from the base of the brain; it is attached by a short stalk of nerve fibers to the hypothalamus, a brain area that controls pituitary function. The pituitary has two lobes, anterior and posterior, which produce a range of hormones. Some pituitary hormones act indirectly, stimulating hormone release in target glands; other pituitary hormones directly affect body functions.

Skin
Melanocytes in skin tissue are stimulated by MSH to produce more melanin, a pigment that darkens skin in response to sunlight.

Artery

Adrenal gland
The steroid hormones produced by these glands influence how the body uses carbohydrates, fats, proteins, and minerals. They also influence the body's response to stress.

Thyroid gland
Hormones produced by this gland have widespread effects on body metabolism.

Bone and general growth
Growth hormone acts on the whole body to promote protein synthesis. This hormone is essential for normal growth and development in children.

Testis and ovary
Male and female hormones released by these glands control sexual development and reproductive function.

ACTH

MSH

TSH

GH

FSH, LH

Anterior lobe
Triggered by the action of the hypothalamus, the anterior lobe of the pituitary gland makes at least six hormones

TESTIS   OVARY

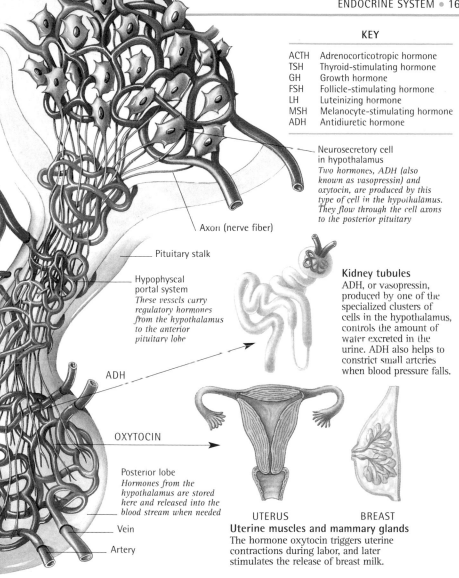

KEY

ACTH	Adrenocorticotropic hormone
TSH	Thyroid-stimulating hormone
GH	Growth hormone
FSH	Follicle-stimulating hormone
LH	Luteinizing hormone
MSH	Melanocyte-stimulating hormone
ADH	Antidiuretic hormone

Neurosecretory cell in hypothalamus
Two hormones, ADH (also known as vasopressin) and oxytocin, are produced by this type of cell in the hypothalamus. They flow through the cell axons to the posterior pituitary

Axon (nerve fiber)

Pituitary stalk

Hypophyseal portal system
These vessels carry regulatory hormones from the hypothalamus to the anterior pituitary lobe

ADH

OXYTOCIN

Posterior lobe
Hormones from the hypothalamus are stored here and released into the blood stream when needed

Vein

Artery

Kidney tubules
ADH, or vasopressin, produced by one of the specialized clusters of cells in the hypothalamus, controls the amount of water excreted in the urine. ADH also helps to constrict small arteries when blood pressure falls.

UTERUS   BREAST
Uterine muscles and mammary glands
The hormone oxytocin triggers uterine contractions during labor, and later stimulates the release of breast milk.

Adrenal glands

The adrenal glands are small, triangular structures located on top of each kidney. They consist of two distinct parts: an outer layer known as the cortex and an inner region called the medulla. The cortex secretes a group of hormones that influence the body's metabolic processes; the medulla releases hormones that affect the body's response to stress.

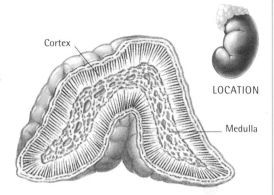

Cortex

LOCATION

Medulla

Adrenal hormones

The adrenal cortex has three zones, each of which produces its own hormones. The adrenal medulla functions almost as a separate endocrine gland. It produces hormones, but its nerve fibers also link it with the sympathetic nervous system and are involved in the "fight-or-flight" response that is triggered by stress.

HORMONES OF THE ADRENAL CORTEX AND ADRENAL MEDULLA	
Aldosterone	Secreted by the outermost zone of the adrenal cortex, this hormone inhibits the amount of sodium excreted in the urine and helps to maintain blood volume and blood pressure.
Cortisol	The middle layer of the cortex produces this hormone. It controls how the body utilizes fat, protein, carbohydrates, and minerals, and helps to reduce inflammation.
Gonado-corticoids	Produced by the inner layer of the adrenal cortex, these sex hormones have only a slight effect on the sex organs. They influence sperm production in males, while in females they mainly influence the distribution of body hair.
Epinephrine and norepinephrine	These hormones from the adrenal medulla affect the body during stress. Epinephrine (also called adrenaline) stimulates heart rate; norepinephrine (also called noradrenaline) helps to maintain constant blood pressure.

Pancreas

The pancreas has a dual function. A major part of its tissues produces digestive enzymes; within these tissues are small clusters of hormone-producing cells known as islets of Langerhans. Each cell cluster contains alpha cells, which produce the hormone glucagon, to increase glucose concentration in the blood; beta cells, which produce insulin to lower blood glucose; and delta cells, which regulate insulin and glucagon secretions.

Pancreatic duct

Tail of pancreas

Head of pancreas

Body of pancreas

Common bile duct

PANCREAS

Duct

Delta cell
The hormone secreted by delta cells regulates the production of insulin and glucagon

Islet of Langerhans
Small clusters (islets) of endocrine cells in pancreatic tissue produce various hormones

Alpha cell
This type of cell secretes the hormone glucagon

Beta cell
Insulin is released by beta cells

Acinar cells
These cells produce digestive enzymes

PANCREATIC CELLS

Thyroid and parathyroid glands

The thyroid gland, which is situated at the front of the neck, partially surrounding the trachea, plays an important role in regulating the body's metabolism. At the back of this gland are the four parathyroid glands, which regulate calcium levels in blood; the hormone they secrete acts on bones to release stored calcium, and on the kidneys to increase production of vitamin D, which in turn increases the body's calcium absorption.

Thyroid gland
The hormones produced by the thyroid gland act on cells to control the body's metabolism, including the use of energy.

Hyoid bone
This bone is the anchor point for a number of important neck muscles

Epiglottis
During swallowing, this flap of cartilage folds back to close the larynx

Thyroid cartilage
The front of this ring of cartilage, which is part of the larynx (voice box), forms the projection known as Adam's apple

Parathyroid gland
The four parathyroid glands are located at the back of the thyroid gland

Cricoid cartilage
This cartilage forms part of the larynx (voice box)

Thyroid gland
This gland is wrapped round the front and sides of the trachea

Trachea (windpipe)

Female sex hormones

In females, the two ovaries secrete the hormones estrogen and progesterone, which are essential to sexual development and fertility. Estrogen is produced by an egg as it develops inside its follicle. At ovulation, when the mature egg is released, the empty follicle forms into a small tissue mass, known as the corpus luteum, which secretes progesterone.

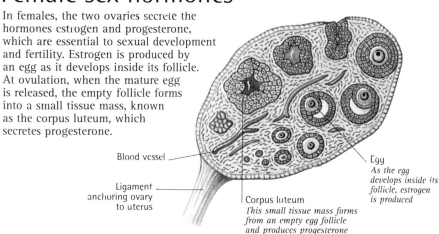

Blood vessel

Ligament anchoring ovary to uterus

Egg
As the egg develops inside its follicle, estrogen is produced

Corpus luteum
This small tissue mass forms from an empty egg follicle and produces progesterone

Male sex hormones

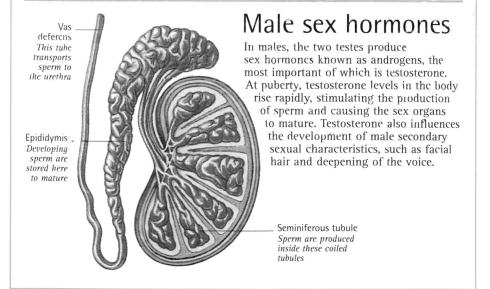

Vas deferens
This tube transports sperm to the urethra

Epididymis
Developing sperm are stored here to mature

In males, the two testes produce sex hormones known as androgens, the most important of which is testosterone. At puberty, testosterone levels in the body rise rapidly, stimulating the production of sperm and causing the sex organs to mature. Testosterone also influences the development of male secondary sexual characteristics, such as facial hair and deepening of the voice.

Seminiferous tubule
Sperm are produced inside these coiled tubules

How hormones work

Hormones are made up of molecules derived from steroids, proteins, or tyrosine (an amino acid). They become active only when bound to a specific receptor on or inside a target cell.

Hormones derived from proteins bind to receptors on the outside of the cell membrane; steroid and tyrosine hormones pass inside the cell and bind to receptors in the cytoplasm or the nucleus.

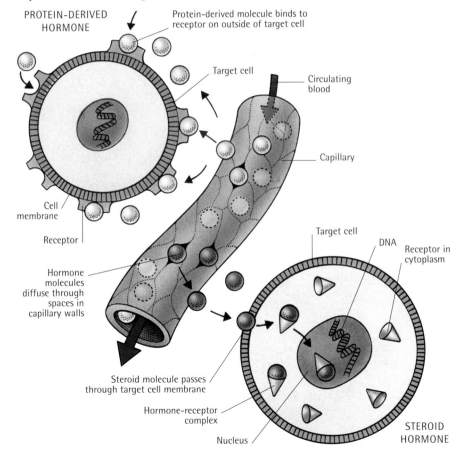

PROTEIN-DERIVED HORMONE

Protein-derived molecule binds to receptor on outside of target cell

Target cell

Circulating blood

Capillary

Cell membrane

Receptor

Hormone molecules diffuse through spaces in capillary walls

Target cell

DNA

Receptor in cytoplasm

Steroid molecule passes through target cell membrane

Hormone-receptor complex

Nucleus

STEROID HORMONE

Feedback mechanisms

A specific mechanism, known as feedback, controls hormone production; it involves the hypothalamus, the pituitary gland, and the target gland. A feedback system can either promote the release of another hormone (positive feedback) or inhibit its release (negative feedback). This involuntary mechanism maintains a balance in the functioning of the body.

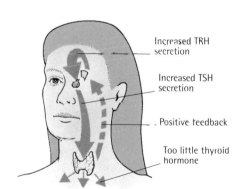

Thyrotropin-releasing hormone (TRH)

Hypothalamus

Pituitary gland

Thyroid-stimulating hormone (TSH)

Thyroid gland

Thyroid hormone

1 Response to hormone levels
Responding to levels of thyroid hormone, the hypothalamus makes TRH. This stimulates the pituitary gland to release TSH. The thyroid gland is then triggered to produce its hormones.

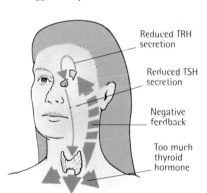

Reduced TRH secretion

Reduced TSH secretion

Negative feedback

Too much thyroid hormone

Increased TRH secretion

Increased TSH secretion

Positive feedback

Too little thyroid hormone

2 Negative feedback
If thyroid hormone levels are too high, negative feedback alerts the hypothalamus so that it produces less TRH. A lower level of TRH results in a reduced level of TSH. The thyroid responds by producing less hormone.

3 Positive feedback
If thyroid hormone levels fall too low, the feedback mechanism is weakened. In response, the hypothalamus makes more TRH; the pituitary gland increases secretion of TSH and the levels of thyroid hormone also rise.

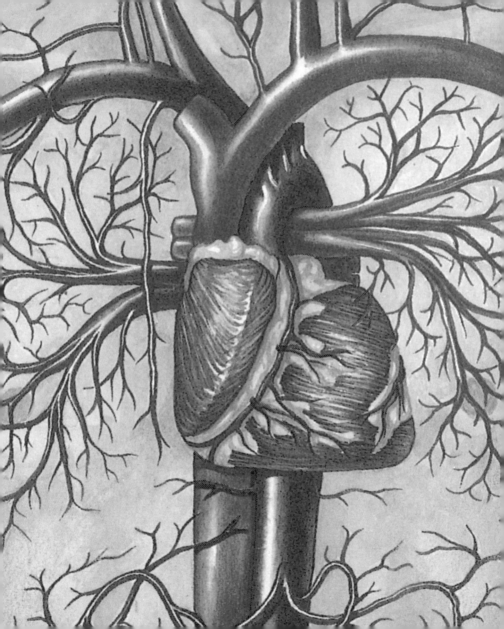

Cardiovascular system

The heart and the body's blood-transporting network of arteries, veins, and smaller vessels form the circulatory, or cardiovascular, system. As blood is continuously pumped out from the heart and around the system in two circuits, it carries oxygen and vital nutrients to all parts of the body and removes harmful waste products from tissues.

Blood circulation
The heart pumps blood through arteries and smaller vessels to all body parts. The blood returns to the heart through the veins.

Heart and circulation

The illustration shows many of the arteries, veins, and smaller blood vessels that together with the heart form the body's cardiovascular system. Red indicates oxygenated blood, which is usually carried by arteries; blue indicates deoxygenated blood, carried by veins. The pulmonary arteries are the only arterial vessels that transport deoxygenated blood. Blood returns to the heart at the same rate at which it is pumped out, making a full circuit of the body in about one minute.

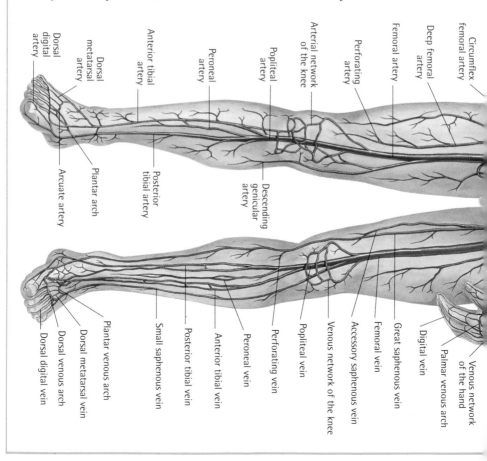

Dorsal digital artery

Dorsal metatarsal artery

Anterior tibial artery

Peroneal artery

Popliteal artery

Arterial network of the knee

Perforating artery

Femoral artery

Deep femoral artery

Circumflex femoral artery

Arcuate artery

Plantar arch

Posterior tibial artery

Descending genicular artery

Venous network of the hand

Dorsal digital vein

Dorsal venous arch

Dorsal metatarsal vein

Plantar venous arch

Small saphenous vein

Posterior tibial vein

Anterior tibial vein

Peroneal vein

Perforating vein

Popliteal vein

Venous network of the knee

Accessory saphenous vein

Femoral vein

Great saphenous vein

Digital vein

Palmar venous arch

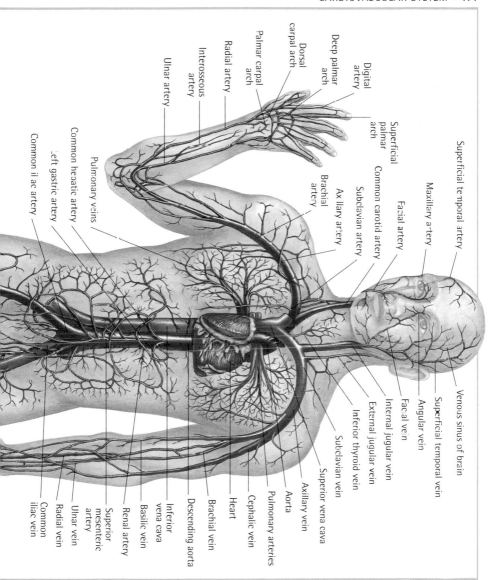

Superficial temporal artery

Maxillary artery

Digital artery

Deep palmar arch

Dorsal carpal arch

Palmar carpal arch

Radial artery

Interosseous artery

Ulnar artery

Superficial palmar arch

Common carotid artery

Facial artery

Brachial artery

Subclavian artery

Axillary artery

Common hepatic artery

Pulmonary veins

Left gastric artery

Common iliac artery

Venous sinus of brain

Superficial temporal vein

Angular vein

Facial vein

Internal jugular vein

External jugular vein

Inferior thyroid vein

Subclavian vein

Superior vena cava

Axillary vein

Aorta

Pulmonary arteries

Cephalic vein

Heart

Brachial vein

Descending aorta

Inferior vena cava

Basilic vein

Renal artery

Superior mesenteric artery

Ulnar vein

Radial vein

Common iliac vein

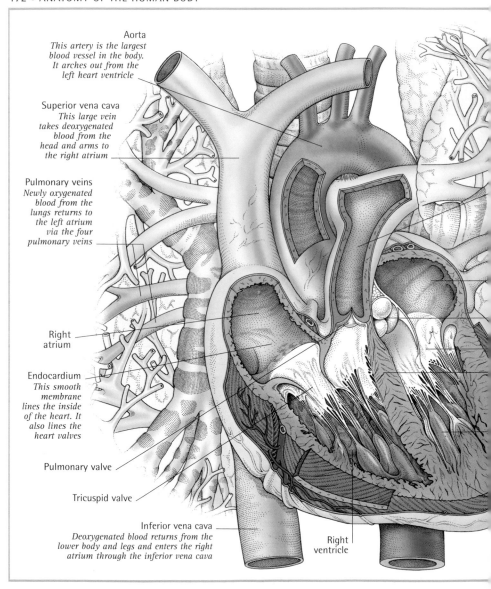

Aorta
This artery is the largest blood vessel in the body. It arches out from the left heart ventricle

Superior vena cava
This large vein takes deoxygenated blood from the head and arms to the right atrium

Pulmonary veins
Newly oxygenated blood from the lungs returns to the left atrium via the four pulmonary veins

Right atrium

Endocardium
This smooth membrane lines the inside of the heart. It also lines the heart valves

Pulmonary valve

Tricuspid valve

Inferior vena cava
Deoxygenated blood returns from the lower body and legs and enters the right atrium through the inferior vena cava

Right ventricle

Heart structure

The heart is a powerful muscle, about the size of a grapefruit, located just left of the center of the chest. It is made of a special type of muscle called myocardium, which occurs nowhere else in the body. Inside, the heart is divided into four chambers: two upper chambers called atria, and two thicker-walled lower chambers known as ventricles. A strong muscular wall, the septum, divides the two sides of the heart. Four one-way valves control blood flow through the heart chambers.

Double heart pump

Deoxygenated blood flows into the right side of the heart, which pumps it to the lungs to pick up oxygen. The oxygenated blood then returns to the left side, which pumps it around the body.

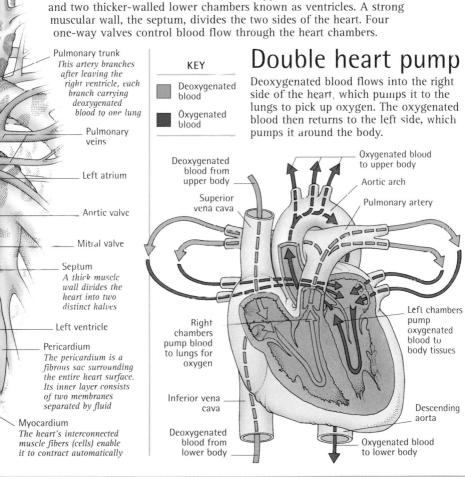

Pulmonary trunk
This artery branches after leaving the right ventricle, each branch carrying deoxygenated blood to one lung

Pulmonary veins

Left atrium

Aortic valve

Mitral valve

Septum
A thick muscle wall divides the heart into two distinct halves

Left ventricle

Pericardium
The pericardium is a fibrous sac surrounding the entire heart surface. Its inner layer consists of two membranes separated by fluid

Myocardium
The heart's interconnected muscle fibers (cells) enable it to contract automatically

KEY
☐ Deoxygenated blood
☐ Oxygenated blood

Deoxygenated blood from upper body

Superior vena cava

Right chambers pump blood to lungs for oxygen

Inferior vena cava

Deoxygenated blood from lower body

Oxygenated blood to upper body

Aortic arch

Pulmonary artery

Left chambers pump oxygenated blood to body tissues

Descending aorta

Oxygenated blood to lower body

Blood supply to the heart

The heart needs a generous supply of oxygen, which is provided by a correspondingly large supply of blood (the brain is the only organ that requires greater blood volume). Blood that flows through the chambers of the heart cannot supply the heart's own muscle cells, so the heart muscle has a separate network of blood vessels called the coronary system.

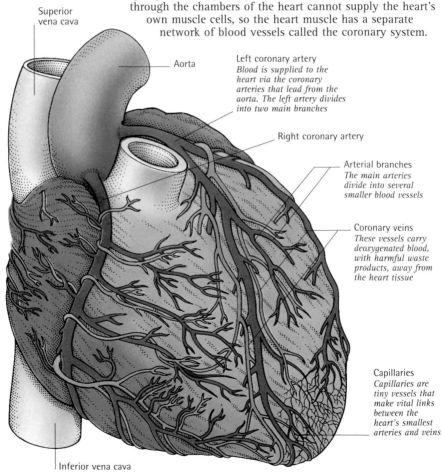

Superior vena cava

Aorta

Left coronary artery
Blood is supplied to the heart via the coronary arteries that lead from the aorta. The left artery divides into two main branches

Right coronary artery

Arterial branches
The main arteries divide into several smaller blood vessels

Coronary veins
These vessels carry deoxygenated blood, with harmful waste products, away from the heart tissue

Capillaries
Capillaries are tiny vessels that make vital links between the heart's smallest arteries and veins

Inferior vena cava

Blood collection

Deoxygenated blood from the heart muscle drains into the coronary veins, most of which in turn drain into the coronary sinus, a large vein at the back of the heart. From this vein, the blood flows back into the right atrium.

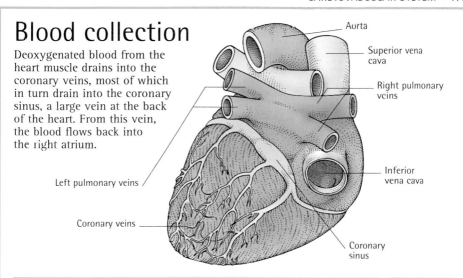

Aorta

Superior vena cava

Right pulmonary veins

Inferior vena cava

Left pulmonary veins

Coronary veins

Coronary sinus

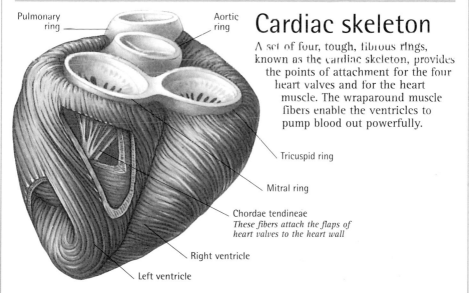

Pulmonary ring

Aortic ring

Cardiac skeleton

A set of four, tough, fibrous rings, known as the cardiac skeleton, provides the points of attachment for the four heart valves and for the heart muscle. The wraparound muscle fibers enable the ventricles to pump blood out powerfully.

Tricuspid ring

Mitral ring

Chordae tendineae
These fibers attach the flaps of heart valves to the heart wall

Right ventricle

Left ventricle

Heart function

The rhythmic, automatically repeated beating of the heart is maintained by electrical impulses originating in the sinoatrial node, which is the body's natural pacemaker. Impulses spread through the atria, stimulating contraction, to the atrioventricular node. After a slight pause at this node, the impulses pass along special conducting muscle fibers through the ventricles, causing them to contract.

ECG recording
Electrocardiography (ECG) records electrical activity in the heart. The colors relate to the impulse pathway (see left).

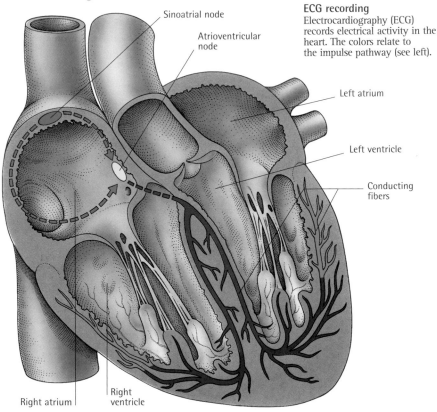

Sinoatrial node

Atrioventricular node

Left atrium

Left ventricle

Conducting fibers

Right atrium

Right ventricle

Nervous system control

Without control by nerves, heart rate would be about 100 beats per minute. The autonomic nerves, especially the vagus nerve of the parasympathetic division, keep the resting rate at about 70 beats by means of impulses from the cardioregulatory center in the medulla. During exercise or stress, the sympathetic cardiac nerves, influenced by the hypothalamus, speed the rate. Heart rate also increases when the adrenal glands release hormones. These changes increase the volume of blood pumped to the muscles.

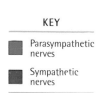

KEY

Parasympathetic nerves

Sympathetic nerves

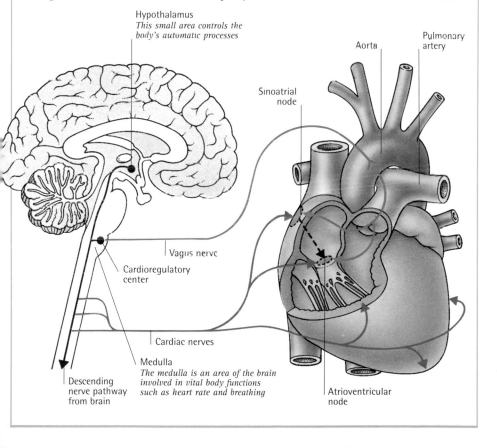

Hypothalamus
This small area controls the body's automatic processes

Aorta

Pulmonary artery

Sinoatrial node

Vagus nerve

Cardioregulatory center

Cardiac nerves

Medulla
The medulla is an area of the brain involved in vital body functions such as heart rate and breathing

Descending nerve pathway from brain

Atrioventricular node

Heart valves

Four valves allow a one-way flow of blood through the heart. The tricuspid and mitral valves are sited between the upper and lower chambers and are attached to the heart walls by fibrous strands called chordae tendineae. The pulmonary and aortic valves are situated at the exits from the ventricles. Valves consist of two or three flaps, or cusps, of tissue that open to let blood through then seal shut to prevent backflow.

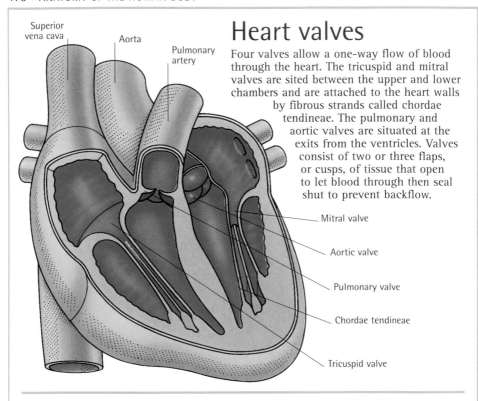

Superior vena cava

Aorta

Pulmonary artery

Mitral valve

Aortic valve

Pulmonary valve

Chordae tendineae

Tricuspid valve

Valve cusps

The thin, fibrous cusps of the heart valves are covered by a smooth membrane called endocardium and reinforced by dense connective tissue. The pulmonary, aortic, and tricuspid valves all have three cusps. The mitral valve has two cusps.

TWO CUSPS

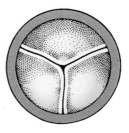

THREE CUSPS

Heartbeat cycle

Three separate and distinct phases make up the sequential beating of the heart. The relaxing of muscle and refilling of the chambers with blood during the first stage is followed by stages when the atria and then the ventricles contract, sending blood through the heart and out into the arteries. The cycle takes about four-fifths of a second; during vigorous exercise, however, or in times of stress, this speed may more than double.

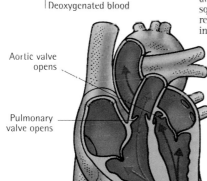

Diastole
In the first phase of the cycle, deoxygenated blood enters the right atrium and oxygenated blood enters the left atrium. The blood then flows through into the ventricles.

Oxygenated blood

Deoxygenated blood

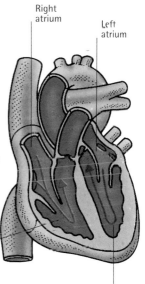

Right atrium

Left atrium

Atrial systole
Impulses from the sinoatrial node initiate the next phase of the cycle, during which the atria contract. This squeezes any blood remaining in the atria into the ventricles.

Right ventricle

Left ventricle

Aortic valve opens

Pulmonary valve opens

Ventricular systole
During the third phase of the heartbeat cycle, the ventricles contract. Valves at the ventricle exits open and blood is forced into the pulmonary artery and the aorta. As this phase ends, diastole starts again.

Blood circulation

Blood circulates through the body in two systems (pulmonary and systemic) through interconnecting pathways of arteries, veins, and smaller blood vessels. In the pulmonary circulation, deoxygenated blood travels from the heart to the lungs, where it is freshly enriched with oxygen before returning to the heart. The oxygenated blood is then transported by the systemic circulation to all body parts.

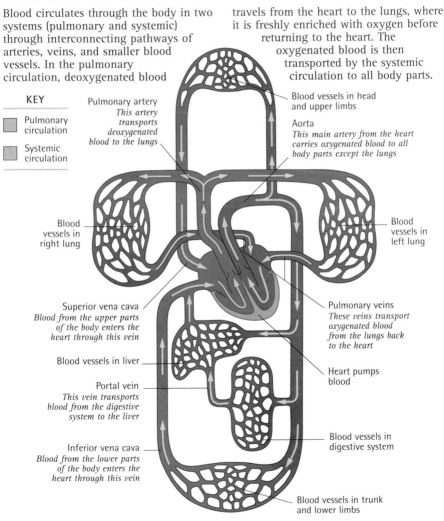

KEY

Pulmonary circulation

Systemic circulation

Pulmonary artery
This artery transports deoxygenated blood to the lungs

Blood vessels in head and upper limbs

Aorta
This main artery from the heart carries oxygenated blood to all body parts except the lungs

Blood vessels in right lung

Blood vessels in left lung

Superior vena cava
Blood from the upper parts of the body enters the heart through this vein

Pulmonary veins
These veins transport oxygenated blood from the lungs back to the heart

Blood vessels in liver

Portal vein
This vein transports blood from the digestive system to the liver

Heart pumps blood

Inferior vena cava
Blood from the lower parts of the body enters the heart through this vein

Blood vessels in digestive system

Blood vessels in trunk and lower limbs

Hepatic portal system

A portal system is an arrangement of blood vessels between two different sets of tissue. Blood from the stomach, spleen, intestines, and pancreas drains into a number of veins, which merge to become the portal vein. This vein transports gastrointestinal blood to the liver, which absorbs and stores nutrients, and also removes toxins and pollutants. Detoxified blood enters the inferior vena cava and returns to the heart and lungs for oxygenation and redistribution.

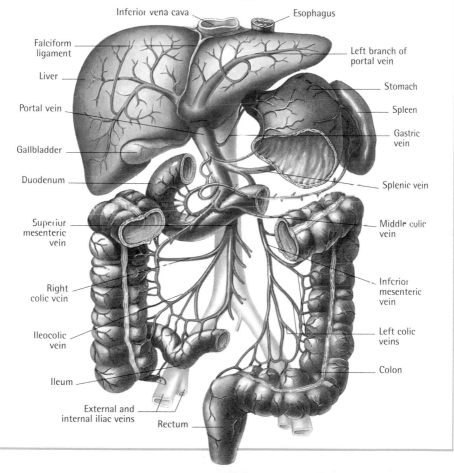

Artery structure

Without oxygen, all cells in the body would quickly die.
Oxygenated blood from the lungs circulates through muscular,
thick-walled arteries, such as the one shown here, into
progressively smaller vessels called arterioles. These vessels
connect with the venous system through capillaries. Blood
also transports nutrients and waste products. Its specialized
cells protect against infection and help to stop bleeding.

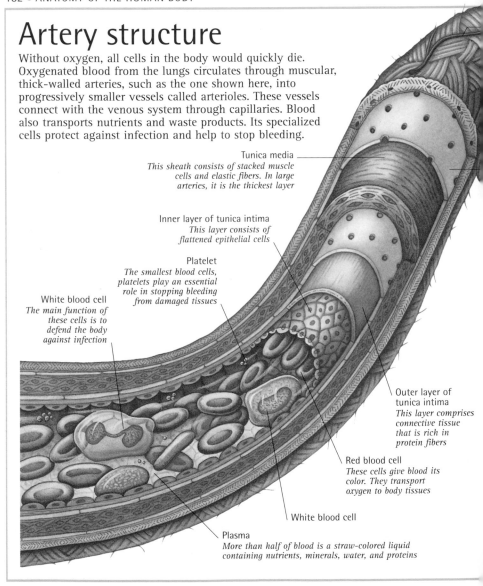

Tunica media
This sheath consists of stacked muscle cells and elastic fibers. In large arteries, it is the thickest layer

Inner layer of tunica intima
This layer consists of flattened epithelial cells

Platelet
The smallest blood cells, platelets play an essential role in stopping bleeding from damaged tissues

White blood cell
The main function of these cells is to defend the body against infection

Outer layer of tunica intima
This layer comprises connective tissue that is rich in protein fibers

Red blood cell
These cells give blood its color. They transport oxygen to body tissues

White blood cell

Plasma
More than half of blood is a straw-colored liquid containing nutrients, minerals, water, and proteins

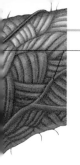

Vasa vasorum
These tiny blood vessels supply nutrients to the artery wall

Elastic layer
In some blood vessels, a flexible membrane separates the tunica adventitia from the tunica media

Linking blood vessel

Tunica adventitia
This protein sheath contains nerves, blood vessels, and lymph vessels

Vein structure

Veins have thin, flexible walls that can expand to hold large volumes of blood. Deoxygenated blood returning to the heart through the veins is circulating at low pressure. Its movement is helped by a succession of one-way valves that prevent the blood from flowing back in the wrong direction.

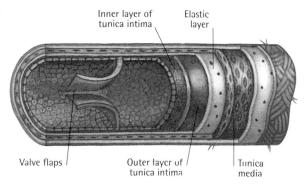

Inner layer of tunica intima

Elastic layer

Valve flaps

Outer layer of tunica intima

Tunica media

Capillary network

The two circulatory routes, arterial and venous, are linked by capillaries. These tiny blood vessels interlink to form networks, the density of which varies with the type of tissue. The thin capillary walls are permeable, allowing the exchange of nutrients, oxygen-bearing fluids, and wastes between the blood and tissue cells.

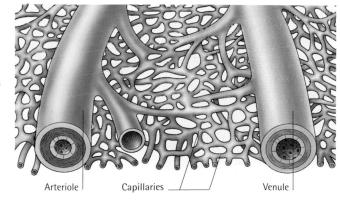

Arteriole

Capillaries

Venule

Components of blood

Blood is composed of tissue cells and a fluid known as plasma, which is mostly water but also contains various substances such as proteins and nutrients. There are several different types of blood cell, by far the most numerous of which are red blood cells. All blood cells have specific roles to play in the efficient functioning of the body. The average volume of blood in an adult human is about 11 pints (5 litres).

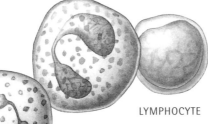

EOSINOPHIL

LYMPHOCYTE

BASOPHIL

MONOCYTE

NEUTROPHIL

White blood cells (leucocytes)
There are five types of white blood cell: neutrophils, eosinophils, lymphocytes, monocytes, and basophils. Their main role is to protect the body against infection from invading organisms.

Platelets (thrombocytes)
The blood contains billions of these minute cells. If a blood vessel is damaged, platelets are activated to start clumping together at the site of the injury. This platelet plug helps to seal the vessel and prevent blood loss.

Red blood cells (erythrocytes)
These cells are the most numerous type of blood cell. Their unusual biconcave shape increases the surface area of the cells, allowing them to absorb oxygen from the lungs with maximum efficiency.

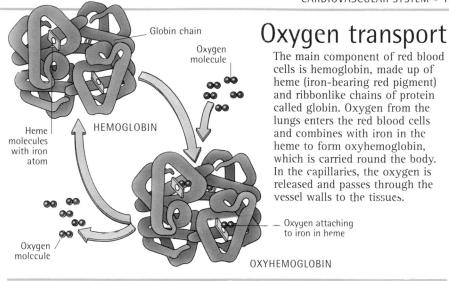

Globin chain

Oxygen molecule

HEMOGLOBIN

Heme molecules with iron atom

Oxygen molecule

Oxygen attaching to iron in heme

OXYHEMOGLOBIN

Oxygen transport

The main component of red blood cells is hemoglobin, made up of heme (iron-bearing red pigment) and ribbonlike chains of protein called globin. Oxygen from the lungs enters the red blood cells and combines with iron in the heme to form oxyhemoglobin, which is carried round the body. In the capillaries, the oxygen is released and passes through the vessel walls to the tissues.

Red blood cell production

Every second, two million oxygen-bearing red blood cells die, but they are replaced at the same rate by new ones generated in the process known as erythropoiesis. This begins in the kidney cells, which are stimulated by low oxygen levels to release a hormone known as erythropoietin. The hormone travels to the body's red bone marrow, where it stimulates increased production of erythrocytes, raising blood oxygen levels.

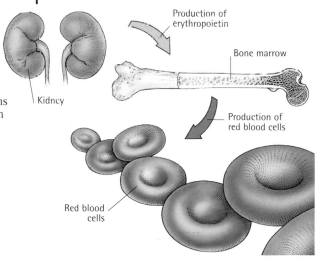

Kidney

Production of erythropoietin

Bone marrow

Production of red blood cells

Red blood cells

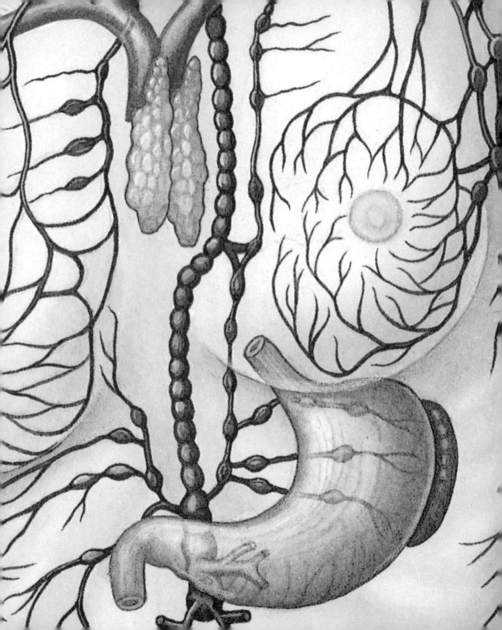

Lymphatic and immune systems

The healthy body has an internal defense mechanism known as the immune system, which guards against invasion by disease-causing organisms and is also activated by cancerous changes in cells. This system is based on specialized white cells called lymphocytes, which respond to infection or cell abnormalities in a number of different ways.

Lymphatic system
The lymphatic system, which comprises various tissues, vessels, and nodes containing white blood cells, is a major part of the immune system.

Lymphatic system

The lymphatic system is part of the body's immune system, which provides defense against disease-causing organisms. Lymph is a watery fluid that leaks out of blood vessels and accumulates in the spaces between the cells of body tissues. This fluid drains into a network of lymph capillaries and then into larger vessels known as lymphatics, which are studded with filters called nodes. The nodes contain specialized white blood cells (lymphocytes and macrophages) that neutralize or destroy microorganisms in lymph before it returns to the bloodstream. Lymph is circulated by the action of body muscles, which pushes the fluid through a series of one-way valves.

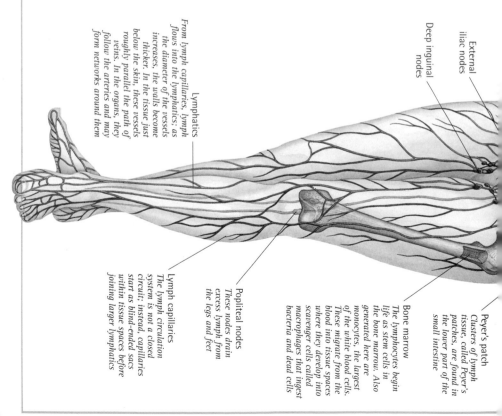

External iliac nodes

Deep inguinal nodes

Lymphatics
From lymph capillaries, lymph flows into the lymphatics; as the diameter of the vessels increases, the walls become thicker. In the tissue just below the skin, these vessels roughly parallel the path of veins. In the organs, they follow the arteries and may form networks around them

Peyer's patch
Clusters of lymph tissue, called Peyer's patches, are found in the lower part of the small intestine

Bone marrow
The lymphocytes begin life as stem cells in the bone marrow. Also generated here are monocytes, the largest of the white blood cells. These migrate from the blood into tissue spaces where they develop into scavenger cells called macrophages that ingest bacteria and dead cells

Popliteal nodes
These nodes drain excess lymph from the legs and feet

Lymph capillaries
The lymph circulation system is not a closed circuit; instead, capillaries start as blind-ended sacs within tissue spaces before joining larger lymphatics

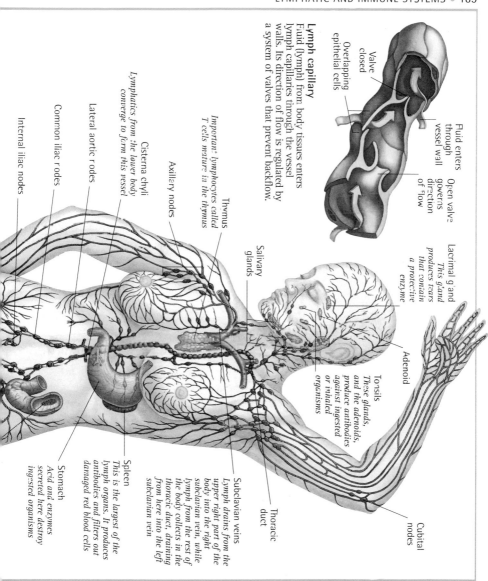

Lymph capillary
Fluid (lymph) from body tissues enters lymph capillaries through the vessel walls. Its direction of flow is regulated by a system of valves that prevent backflow.

Valve closed

Fluid enters through vessel wall

Overlapping epithelial cells

Open valve governs direction of flow

Thymus
Important lymphocytes called T cells mature in the thymus

Cisterna chyli
Lymphatics from the lower body converge to form this vessel

Axillary nodes

Lateral aortic nodes

Common iliac nodes

Internal iliac nodes

Salivary glands

Lacrimal gland
This gland produces tears that contain a protective enzyme

Adenoid

Tonsils
These glands, and the adenoids, produce antibodies against ingested or inhaled organisms

Cubital nodes

Thoracic duct

Subclavian veins
Lymph drains from the upper right part of the body into the right subclavian vein, while lymph from the rest of the body collects in the thoracic duct, draining from here into the left subclavian vein

Spleen
This is the largest of the lymph organs. It produces antibodies and filters out damaged red blood cells

Stomach
Acid and enzymes secreted here destroy ingested organisms

Lymph node structure

Lymph nodes are small masses of tissue enclosed in a fibrous capsule. They range in size from 0.04–0.8in (1–20mm), and within each node are cavities (sinuses) containing two types of white blood cells: lymphocytes and macrophages. These cells play a major role in defending the body against infection. Lymph from most tissues filters through at least one node before returning to the bloodstream.

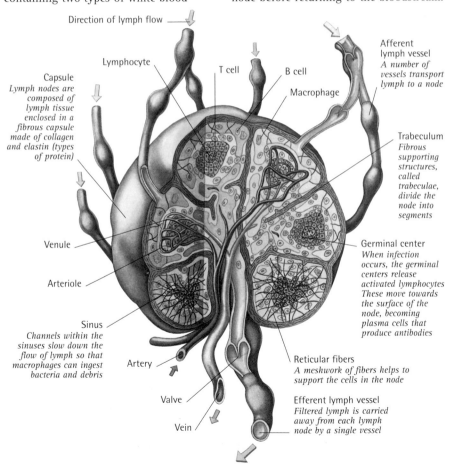

Direction of lymph flow

Lymphocyte

T cell

B cell

Macrophage

Afferent lymph vessel
A number of vessels transport lymph to a node

Capsule
Lymph nodes are composed of lymph tissue enclosed in a fibrous capsule made of collagen and elastin (types of protein)

Trabeculum
Fibrous supporting structures, called trabeculae, divide the node into segments

Venule

Arteriole

Germinal center
When infection occurs, the germinal centers release activated lymphocytes These move towards the surface of the node, becoming plasma cells that produce antibodies

Sinus
Channels within the sinuses slow down the flow of lymph so that macrophages can ingest bacteria and debris

Artery

Reticular fibers
A meshwork of fibers helps to support the cells in the node

Valve

Vein

Efferent lymph vessel
Filtered lymph is carried away from each lymph node by a single vessel

Inflammatory response

Some disease organisms trigger an inflammatory response in affected tissues. This type of defense is not specially tailored to destroy a specific organism, but attacks all invading organisms in the same way. The body's inflammatory response increases blood flow and brings special cells called neutrophils to the infected area (shown here are bronchi) to ingest and destroy the organisms.

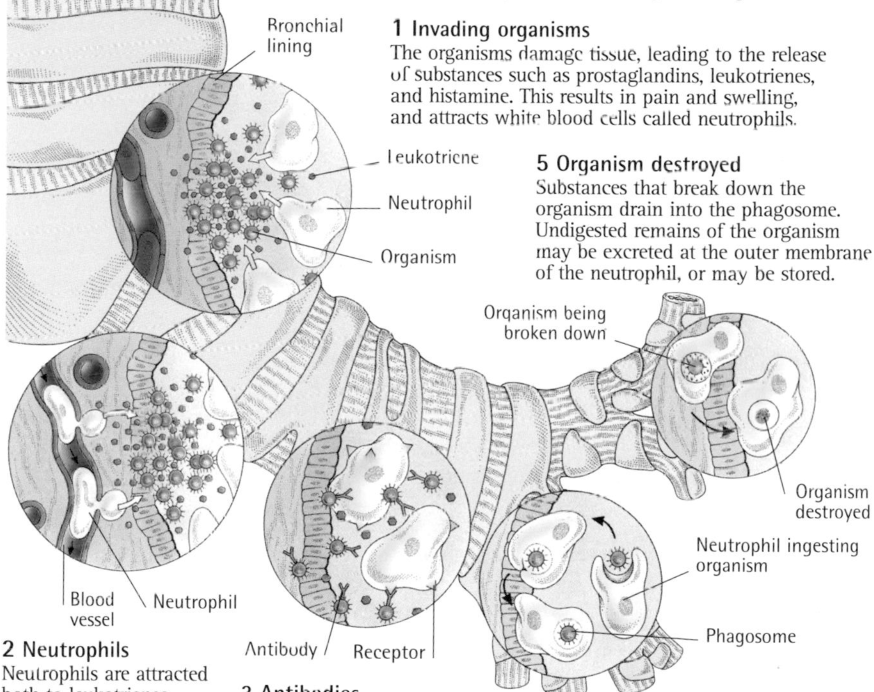

Bronchial lining

Leukotriene

Neutrophil

Organism

1 Invading organisms

The organisms damage tissue, leading to the release of substances such as prostaglandins, leukotrienes, and histamine. This results in pain and swelling, and attracts white blood cells called neutrophils.

5 Organism destroyed

Substances that break down the organism drain into the phagosome. Undigested remains of the organism may be excreted at the outer membrane of the neutrophil, or may be stored.

Organism being broken down

Organism destroyed

Neutrophil ingesting organism

Phagosome

Blood vessel

Neutrophil

Antibody

Receptor

2 Neutrophils

Neutrophils are attracted both to leukotrienes and to toxins produced by the organisms. They pass through spaces in the blood vessel walls to reach the damaged tissues.

3 Antibodies

Specifically created proteins called antibodies attach to the invading organisms. Both antibodies and disease organisms then attach to receptors on the neutrophils.

4 Phagocytosis

The neutrophils engulf the organisms. After ingestion, the organism is isolated within a small structure called a phagosome. The process is known as phagocytosis.

Specific immune responses

If infection is not overcome by the inflammatory response, two types of specific defense – an antibody and a cellular defense – may be activated. These defenses are called immune responses. They depend on the action of white blood cells known as B and T lymphocytes, and provide protection against future infections.

Antibody defenses
B lymphocytes recognize foreign molecules from disease organisms, called antigens, that are different from natural body proteins. Antigens trigger B cells to multiply. Some develop into plasma cells, which secrete antibodies (special proteins that inactivate the antigens).

Antigen

B lymphocyte
These cells begin life as stem cells in bone marrow

Plasma cell

Killer T cell

Invading organism with antigen

T cells multiply

Memory T cell
These cells may survive for many years to respond to an attempted second invasion by the same antigen. They mobilize very quickly

Infected cell

Lymphokine

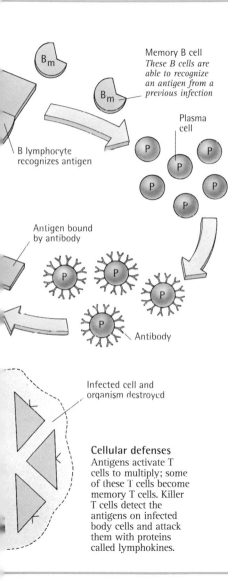

Memory B cell
These B cells are able to recognize an antigen from a previous infection

B lymphocyte recognizes antigen

Plasma cell

Antigen bound by antibody

Antibody

Infected cell and organism destroyed

Cellular defenses
Antigens activate T cells to multiply; some of these T cells become memory T cells. Killer T cells detect the antigens on infected body cells and attack them with proteins called lymphokines.

Complement system

Circulating in the blood are about 20 inactive proteins. These "complement" proteins are activated when they come into contact with foreign particles or when the immune system creates specific antibodies against disease organisms. They help to destroy bacteria and neutralize their toxins.

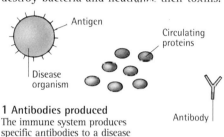

Antigen

Circulating proteins

Disease organism

Antibody

1 Antibodies produced
The immune system produces specific antibodies to a disease organism; these antibodies attach to the organism.

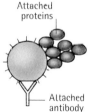

Attached proteins

2 Proteins activated
Activated complement proteins also attach to the disease organism, making a hole in the organism's cell membrane.

Attached antibody

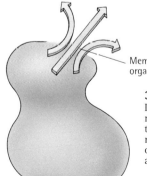

Membrane of disease organism ruptures

3 Organism dies
Intracellular fluid rushes through the hole in the membrane; the organism bursts and dies.

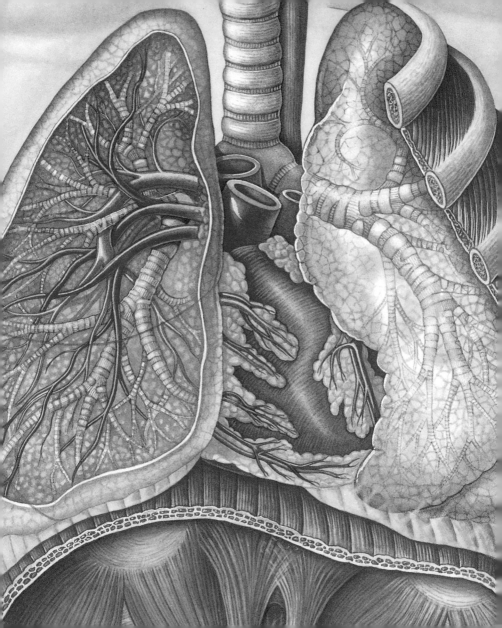

Respiratory system

To function, body cells need oxygen. The respiratory system, which consists of air passages, pulmonary vessels, and the lungs, as well as breathing muscles, supplies fresh oxygen to the blood for distribution to the rest of the body tissues. In addition, respiration removes carbon dioxide, a waste product of body processes.

Breathing
Breathing, when air passes into and out of the lungs, is usually an involuntary process but can be controlled consciously.

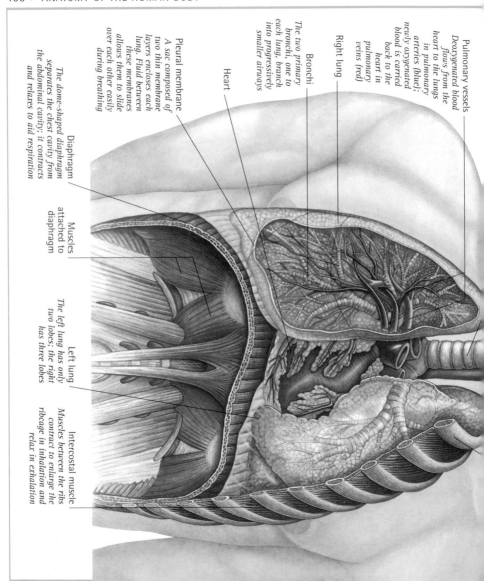

Pulmonary vessels
Deoxygenated blood flows from the heart to the lungs in pulmonary arteries (blue); newly oxygenated blood is carried back to the heart in pulmonary veins (red)

Right lung

Bronchi
The two primary bronchi, one to each lung, branch into progressively smaller airways

Heart

Pleural membrane
A sac composed of two thin membrane layers encloses each lung. Fluid between these membranes allows them to slide over each other easily during breathing

Diaphragm
The dome-shaped diaphragm separates the chest cavity from the abdominal cavity; it contracts and relaxes to aid respiration

Muscles attached to diaphragm

Left lung
The left lung has only two lobes; the right has three lobes

Intercostal muscle
Muscles between the ribs contract in inhalation and relax in exhalation

Air passages

As air is inhaled and passes through the nasal passages, it is filtered, heated, and humidified. The filtering process continues as air flows down through the pharynx, larynx, trachea, and bronchi to the lungs. Each lung contains a tree of branching tubes that end in tiny air sacs, or alveoli, where gases diffuse into and out of the blood stream through tiny vessels.

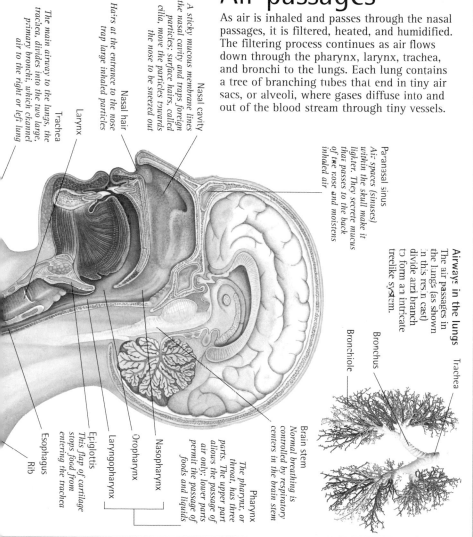

Nasal cavity
A sticky mucous membrane lines the nasal cavity and traps foreign particles; surface hairs, called cilia, move the particles towards the nose to be sneezed out

Nasal hair
Hairs at the entrance to the nose trap large inhaled particles

Larynx

Trachea
The main airway to the lungs, the trachea, divides into the two large, primary bronchi, which channel air to the right or left lung

Paranasal sinus
Air spaces (sinuses) within the skull make it lighter. They secrete mucus that passes to the back of the nose and moistens inhaled air

Airways in the lungs
The air passages in the lungs (as shown in this resin cast) divide and branch to form an intricate treelike system.

Trachea

Bronchus

Bronchiole

Brain stem
Normal breathing is controlled by respiratory centers in the brain stem

Pharynx
The pharynx, or throat, has three parts. The upper part allows the passage of air only; lower parts permit the passage of foods and liquids

Esophagus

Rib

Epiglottis
This flap of cartilage stops food from entering the trachea

Laryngopharynx

Oropharynx

Nasopharynx

Lung structure

Each lung is a cone-shaped organ of spongy tissue that contains millions of tiny air sacs called alveoli. Within the lungs is an intricate network of air passages that starts at the trachea below the larynx. The trachea bifurcates to form two primary (main) bronchi, which enter each lung and subdivide into increasingly smaller branches called bronchioles. The bronchioles in turn lead to the alveoli, which they supply with air.

Trachea

Right primary bronchus

Lobes of the lung
The right lung is separated by surface fissures into three lobes, while the smaller left lung is divided into only two lobes. Each lobe is subdivided into segments

Rib

Tertiary bronchus
These branches of the five lobar (secondary) bronchi are also called segmental bronchi because each one aerates an individual segment within each lobe. They may subdivide further into 50–80 terminal bronchioles

Secondary bronchus
The five secondary, or lobar, bronchi are branches of the primary bronchi. Each supplies an individual lobe

Diaphragm
This dome-shaped muscle separates the chest cavity from the abdominal cavity. It contracts and flattens during inhalation, creating space for air intake. It relaxes as air is expelled

Terminal bronchiole
These tiny bronchioles – about 30,000 in each lung – are the terminal ends of the segmental bronchi. They divide into two or more respiratory bronchioles that lead into the alveoli via alveolar ducts

Alveolus

Pleural membranes
Each lung is covered by two membrane layers separated by a lubricating fluid that allows them to slide easily during breathing

Left primary bronchus

Surfactant

The lungs remain partly inflated even after exhalation because of a vital substance, known as surfactant, secreted inside the alveoli. This surfactant, which is composed mainly of special lipids (fats), is produced by specialized cells.

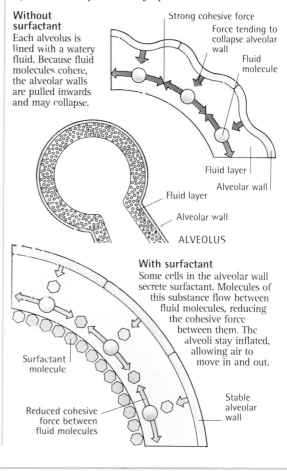

Without surfactant
Each alveolus is lined with a watery fluid. Because fluid molecules cohere, the alveolar walls are pulled inwards and may collapse.

Strong cohesive force

Force tending to collapse alveolar wall

Fluid molecule

Fluid layer

Alveolar wall

Fluid layer

Alveolar wall

ALVEOLUS

With surfactant
Some cells in the alveolar wall secrete surfactant. Molecules of this substance flow between fluid molecules, reducing the cohesive force between them. The alveoli stay inflated, allowing air to move in and out.

Surfactant molecule

Reduced cohesive force between fluid molecules

Stable alveolar wall

Alveoli

The air sacs of the lungs, called alveoli, are elastic, thin-walled structures that are supplied with air by respiratory bronchioles. Tiny blood capillaries surrounding the alveolar walls allow oxygen to be carried into the bloodstream. In exchange, carbon dioxide waste diffuses from blood into the alveoli, from where it is exhaled. White blood cells, called macrophages, on the inner surface of each alveolus trap and destroy inhaled bacteria and fine particles.

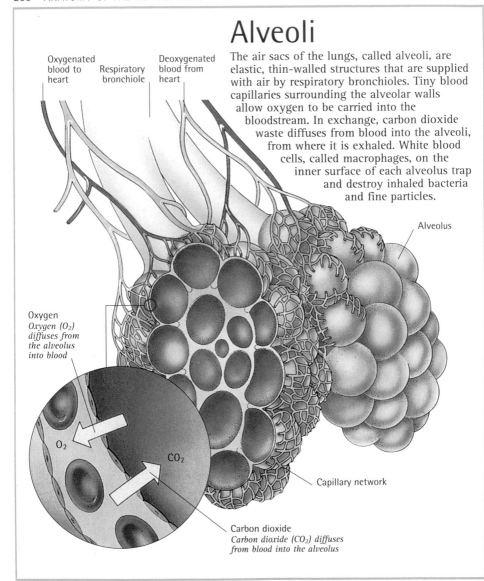

Oxygenated blood to heart

Respiratory bronchiole

Deoxygenated blood from heart

Alveolus

Oxygen
Oxygen (O$_2$) diffuses from the alveolus into blood

O$_2$

CO$_2$

Capillary network

Carbon dioxide
Carbon dioxide (CO$_2$) diffuses from blood into the alveolus

Respiration

The process by which oxygen is carried to body cells and carbon dioxide is removed is called respiration. This process takes place in three parts: pulmonary ventilation, when air is taken into the lungs; external respiration, the exchange of respiratory gases between the lungs and the blood; and internal respiration, the exchange of respiratory gases between blood and body tissues.

Gas exchange

In the lungs, carbon dioxide (CO_2) from the blood passes into the alveoli through the respiratory membrane, a thin barrier that has several layers. Oxygen (O_2) crosses the membrane in the opposite direction, from the alveoli to the blood capillaries.

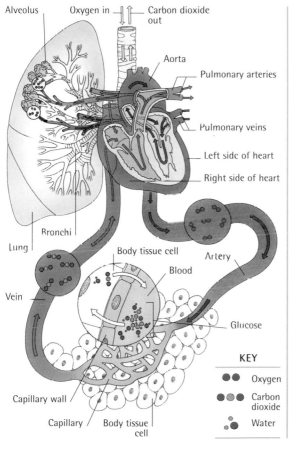

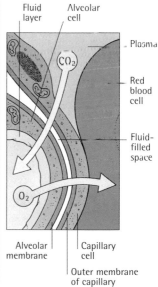

KEY

- Oxygen
- Carbon dioxide
- Water

Breathing

Movement of air in and out of the lungs is generated by differences in pressure inside and outside the body. The most important muscle used in breathing is the diaphragm, a muscular sheet between the base of the lungs and the abdominal cavity. The diaphragm is assisted by the internal and external intercostal muscles (which lie between the ribs) and by the neck and abdominal muscles. A person normally breathes in and out about 1pt (500ml) of air 12–17 times a minute.

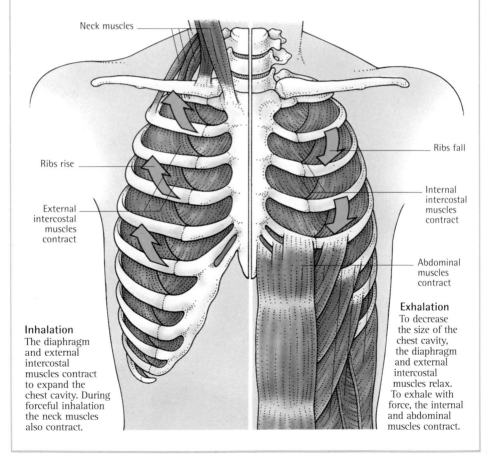

Neck muscles

Ribs rise

External intercostal muscles contract

Ribs fall

Internal intercostal muscles contract

Abdominal muscles contract

Inhalation
The diaphragm and external intercostal muscles contract to expand the chest cavity. During forceful inhalation the neck muscles also contract.

Exhalation
To decrease the size of the chest cavity, the diaphragm and external intercostal muscles relax. To exhale with force, the internal and abdominal muscles contract.

Pressure changes

The human body is adapted to breathing in an average air pressure of about 760mmHg. During an intake of breath, the diaphragm contracts and increases the size of the chest cavity; pressure drops within the lungs and in the space between the two layers of the membrane around the lungs (the pleural space). Air rushes into the lungs to equalize pressure. As the diaphragm relaxes, pressure rises in the decreased area of the chest cavity. To equalize pressure, air is exhaled.

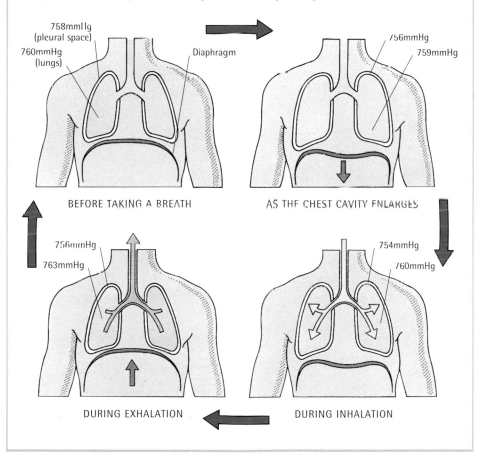

758mmHg (pleural space)

760mmHg (lungs)

Diaphragm

BEFORE TAKING A BREATH

756mmHg

759mmHg

AS THE CHEST CAVITY ENLARGES

756mmHg

763mmHg

DURING EXHALATION

754mmHg

760mmHg

DURING INHALATION

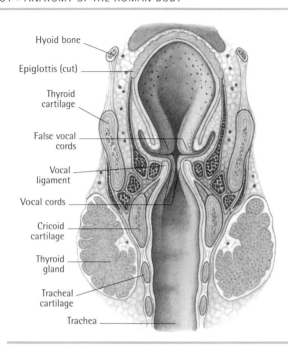

Hyoid bone

Epiglottis (cut)

Thyroid cartilage

False vocal cords

Vocal ligament

Vocal cords

Cricoid cartilage

Thyroid gland

Tracheal cartilage

Trachea

Larynx

The larynx, or voice box, lies between the pharynx and the trachea. It is composed of areas of cartilage and connective tissue. At the entrance to the larynx, there is a leaf-shaped flap of cartilage called the epiglottis. This structure remains upright to allow the passage of air but tips back during swallowing to close the larynx and prevent food from entering the trachea. The vocal cords that stretch across the larynx are responsible for the production of speech.

Vocal cords

The vocal cords are two fibrous bands at the base of the larynx. When exhaled air passes between them, the cords vibrate, making it possible to generate sounds. Depending on the tension in the cords, the sound varies. An upper pair of bands, the false vocal cords, do not produce sound, but move together during swallowing to close the larynx.

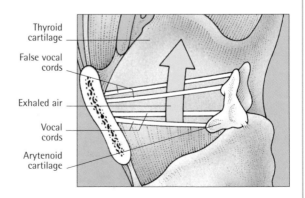

Thyroid cartilage

False vocal cords

Exhaled air

Vocal cords

Arytenoid cartilage

Coughing reflex

If the airways are irritated, for example by foreign particles, chemical fumes, or excess mucus, this stimulates nerve cell receptors in the larynx, trachea, and bronchi. Nerve signals are transmitted to the brain stem, which then relays a response that triggers coughing. This action (illustrated below) is often successful in expelling irritants and blockages from the airways.

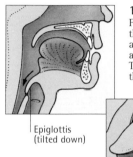

1 Larynx closes
Following a deep intake of air, the epiglottis, a flap of cartilage at the top of the larynx, tilts down and the false vocal cords close. This shuts off the larynx, trapping the air in the lungs.

Epiglottis
(tilted down)

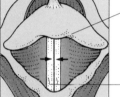

Epiglottis

VIEW OF VOCAL
CORDS FROM ABOVE

Vocal cords
(closed)

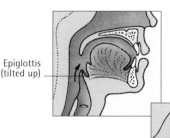

Epiglottis
(tilted up)

Epiglottis

3 Air expelled
When air pressure reaches its highest point, the epiglottis tilts up and the vocal cords move apart. Air is forced up the airway and propelled out as a cough.

Diaphragm

2 Diaphragm rises
The diaphragm rises and the abdominal muscles contract so that the lungs are increasingly compressed. Air in the lungs is held under greater pressure.

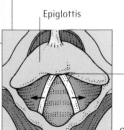

Vocal cords
(open)

VIEW OF VOCAL
CORDS FROM ABOVE

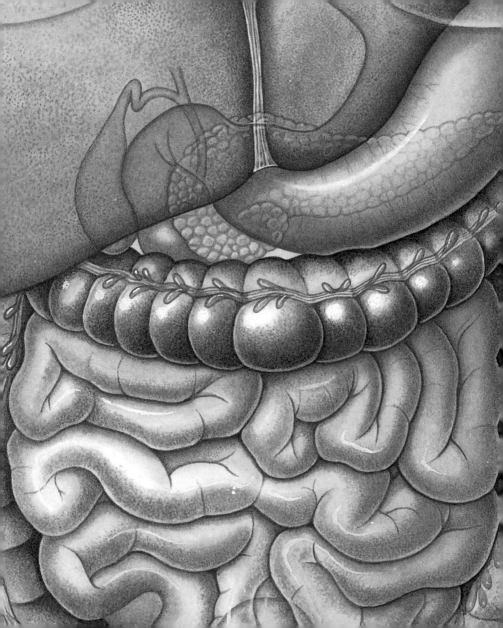

Digestive system

The task of the digestive system is the physical and chemical breakdown of food. Following ingestion, food and fluids are processed by the digestive organs so that nutrients can be absorbed from the intestines and circulated around the body. Any residue of food that is not digested is solidified and eliminated from the body in the form of feces.

Digestive tract
The process of digestion begins when food enters the mouth and is largely complete at the end of the small intestine.

Digestive organs

The mouth, pharynx, esophagus, stomach, small intestine, large intestine, rectum, and anus make up the digestive tract, which is basically a food-processing pipe about 30ft (9m) long. Associated digestive structures include three pairs of salivary glands, the pancreas, the liver, and the gallbladder, each of which has an important role. The appendix – a short, blind-ended tube attached to the large intestine – has no known function. Food is moved through the digestive tract by muscular contractions called peristalsis.

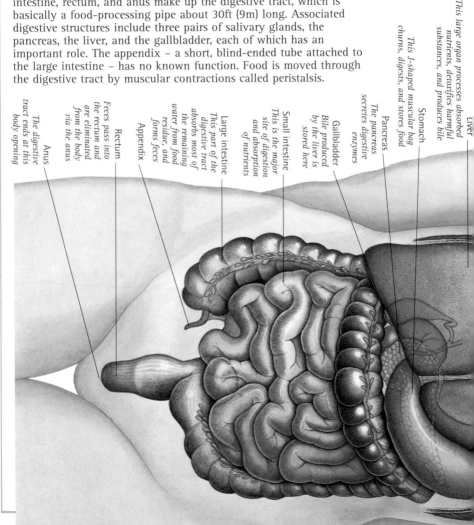

Liver
This large organ processes absorbed nutrients, detoxifies harmful substances, and produces bile

Stomach
This J-shaped muscular bag churns, digests, and stores food

Pancreas
The pancreas secretes digestive enzymes

Gallbladder
Bile produced by the liver is stored here

Small intestine
This is the major site of digestion and absorption of nutrients

Large intestine
This part of the digestive tract absorbs most of the remaining water from food residue, and forms feces

Appendix

Rectum
Feces pass into the rectum and are eliminated from the body via the anus

Anus
The digestive tract ends at this body opening

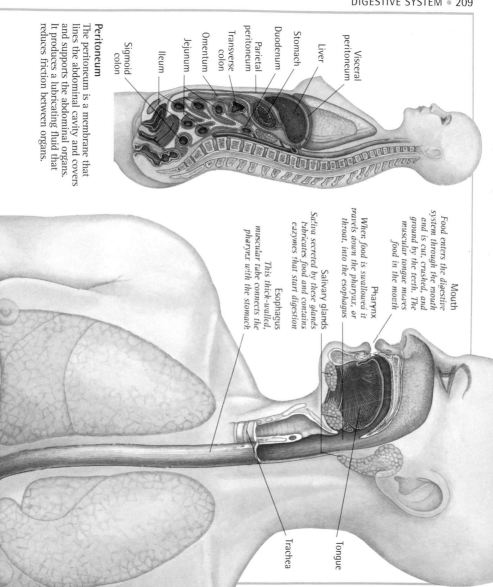

Peritoneum

The peritoneum is a membrane that lines the abdominal cavity and covers and supports the abdominal organs. It produces a lubricating fluid that reduces friction between organs.

Visceral peritoneum

Liver

Stomach

Duodenum

Parietal peritoneum

Transverse colon

Omentum

Jejunum

Ileum

Sigmoid colon

Mouth
Food enters the digestive system through the mouth and is cut, crushed, and ground by the teeth. The muscular tongue moves food in the mouth

Pharynx
When food is swallowed it travels down the pharynx, or throat, into the esophagus

Salivary glands
Saliva secreted by these glands lubricates food and contains enzymes that start digestion

Esophagus
This thick-walled, muscular tube connects the pharynx with the stomach

Tongue

Trachea

Breakdown of food

The digestive process breaks down food by chemical and mechanical action into substances that can pass into the bloodstream and be distributed to body cells. Certain nutrients, such as salts and minerals, can be absorbed directly into the circulation. Fats, complex carbohydrates, and proteins are broken down into smaller molecules before being absorbed. Fats are split into glycerol and fatty acids; carbohydrates are split into monosaccharide sugars; and proteins are split into linked amino acids called peptides, and then into individual amino acids.

3 Duodenum
Lipase, a pancreatic enzyme, breaks down fats into glycerol and fatty acids. Amylase, another enzyme produced by the pancreas, breaks down starch into maltose, a disaccharide sugar. Trypsin and chymotrypsin are powerful pancreatic enzymes that split proteins into polypeptides and peptides.

Small intestine

Large intestine

4 Small intestine
Maltase, sucrase, and lactase are enzymes produced by the lining of the small intestine. They convert disaccharide sugars into monosaccharide sugars. Peptidase, another enzyme produced in the intestine, splits large peptides into smaller peptides and then into amino acids.

5 Large intestine
Undigested food enters the large intestine, where water and salt are absorbed by the intestinal lining. The residue, together with waste pigments, dead cells, and bacteria, is pressed into feces and stored for excretion.

KEY

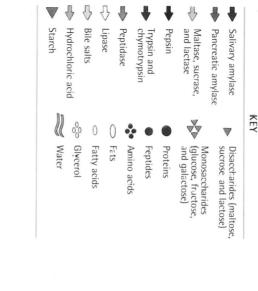

Salivary amylase	Disaccharides (maltose, sucrose and lactose)
Pancreatic amylase	Monosaccharides (glucose, fructose, and galactose)
Maltase, sucrase, and lactase	
Pepsin	Proteins
Trypsin and chymotrypsin	Peptides
Peptidase	Amino acids
Lipase	Fats
Bile salts	Fatty acids
Hydrochloric acid	Glycerol
Starch	Water

1 Mouth and Esophagus

Food is chewed with the teeth and mixed with saliva. The enzyme amylase, present in saliva, begins the breakdown of starch into sugar. Each lump of soft food, called a bolus, is swallowed and propelled by contractions down the esophagus to the stomach.

2 Stomach

Pepsin is an enzyme produced when pepsinogen, a substance secreted by the stomach lining, is modified by hydrochloric acid (also produced by the stomach lining). Pepsin breaks down proteins into smaller units called polypeptides and peptides. Lipase is a stomach enzyme that breaks down fats into glycerol and fatty acids. The acid produced by the stomach also kills bacteria.

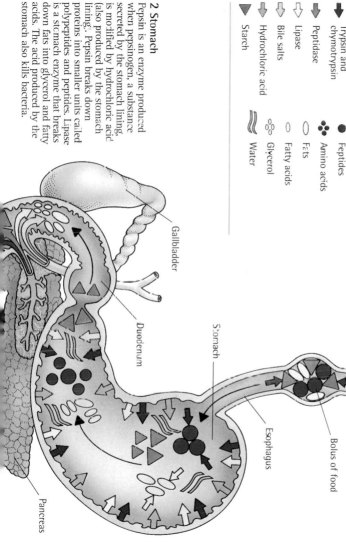

Gallbladder

Duodenum

Stomach

Esophagus

Bolus of food

Pancreas

Digestive enzymes

Enzymes are protein molecules that increase the speed of chemical reactions in the body. They work by combining with and altering the molecules of other chemical substances. There are thousands of different types of enzyme, with varied structures that determine their particular activity. The digestive enzymes secreted in the digestive tract split large molecules of food into smaller units for absorption.

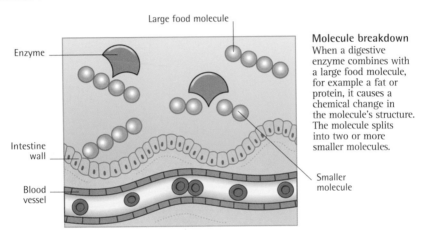

Large food molecule

Enzyme

Intestine wall

Blood vessel

Smaller molecule

Molecule breakdown
When a digestive enzyme combines with a large food molecule, for example a fat or protein, it causes a chemical change in the molecule's structure. The molecule splits into two or more smaller molecules.

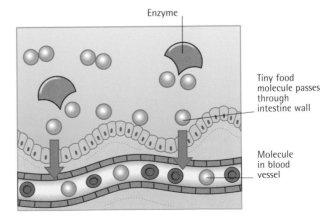

Enzyme

Tiny food molecule passes through intestine wall

Molecule in blood vessel

Absorption
The smaller food molecules separate from the enzyme and can now be absorbed. They pass through the wall of the intestine and into the bloodstream for distribution around the body. The digestive enzyme remains unaltered and will continue the process with further food molecules.

Components of food

One of the major components of food is water. Food also contains carbohydrates, fats, and proteins as well as vitamins, minerals, and fiber. Starchy and sugary foods are rich in carbohydrates, which, along with fats, are the body's main source of energy. Fats and protein are used for cell growth and repair.

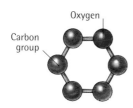

Oxygen

Carbon group

Monosaccharides
These single sugar units usually have a ring-shaped structure. They form the building blocks of the more complex carbohydrates.

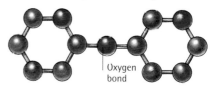

Oxygen bond

Disaccharides
Disaccharide molecules are formed from two monosaccharide units chemically bonded together. Sucrose, maltose, and lactose are the main disaccharides.

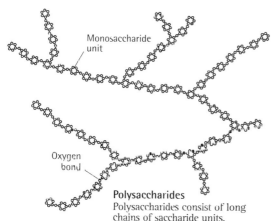

Monosaccharide unit

Oxygen bond

Polysaccharides
Polysaccharides consist of long chains of saccharide units. Starch and glycogen, which are both carbohydrates, are examples of polysaccharides.

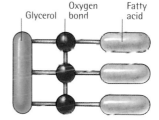

Glycerol

Oxygen bond

Fatty acid

Fats
Most dietary fats consist of three fatty acids linked by oxygen bonds to a glycerol molecule. Fatty acids are either saturated or unsaturated.

Proteins
Proteins are complex molecules consisting of long chains of amino acids. These acids link in various ways to form many different proteins.

Amino acid

How food provides energy

Energy is needed for the functioning of body cells. This energy is produced by breaking down food molecules in various chemical reactions, in particular the Krebs cycle that takes place within a cell's mitochondria. Because the energy cannot be used directly by the cell, it is stored in a high-energy chemical bond, which links a molecule of ADP (adenosine diphosphate) to a phosphate group, forming ATP (adenosine triphosphate). The bond breaks, releasing the energy and converting ATP back into ADP.

Outer membrane of a mitochondrion

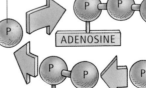

Mitochondria
Many energy-producing processes in body cells take place inside small structures called mitochondria.

Fuel for energy
Glucose and fatty acids are the main substances that are broken down to provide the fuel used by the Krebs cycle to produce energy.

Glucose and other fuels

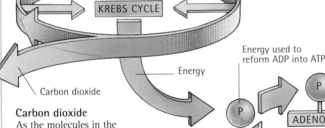

KREBS CYCLE

Energy used to reform ADP into ATP

Energy

Carbon dioxide

Carbon dioxide
As the molecules in the Krebs cycle are altered to produce energy, carbon dioxide is released as a by-product.

ATP splits, forming ADP and releasing energy

Energy release
Energy is released when ATP, the body's main energy-carrying chemical, converts to ADP. ADP continuously converts back into ATP, using the energy that has been released from glucose and fatty acids.

ENERGY

KEY

 Phosphate groups

Role of fiber

Dietary fiber consists of the parts of foods that are not broken down during digestion. It adds bulk to feces and increases the speed of their passage through the bowel. By delaying the absorption of sugar, fiber helps to control its level in the blood. Fiber also binds with cholesterol and bile salts, which are derived from cholesterol, and may reduce the amount of cholesterol in blood.

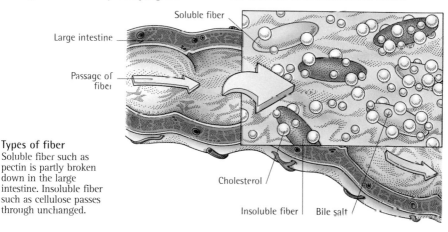

Soluble fiber

Large intestine

Passage of fiber

Cholesterol

Insoluble fiber | Bile salt

Types of fiber
Soluble fiber such as pectin is partly broken down in the large intestine. Insoluble fiber such as cellulose passes through unchanged.

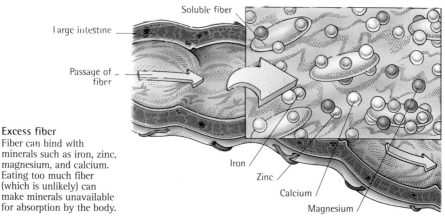

Soluble fiber

Large intestine

Passage of fiber

Iron

Zinc

Calcium

Magnesium

Excess fiber
Fiber can bind with minerals such as iron, zinc, magnesium, and calcium. Eating too much fiber (which is unlikely) can make minerals unavailable for absorption by the body.

Using the jaws

The processs of digestion starts as soon as food enters the mouth. Food is chewed, lubricated by saliva, and pressed by the tongue. The flexible temporomandibular joint, which attaches the jaw bone to the skull, allows the jaw to move in several different directions: from side to side, up and down, and backwards and forwards. With the assistance of the tongue, this allows food to be moved in the mouth and ground by the teeth until it is formed into a soft lump, called a bolus, for swallowing.

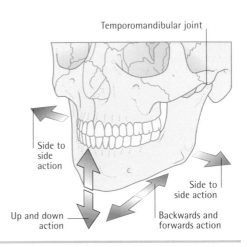

Temporomandibular joint

Side to side action

Side to side action

Up and down action

Backwards and forwards action

Role of the teeth

There are 20 teeth in the first (primary) set and 32 in the adult (secondary) set; they vary in type, according to their role. The incisors are chisel-shaped with sharp edges for cutting, while the pointed canines are designed for tearing. The ridged premolars and the flatter molars, the largest teeth, crush and grind food.

UPPER

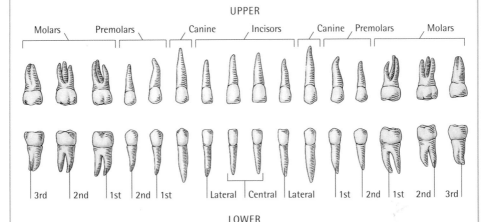

Molars | Premolars | Canine | Incisors | Canine | Premolars | Molars

3rd | 2nd | 1st | 2nd | 1st | Lateral | Central | Lateral | 1st | 2nd | 1st | 2nd | 3rd

LOWER

Tooth structure

The teeth are encased in sockets in the jaw bone and held in place by ligaments and shock-absorbent gums. At the center of each tooth there is a soft pulp which contains blood vessels and nerves; this is surrounded by a layer of sensitive tissue called dentine. Above the gum, the tooth has an outer covering of hard enamel. Below the gum, a bonelike tissue called cementum forms the tooth's outer layer.

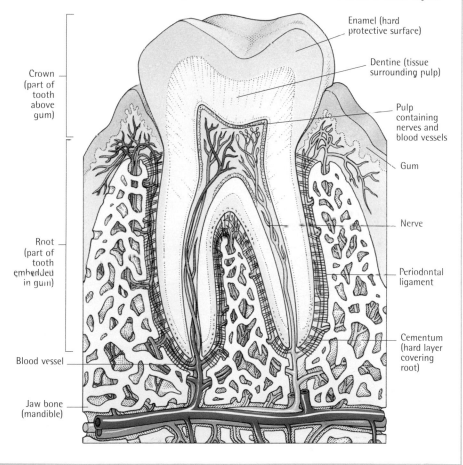

Enamel (hard protective surface)

Dentine (tissue surrounding pulp)

Pulp containing nerves and blood vessels

Gum

Nerve

Periodontal ligament

Cementum (hard layer covering root)

Crown (part of tooth above gum)

Root (part of tooth embedded in gum)

Blood vessel

Jaw bone (mandible)

Salivary glands

Saliva is produced by three pairs of salivary glands: the parotid, sublingual, and submandibular. There are also numerous small accessory glands in the mucous membrane lining the mouth and tongue. Saliva, which contains a digestive enzyme called amylase, is carried by ducts into the mouth. It moistens and softens food, making chewing and swallowing easier.

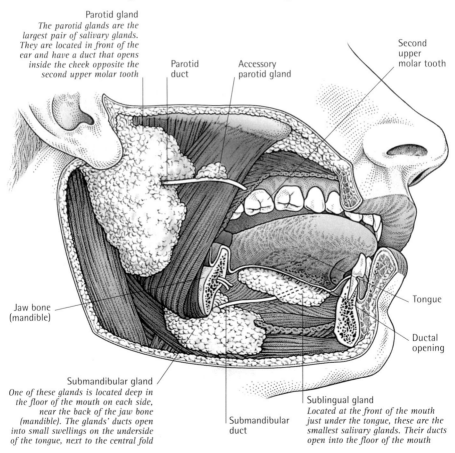

Parotid gland
The parotid glands are the largest pair of salivary glands. They are located in front of the ear and have a duct that opens inside the cheek opposite the second upper molar tooth

Parotid duct

Accessory parotid gland

Second upper molar tooth

Jaw bone (mandible)

Tongue

Ductal opening

Submandibular gland
One of these glands is located deep in the floor of the mouth on each side, near the back of the jaw bone (mandible). The glands' ducts open into small swellings on the underside of the tongue, next to the central fold

Submandibular duct

Sublingual gland
Located at the front of the mouth just under the tongue, these are the smallest salivary glands. Their ducts open into the floor of the mouth

Pharynx

The pharynx (throat) is a channel for both air and food. Its upper parts connect with the nose and mouth; the lower part connects with the larynx (voice box) and leads into the esophagus for swallowing. The epiglottis, and reflexes controlled by the brain, prevent food from being swallowed the wrong way. If food enters the larynx, the coughing reflex expels inhaled particles.

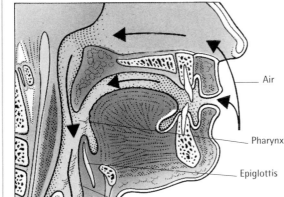

Air

Pharynx

Epiglottis

Larynx

Trachea

Esophagus

Breathing

During breathing, the vocal cords at the entrance to the larynx are relaxed and open, creating a space between them called the glottis. Air passes from the pharynx through the glottis into the trachea when it is inhaled, and passes from the trachea into the pharynx when it is exhaled.

Swallowing

During swallowing, the flap of cartilage known as the epiglottis tilts back and the larynx rises up. The vocal cords are pressed together, which closes the glottis and seals off the entrance to the trachea. As soon as food has entered the esophagus, the glottis reopens.

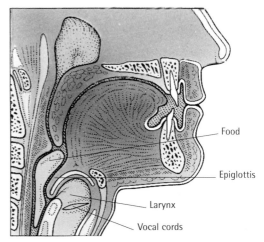

Food

Epiglottis

Larynx

Vocal cords

Esophagus

The esophagus is a thick-walled tube, consisting of layers of longitudinal and circular muscle fibers, which runs between the pharynx and the stomach. At its top end, the narrowest section of the entire digestive tract, a sphincter muscle controls the passage of food into the esophagus from the pharynx. The lower esophageal sphincter opens and closes the entrance to the stomach.

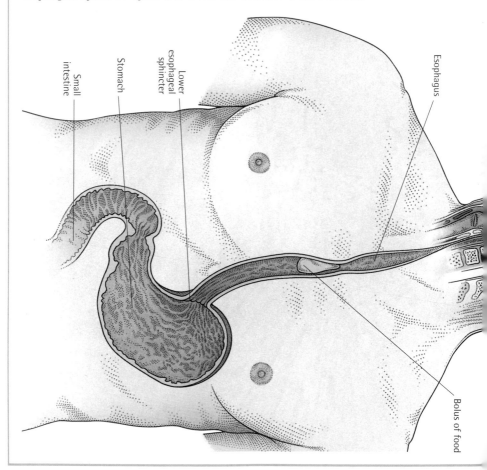

Small intestine

Stomach

Lower esophageal sphincter

Esophagus

Bolus of food

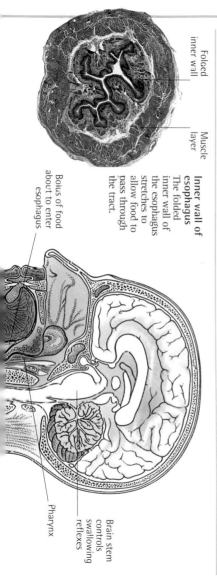

Folded
inner wall

Muscle
layer

**Inner wall of
esophagus**
The folded
inner wall of
the esophagus
stretches to
allow food to
pass through
the tract.

Bolus of food
about to enter
esophagus

Pharynx

Brain stem
controls
swallowing
reflexes

Peristalsis in the Esophagus

Peristalsis, the sequence of involuntary muscle contractions that occurs throughout the digestive tract, carries food down the esophagus to the stomach in about 4–8 seconds. As muscles in front of a bolus of food widen the esophagus, muscles behind it tighten the tube to push the food along. Controlled by the brain stem, the process makes it possible for food to be propelled along, even if the body is upside down.

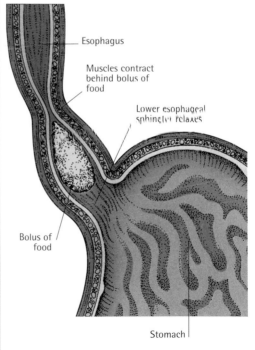

Esophagus

Muscles contract
behind bolus of
food

Lower esophageal
sphincter relaxes

Bolus of
food

Stomach

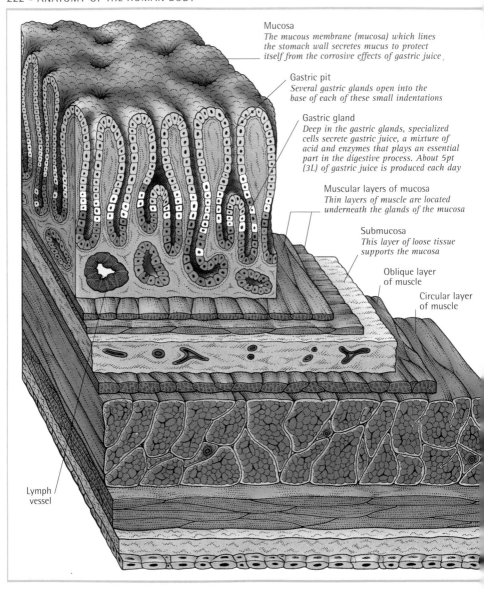

Mucosa
The mucous membrane (mucosa) which lines the stomach wall secretes mucus to protect itself from the corrosive effects of gastric juice

Gastric pit
Several gastric glands open into the base of each of these small indentations

Gastric gland
Deep in the gastric glands, specialized cells secrete gastric juice, a mixture of acid and enzymes that plays an essential part in the digestive process. About 5pt (3L) of gastric juice is produced each day

Muscular layers of mucosa
Thin layers of muscle are located underneath the glands of the mucosa

Submucosa
This layer of loose tissue supports the mucosa

Oblique layer of muscle

Circular layer of muscle

Lymph vessel

Stomach

The stomach is an elastic sac that forms the widest part of the digestive tract. It consists of the serosa, a membranous coating on the outer surface; layers of longitudinal, circular, and oblique muscle; the submucosa, composed of loose connective tissue; and the mucosa, the inner membrane, containing cells that produce mucus, acid, digestive enzymes, and hormones. Food is churned by the muscles and mixed with stomach secretions.

KEY

- Acid-secreting cell (secretes hydrochloric acid)
- Enzyme-secreting cell
- Hormone-secreting cell
- Stem cell (can develop into any type of cell)
- Mucus-secreting cell

Stomach muscles

The stomach contains longitudinal, circular, and oblique layers of smooth muscle. Where the esophagus joins the stomach, a muscular area known as the lower esophageal sphincter controls the entry of food. At the exit from the stomach, another area of muscle, the pyloric sphincter, opens and closes to allow food to move into the duodenum.

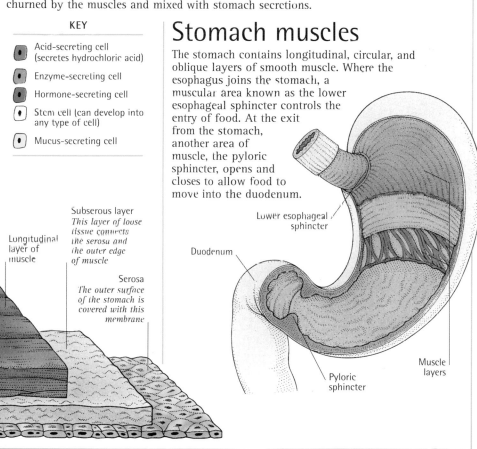

Longitudinal layer of muscle

Subserous layer
This layer of loose tissue connects the serosa and the outer edge of muscle

Serosa
The outer surface of the stomach is covered with this membrane

Lower esophageal sphincter

Duodenum

Pyloric sphincter

Muscle layers

Small intestine

The small intestine has three sections: the duodenum, jejunum, and ileum, all of which produce digestive enzymes and contribute to the breakdown and absorption of food. The duodenum is a short, curved tube attached to the stomach that receives secretions from the liver and pancreas. Joined to this section is the long, coiled jejunum. The main role of the ileum, the final and longest section of the small intestine, is to complete the absorption of nutrients from digested food passed along from the other two sections.

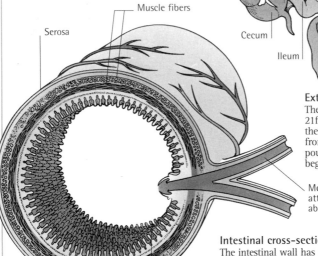

Pyloric sphincter

Duodenum

Serosa

Muscle fibers

Cecum

Ileum

Jejunum

Extent of small intestine
The small intestine is approximately 21ft (6.5m) in length. It starts from the pyloric sphincter at the exit from the stomach and ends at the pouchlike cecum, which is the beginning of the large intestine.

Mesentery (membrane attaching intestine to abdominal wall)

Mucosa

Submucosa

Intestinal cross-section
The intestinal wall has four layers. The outermost, protective coat is known as the serosa. Next is the muscle layer, containing outer longitudinal and inner circular muscle fibers. Adjoining this is the submucosa, a loose layer carrying vessels and nerves. The innermost layer is a mucus-secreting and absorptive membrane called the mucosa.

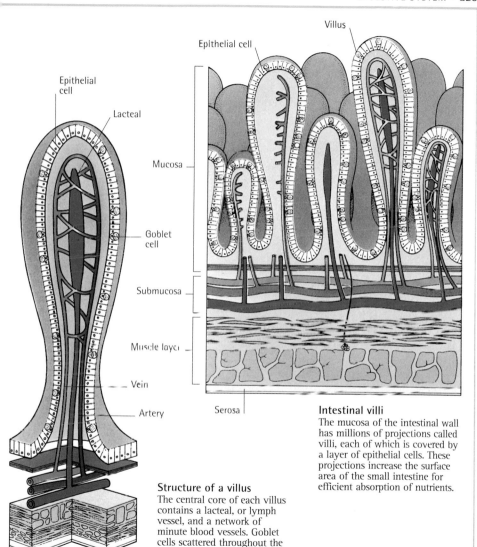

Villus

Epithelial cell

Epithelial cell

Lacteal

Mucosa

Goblet cell

Submucosa

Muscle layer

Vein

Artery

Serosa

Intestinal villi
The mucosa of the intestinal wall has millions of projections called villi, each of which is covered by a layer of epithelial cells. These projections increase the surface area of the small intestine for efficient absorption of nutrients.

Structure of a villus
The central core of each villus contains a lacteal, or lymph vessel, and a network of minute blood vessels. Goblet cells scattered throughout the epithelium secrete mucus.

Function of the stomach

Swallowing triggers relaxation of the lower esophageal sphincter, the muscle that controls the passage of food into the stomach. Waves of muscular contractions, called peristalsis, churn and mix food with gastric juice, producing a substance called chyme. The peristaltic action also moves food through the stomach towards the small intestine. Opening and closing of the pyloric sphincter at the exit from the stomach allows food to pass into the duodenum, a small amount at a time.

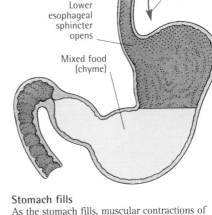

Direction of food

Lower esophageal sphincter opens

Mixed food (chyme)

Stomach fills
As the stomach fills, muscular contractions of the stomach wall start to churn and mix the food with gastric juice; this action produces a thick, creamy substance called chyme.

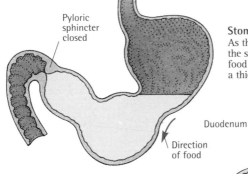

Pyloric sphincter closed

Direction of food

Movement of stomach contents
Peristaltic waves are most marked in the lower half of the stomach. These contractions move the stomach contents towards the still-closed pyloric sphincter.

Duodenum

Pyloric sphincter opens

Direction of food

Food exits stomach
The valvelike pyloric sphincter, stimulated by chyme, opens to let food pass into the duodenum in small quantities at a time.

Intestinal movement

The small intestine moves food by peristalsis and mixes it by segmentation. Peristalsis propels food through the intestine by waves of contractions. Segmentation (illustrated below) is the main activity of the small intestine. This is a sequence of muscular movements in which segments of the intestine alternately contract and relax. The action mixes chyme up to 16 times a minute.

Contraction 1
Short segments of the intestine contract for a few seconds, while others segments relax. This action mixes the chyme with digestive enzymes.

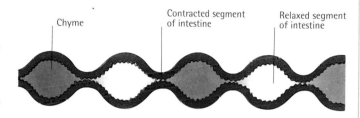

Chyme

Contracted segment of intestine

Relaxed segment of intestine

Contraction 2
As a contracted segment relaxes another series of contractions begins at a point between the previous contractions.

Another series of contractions begins

Chyme mixes

Contraction 3
Segmentation continues in a repeating pattern up to 16 times a minute. The chyme becomes mixed thoroughly.

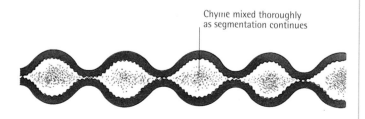

Chyme mixed thoroughly as segmentation continues

Liver structure

The wedge-shaped liver, which is located in the upper right-hand side of the abdominal cavity, is one of the largest organs in the body. It is divided by the falciform ligament into two lobes, with the left lobe smaller than the right. Each lobe is composed of thousands of hexagonal lobules made up of billions of cells. Tiny tubes known as bile ducts form a network throughout the liver.

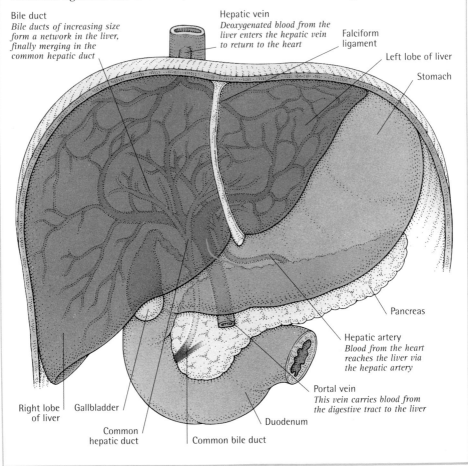

Bile duct
Bile ducts of increasing size form a network in the liver, finally merging in the common hepatic duct

Hepatic vein
Deoxygenated blood from the liver enters the hepatic vein to return to the heart

Falciform ligament

Left lobe of liver

Stomach

Pancreas

Hepatic artery
Blood from the heart reaches the liver via the hepatic artery

Portal vein
This vein carries blood from the digestive tract to the liver

Duodenum

Common bile duct

Common hepatic duct

Gallbladder

Right lobe of liver

Liver function

The liver is like a chemical processing factory and has many functions. It secretes a digestive fluid called bile, largely produced from the breakdown products of dietary fat and old red blood cells. The liver also produces proteins, and stores glycogen, iron, and some vitamins. It removes toxins (poisons) and wastes from the blood and converts them into less harmful substances.

Hepatocyte
The liver's chemical activities take place within millions of these cells, which are packed with organelles and enzymes, as well as storage particles. Hepatocytes manufacture bile and secrete it into tiny channels called canaliculi, which drain into bile ducts

Sinusoid

Lobule
Thousands of hexagonal units, called lobules, make up the liver. Each lobule is about 0.04in (1mm) in diameter, and contains a central vein from which rows of hepatocytes radiate

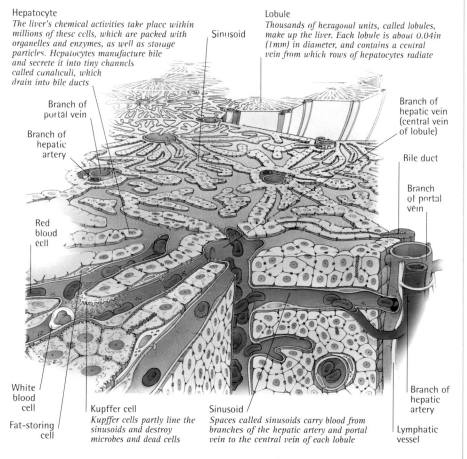

Branch of portal vein

Branch of hepatic artery

Red blood cell

Branch of hepatic vein (central vein of lobule)

Bile duct

Branch of portal vein

White blood cell

Kupffer cell
Kupffer cells partly line the sinusoids and destroy microbes and dead cells

Fat-storing cell

Sinusoid
Spaces called sinusoids carry blood from branches of the hepatic artery and portal vein to the central vein of each lobule

Branch of hepatic artery

Lymphatic vessel

Pancreas

The pancreas is an elongated gland lying behind the stomach and partly within the curve of the duodenum. In response to the entry of food into the upper digestive tract, the pancreas secretes digestive juice containing enzymes that break down fats, nucleic acids, proteins, and carbohydrates. The juice also contains sodium bicarbonate to neutralize stomach acid. The enzymes are secreted into ducts that converge to form the pancreatic duct, which transports the enzymes to the duodenum.

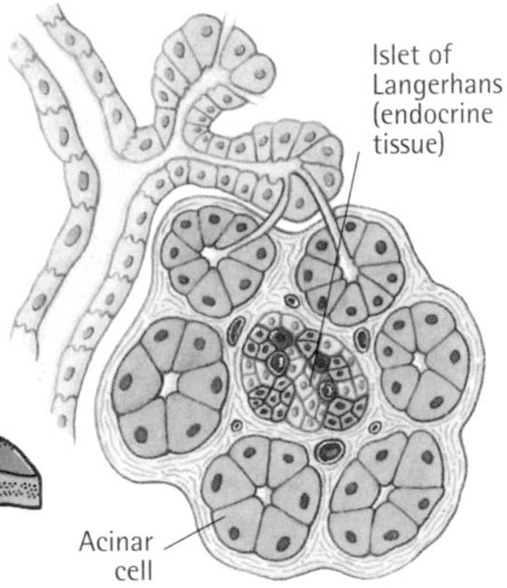

Islet of Langerhans (endocrine tissue)

Acinar cell

Acinar cells of the pancreas
Grapelike clusters of cells in the pancreas, known as acini, secrete pancreatic juice.

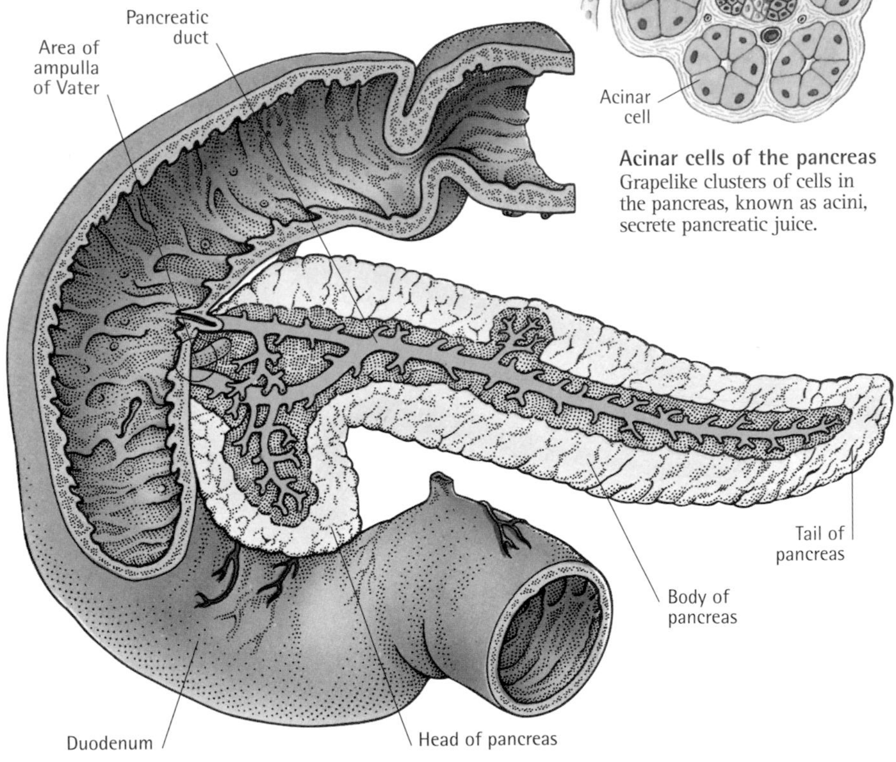

Pancreatic duct

Area of ampulla of Vater

Tail of pancreas

Body of pancreas

Duodenum

Head of pancreas

Biliary system

The gallbladder and a network of ducts carrying bile form the biliary system. Bile is a greenish-brown fluid that is partly a waste product of the liver's chemical processes, but it also plays a vital part in the digestion of fats. It passes from the liver either directly to the duodenum or to the gallbladder, where it is stored and concentrated. The fluid is then carried down the common bile duct and into an expanded area called the ampulla of Vater, before entering the duodenum through an opening called the sphincter of Oddi.

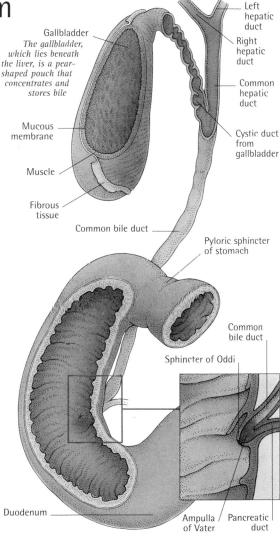

Gallbladder
The gallbladder, which lies beneath the liver, is a pear-shaped pouch that concentrates and stores bile

Left hepatic duct

Right hepatic duct

Common hepatic duct

Cystic duct from gallbladder

Mucous membrane

Muscle

Fibrous tissue

Common bile duct

Pyloric sphincter of stomach

Common bile duct

Sphincter of Oddi

Duodenum

Ampulla of Vater

Pancreatic duct

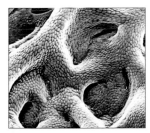

Gallbladder wall
The gallbladder has a muscular wall lined with a folded mucous membrane (shown above, highly magnified).

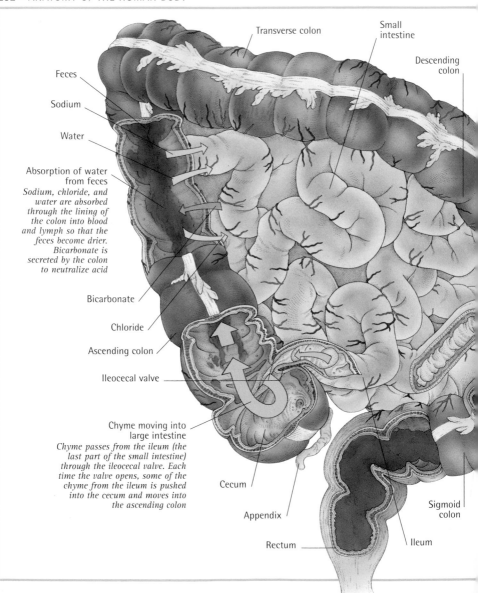

Transverse colon

Small intestine

Descending colon

Feces

Sodium

Water

Absorption of water from feces
Sodium, chloride, and water are absorbed through the lining of the colon into blood and lymph so that the feces become drier. Bicarbonate is secreted by the colon to neutralize acid

Bicarbonate

Chloride

Ascending colon

Ileocecal valve

Chyme moving into large intestine
Chyme passes from the ileum (the last part of the small intestine) through the ileocecal valve. Each time the valve opens, some of the chyme from the ileum is pushed into the cecum and moves into the ascending colon

Cecum

Appendix

Rectum

Sigmoid colon

Ileum

Large intestine: colon

By the time digested food (chyme) reaches the large intestine, most of the nutrient content has been absorbed. The primary role of the colon, the major section of the large intestine, is to convert chyme into feces for excretion. During this process, the colon absorbs water from the chyme, changing it from liquid to solid. Billions of bacteria within the colon synthesize the vitamins K and B, as well as the gases hydrogen, carbon dioxide, hydrogen sulphide, and methane.

Lining of large intestine   Lining of small intestine

Intestinal lining
The junction between the large and small intestines illustrates the structural difference in their linings. The lining of the small intestine is folded for better absorption of nutrients; that of the large intestine is smoother.

Chyme in small intestine

Consolidation of feces
Billions of bacteria live in the intestinal tract, and are normally harmless provided they do not spread to other body parts. They feed on undigested fiber and other substances, such as proteins, in fecal material, helping to reduce the amount of feces produced

Transit times

The illustration below shows the approximate time food takes to pass through each part of the digestive system. Food normally spends longer in the colon than anywhere else in the digestive tract, but the timing of its passage depends on its type and on the individual person.

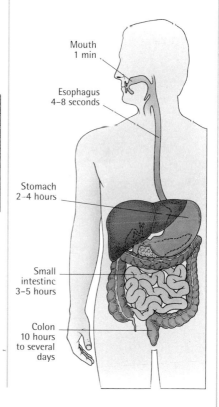

Mouth
1 min

Esophagus
4–8 seconds

Stomach
2–4 hours

Small intestine
3–5 hours

Colon
10 hours to several days

Colonic movement

Muscular movement in the wall of the colon mixes and propels feces along the colon towards the rectum. The movement of feces varies in rate, intensity, and nature. Types of movement are Haustral churning, peristaltic contractions, and mass movements. As fecal material passes through the colon, approximately 2½pt (1.4L) of water is reabsorbed every day. The lining of the colon secretes mucus to lubricate the intestine and ease the passage of feces.

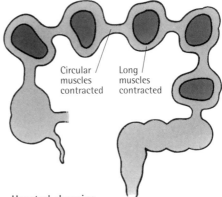

Circular muscles contracted

Long muscles contracted

Haustral churning

In a process similar to segmentation, contractions occur at regular intervals. This mixes the feces but propels them along only slowly.

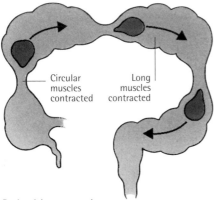

Circular muscles contracted

Long muscles contracted

Peristaltic contractions

Waves of peristaltic contractions propel feces towards the rectum. Circular muscles behind fecal matter and longitudinal muscles in front all contract. Peristalsis occurs more slowly in the colon than it does in the small intestine.

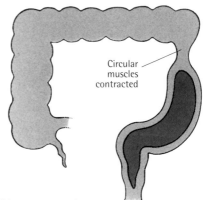

Circular muscles contracted

Mass movements

Mass movements are strong peristaltic waves that propel feces relatively long distances; these movements occur about two or three times a day.

Large intestine: rectum and anus

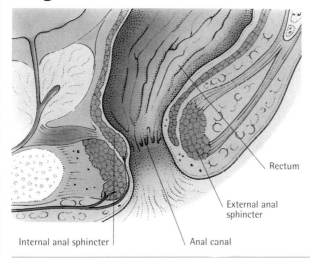

The rectum is the final section of the large intestine. It is about 5in (12cm) long, and is normally empty except just before and during defecation. Below the rectum lies the anal canal, or anus, which is about 1.5in (4cm) long and lined with vertical ridges called anal columns. In the walls of the anal canal are two strong, flat sheets of muscles called the internal and external sphincters, which act like valves and relax during defecation.

Rectum

External anal sphincter

Internal anal sphincter

Anal canal

Defecation

Peristaltic waves in the colon push feces into the rectum, which triggers the defecation reflex. Muscle contractions push the feces along into the anal canal, and the anal sphincters relax to allow feces to pass out of the body. The defecation reflex may be aided by voluntary contraction of the abdominal muscles, or overridden by conscious control that keeps the external anal sphincter closed.

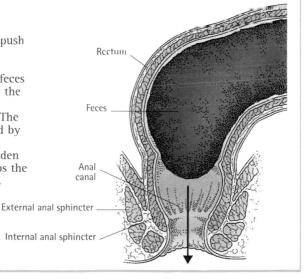

Rectum

Feces

Anal canal

External anal sphincter

Internal anal sphincter

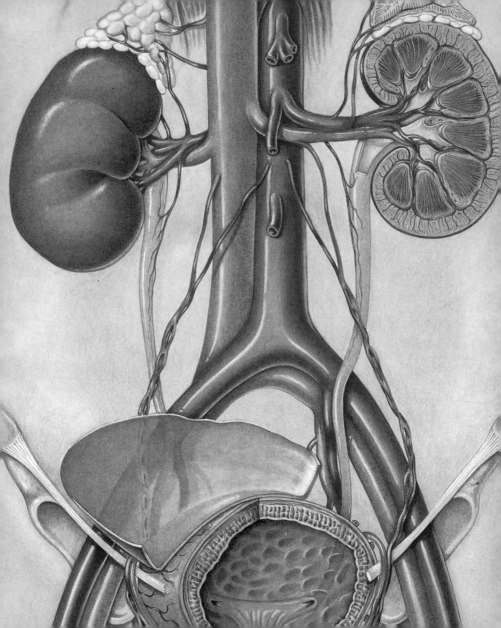

Urinary system

The urinary tract is the body's filtering system. As blood passes through the kidneys, waste products are removed and, together with excess fluids, are excreted as urine. The urinary system also regulates the volume and composition of fluids in the body, keeping an internal chemical balance.

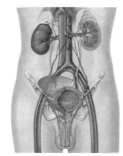

Urinary tract
The function of the urinary tract, comprising the kidneys, ureters, bladder, and urethra, is to make and excrete urine.

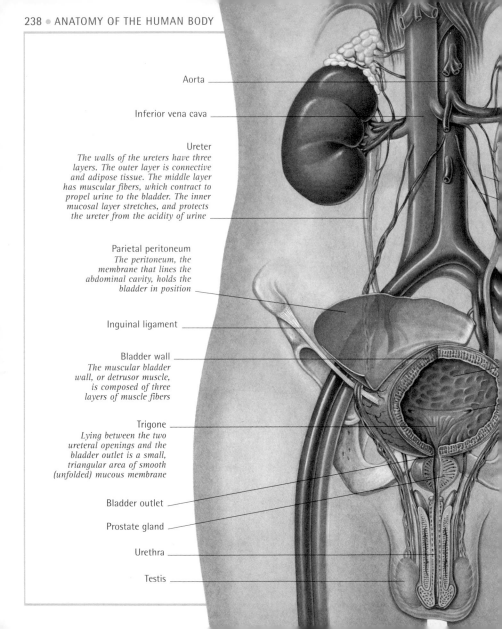

Aorta

Inferior vena cava

Ureter
The walls of the ureters have three layers. The outer layer is connective and adipose tissue. The middle layer has muscular fibers, which contract to propel urine to the bladder. The inner mucosal layer stretches, and protects the ureter from the acidity of urine

Parietal peritoneum
The peritoneum, the membrane that lines the abdominal cavity, holds the bladder in position

Inguinal ligament

Bladder wall
The muscular bladder wall, or detrusor muscle, is composed of three layers of muscle fibers

Trigone
Lying between the two ureteral openings and the bladder outlet is a small, triangular area of smooth (unfolded) mucous membrane

Bladder outlet

Prostate gland

Urethra

Testis

Adrenal gland

Adipose capsule
*A layer of adipose tissue
(fat) surrounds the
kidney and protects it
from damage*

Renal artery

Renal vein

Kidney
*Each kidney is about
4–5in (10–12.5cm)
long, and contains
about one million
filtering units*

Testicular vein

Testicular artery

Bladder lining
*When the bladder is empty,
the mucosal lining is folded;
these folds smooth out as
the bladder fills*

Opening of
ureter

Femoral vein

Femoral
artery

Urinary system

The urinary system, also known as
the urinary tract, consists of two
kidneys, the ureters, the bladder, and
the urethra. Waste products are
filtered from the blood by the kidneys
for excretion in the urine. From the
kidneys, urine passes down two
tubes, the ureters, to the hollow
bladder. Urine is stored in the bladder
until a convenient time, when the
muscles at the bladder outlet relax,
allowing the urine to be expelled
from the body through the urethra.

Aorta

Superior
mesenteric
artery

Celiac
trunk

Right
renal
artery

Left
renal
artery

Right
ureter

Left
ureter

Blood supply to the kidneys
The renal arteries that supply blood to the kidneys
branch directly from the aorta, the body's main
blood vessel. Inside the kidneys, the arteries
branch into progressively smaller blood vessels.

Female bladder and urethra

In females, the bladder is situated lower in the pelvic cavity than in males. The uterus rests just above the bladder and is pushed up and backwards when the bladder fills; enlargement of the uterus in pregnancy compresses the bladder, commonly causing frequent urination. The female urethra is short, about 1$\frac{1}{2}$in (4cm) long, which causes more frequent urinary tract infections in females than in males. The outlet of the urethra is just in front of the vagina.

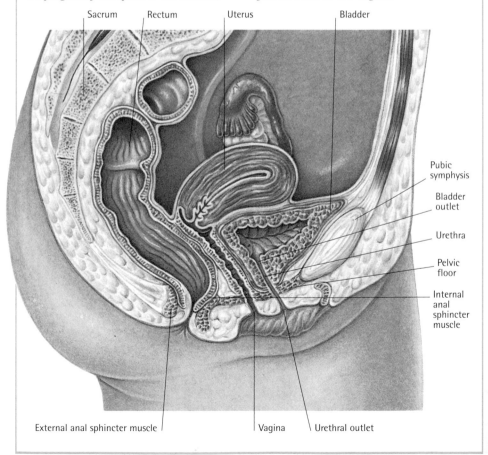

Sacrum Rectum Uterus Bladder

Pubic symphysis

Bladder outlet

Urethra

Pelvic floor

Internal anal sphincter muscle

External anal sphincter muscle   Vagina   Urethral outlet

Male bladder and urethra

The male urethra is about 8in (20cm) long, and is made up of three sections: the spongy urethra, the membranous urethra, and the prostatic urethra. It transports both urine and semen out of the body. The upper end of the urethra, where it leaves the bladder, is encircled by the prostate gland. In older men, enlargement of the prostate gland may compress the urethra, causing problems with urination. The urethra runs through the penis to an outlet at its tip.

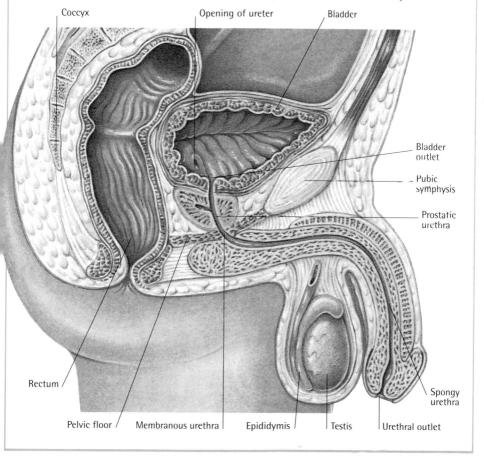

Coccyx

Opening of ureter

Bladder

Bladder outlet

Pubic symphysis

Prostatic urethra

Rectum

Pelvic floor

Membranous urethra

Epididymis

Testis

Urethral outlet

Spongy urethra

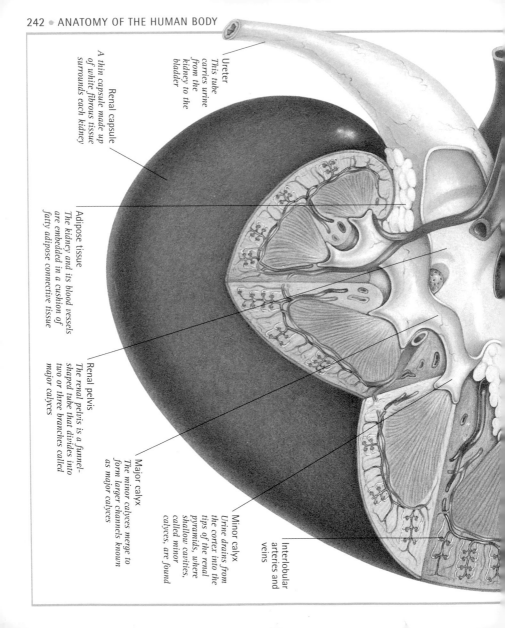

Ureter
This tube carries urine from the kidney to the bladder

Renal capsule
A thin capsule made up of white fibrous tissue surrounds each kidney

Adipose tissue
The kidney and its blood vessels are embedded in a cushion of fatty adipose connective tissue

Renal pelvis
The renal pelvis is a funnel-shaped tube that divides into two or three branches called major calyces

Major calyx
The minor calyces merge to form larger channels known as major calyces

Minor calyx
Urine drains from the cortex into the tips of the renal pyramids, where shallow cavities, called minor calyces, are found

Interlobular arteries and veins

Kidney structure

Each kidney has an outer rim, the renal cortex; this rim surrounds an inner region, the renal medulla, which is composed of many conical segments known as renal pyramids. Kidneys contain numerous urine-making units, known as nephrons, and urine-collecting tubules. Urine drains from these small tubules into wider tubules called collecting ducts. These ducts open at the tips of the renal pyramids into calyces (cavities).

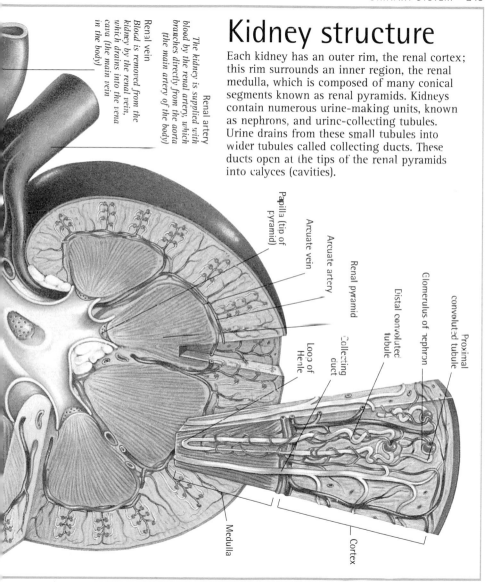

Renal artery
The kidney is supplied with blood by the renal artery, which branches directly from the aorta (the main artery of the body)

Renal vein
Blood is removed from the kidney by the renal vein, which drains into the vena cava (the main vein in the body)

Papilla (tip of pyramid)

Arcuate vein

Arcuate artery

Renal pyramid

Glomerulus of nephron

Proximal convoluted tubule

Distal convoluted tubule

Loop of Henle

Collecting duct

Medulla

Cortex

Nephron structure

Each kidney contains more than one million filtering units called nephrons. Within each tiny unit is a glomerulus (a cluster of blood capillaries), where blood is filtered. The glomerulus is enveloped in a cup-shaped membrane known as Bowman's capsule. This capsule forms the end of a long, convoluted renal tubule, which converts filtrate from the glomerulus into urine.

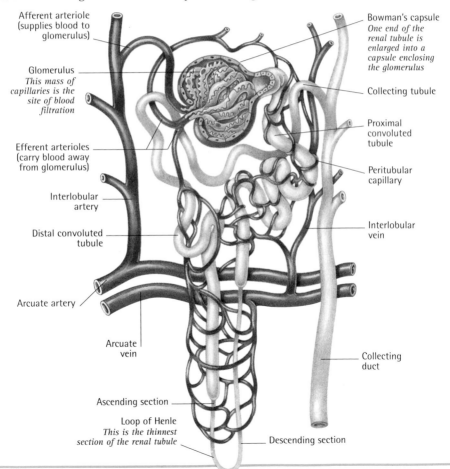

Afferent arteriole (supplies blood to glomerulus)

Glomerulus *This mass of capillaries is the site of blood filtration*

Efferent arterioles (carry blood away from glomerulus)

Interlobular artery

Distal convoluted tubule

Arcuate artery

Arcuate vein

Ascending section

Loop of Henle *This is the thinnest section of the renal tubule*

Bowman's capsule *One end of the renal tubule is enlarged into a capsule enclosing the glomerulus*

Collecting tubule

Proximal convoluted tubule

Peritubular capillary

Interlobular vein

Collecting duct

Descending section

Glomerular filtration

Blood passing through the glomerular capillaries is filtered under pressure into Bowman's capsule. This filtered fluid, or filtrate, contains water, potassium, bicarbonate, sodium, glucose, and amino acids, as well as waste products such as urea and uric acid. Large particles, such as blood cells, stay in the capillaries.

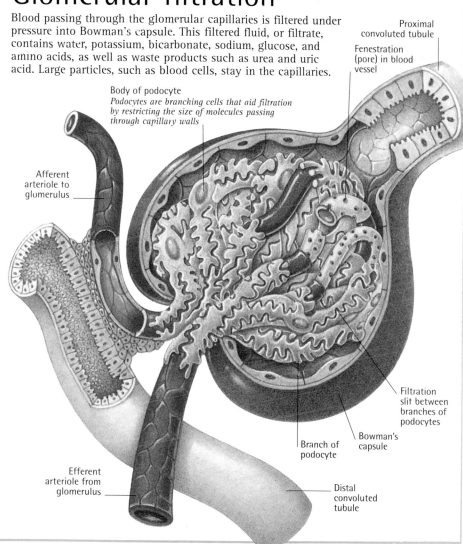

Proximal convoluted tubule

Fenestration (pore) in blood vessel

Body of podocyte
Podocytes are branching cells that aid filtration by restricting the size of molecules passing through capillary walls

Afferent arteriole to glomerulus

Filtration slit between branches of podocytes

Branch of podocyte

Bowman's capsule

Efferent arteriole from glomerulus

Distal convoluted tubule

Urine formation

After blood has been filtered in the glomerulus, the resulting filtrate passes along the coiled renal tubule. Here, water and other substances from the fluid are reabsorbed into the blood. The remainder of the filtrate is eliminated from the body as urine. The volume and composition of urine are regulated by hormones acting on the kidneys; this ensures that the body's fluid balance is maintained.

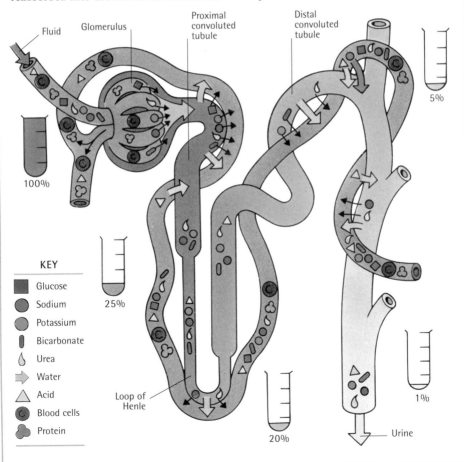

Fluid

Glomerulus

Proximal convoluted tubule

Distal convoluted tubule

5%

100%

25%

KEY

- Glucose
- Sodium
- Potassium
- Bicarbonate
- Urea
- Water
- Acid
- Blood cells
- Protein

Loop of Henle

20%

1%

Urine

Bladder

The bladder is a hollow organ that can hold up to 1¾ pints (800ml) of urine. Its walls consist of layers of muscle, with an elastic inner lining of tissue called transitional epithelium. Urine is transported into the bladder by the two ureters that run from the kidneys. The bladder's outlet into the urethra is kept closed by a strong ring of muscle, the sphincter, until urine is passed.

Lining of empty bladder
While the bladder is empty, the cells of the transitional epithelium that lines the bladder walls are rounded and packed tightly together.

Lining of full bladder
As the bladder fills with urine and expands, the cells of the transitional epithelium are stretched and become thin and flattened.

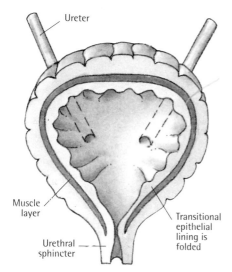

Empty bladder
When the bladder is empty, its walls are thickened. The transitional epithelial tissue lining the bladder walls is pulled into folds as the underlying layer of muscle contracts.

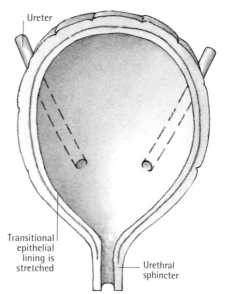

Full bladder
The bladder expands as it fills, and the walls become thinner; the folds in the lining are stretched out. When the sphincter muscle is relaxed, urine passes out of the bladder.

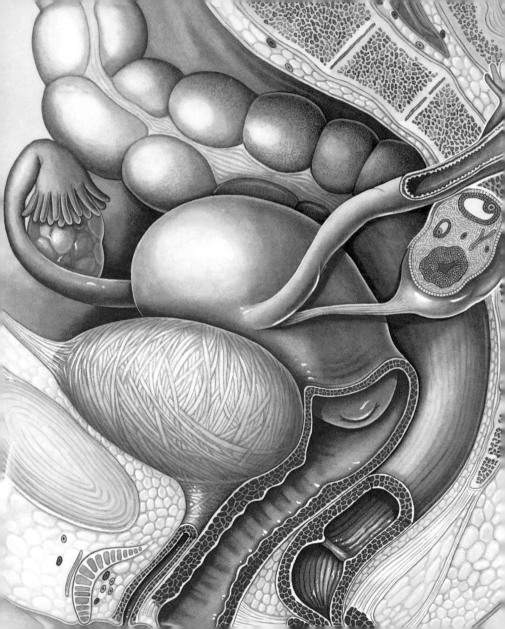

Reproductive system

In biological terms, the primary function of the reproductive system in both sexes is to produce offspring by generating and bringing together male and female sex cells. Hormones manufactured by the ovaries in women and the testes in men influence the development of sexual characteristics, such as body shape.

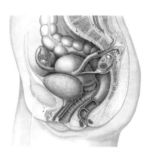

Female reproductive system
The female reproductive organs lie within the pelvis. Eggs travel from the ovaries through the fallopian tubes to the uterus.

Male reproductive organs

The male genitals are the penis; the testes, a pair of glands that produce sperm and male sex hormones; and the scrotum, the sac that contains the testes. From each testis, sperm pass into a long coiled tube, the epididymis, where they mature and are stored. During sexual intercourse, sperm are propelled along a duct called the vas deferens. Fluid secreted by the seminal vesicles and the prostate gland is added to sperm to produce semen, which is ejaculated from the erect penis through the urethra.

Testes

The testes hang suspended outside the pelvic cavity in a pouch called the scrotum. This arrangement maintains sperm at slightly below body temperature, which is necessary for them to survive.

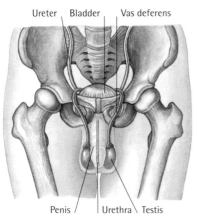

Ureter   Bladder   Vas deferens

Penis / Urethra \ Testis

Penis
The penis is composed of columns of erectile tissue. When this tissue fills with blood, the penis becomes erect

Corpus cavernosum
This column of erectile tissue is one of two that lie side by side

Corpus spongiosum
This spongy, erectile tissue surrounds the urethra, and widens to form the glans (tip)

Urethra

Spermatic cord

Seminiferous tubules
The testes contain many of these tightly coiled tubules, which produce millions of sperm per day

Epididymis
Sperm mature and are stored inside this coiled tube

Glans penis

Testis

Scrotum

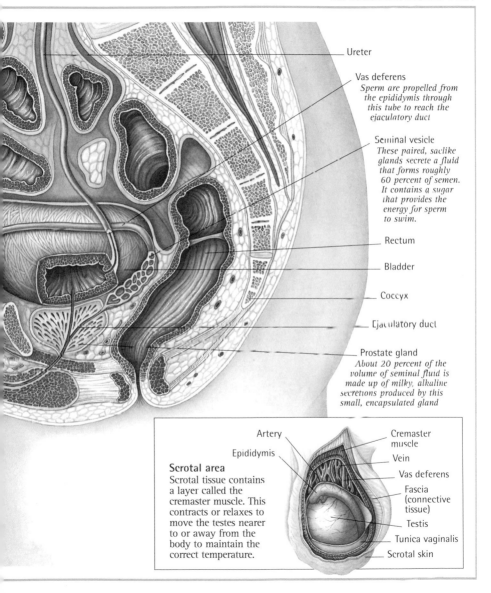

Ureter

Vas deferens
Sperm are propelled from the epididymis through this tube to reach the ejaculatory duct

Seminal vesicle
These paired, saclike glands secrete a fluid that forms roughly 60 percent of semen. It contains a sugar that provides the energy for sperm to swim.

Rectum

Bladder

Coccyx

Ejaculatory duct

Prostate gland
About 20 percent of the volume of seminal fluid is made up of milky, alkaline secretions produced by this small, encapsulated gland

Artery

Epididymis

Cremaster muscle

Vein

Vas deferens

Fascia (connective tissue)

Testis

Tunica vaginalis

Scrotal skin

Scrotal area
Scrotal tissue contains a layer called the cremaster muscle. This contracts or relaxes to move the testes nearer to or away from the body to maintain the correct temperature.

Female reproductive organs

The female sex glands are the ovaries. From puberty, these paired glands release ova, the female egg cells; they also manufacture the sex hormones, notably estrogen, that influence the menstrual cycle and the development of female sexual characteristics. Each month an egg is released and travels down one of the two fallopian tubes to the uterus, a hollow structure in the center of the pelvis; if the egg is fertilized by a sperm it may develop into an embryo.

Ovaries

Each ovary contains numerous cell clusters called follicles, in which egg cells develop. At ovulation, the follicle ruptures, releasing the mature egg into the fallopian tube. The empty follicle develops into a small tissue mass, the corpus luteum, which secretes the female sex hormones progesterone and estrogen.

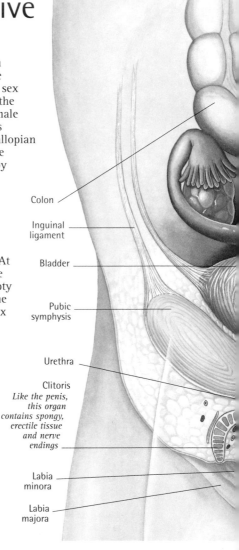

Colon

Inguinal ligament

Bladder

Pubic symphysis

Urethra

Clitoris
Like the penis, this organ contains spongy, erectile tissue and nerve endings

Labia minora

Labia majora

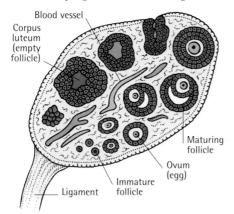

Blood vessel

Corpus luteum (empty follicle)

Maturing follicle

Ovum (egg)

Immature follicle

Ligament

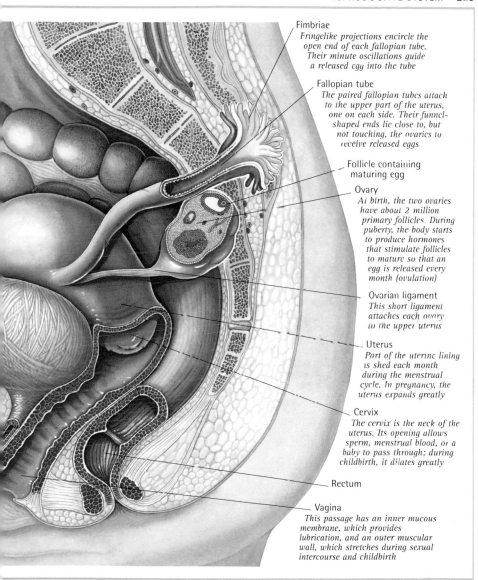

Fimbriae
Fringelike projections encircle the open end of each fallopian tube. Their minute oscillations guide a released egg into the tube

Fallopian tube
The paired fallopian tubes attach to the upper part of the uterus, one on each side. Their funnel-shaped ends lie close to, but not touching, the ovaries to receive released eggs

Follicle containing maturing egg

Ovary
At birth, the two ovaries have about 2 million primary follicles. During puberty, the body starts to produce hormones that stimulate follicles to mature so that an egg is released every month (ovulation)

Ovarian ligament
This short ligament attaches each ovary to the upper uterus

Uterus
Part of the uterine lining is shed each month during the menstrual cycle. In pregnancy, the uterus expands greatly

Cervix
The cervix is the neck of the uterus. Its opening allows sperm, menstrual blood, or a baby to pass through; during childbirth, it dilates greatly

Rectum

Vagina
This passage has an inner mucous membrane, which provides lubrication, and an outer muscular wall, which stretches during sexual intercourse and childbirth

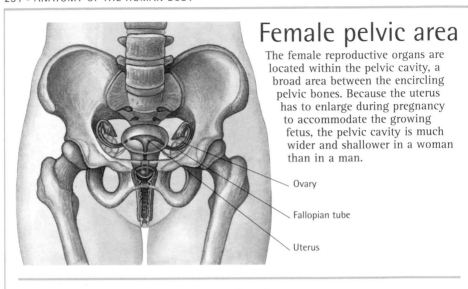

Female pelvic area

The female reproductive organs are located within the pelvic cavity, a broad area between the encircling pelvic bones. Because the uterus has to enlarge during pregnancy to accommodate the growing fetus, the pelvic cavity is much wider and shallower in a woman than in a man.

Ovary

Fallopian tube

Uterus

External female genitals

The clitoris and the openings of the vagina and the urethra are surrounded by folds of skin called labia. During sexual arousal, the clitoris swells with blood, becoming erect and highly sensitive; two tissue masses (vestibular bulbs) on either side of the vagina also become erect, while glands lubricate the mucous membranes.

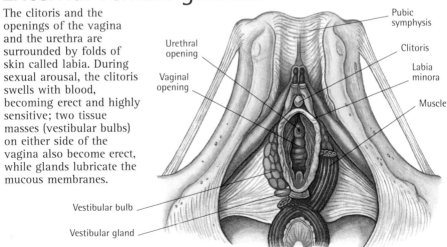

Pubic symphysis

Urethral opening

Clitoris

Labia minora

Vaginal opening

Muscle

Vestibular bulb

Vestibular gland

Anatomy of the breasts

The female breast is composed of lobules of milk-producing glands surrounded by fatty tissue. Ducts run from these glands to outlets at the nipple. Around the nipple is a circular pigmented area called the areola, which contains sweat glands and sebaceous glands. The breasts have no muscles but are held in place by ligaments. A major network of lymph vessels surrounds the breasts and drains into lymph nodes in the armpit area (axilla).

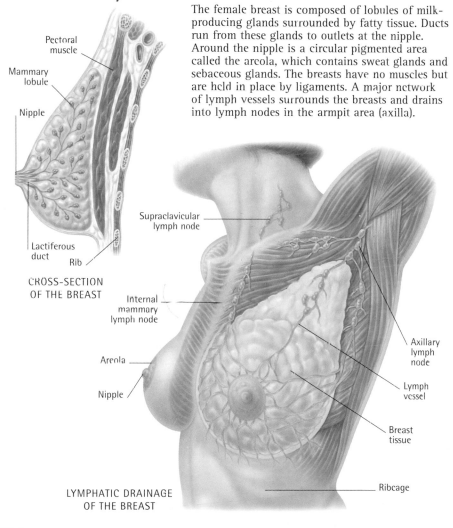

Pectoral muscle

Mammary lobule

Nipple

Lactiferous duct

Rib

CROSS-SECTION OF THE BREAST

Supraclavicular lymph node

Internal mammary lymph node

Areola

Nipple

Axillary lymph node

Lymph vessel

Breast tissue

Ribcage

LYMPHATIC DRAINAGE OF THE BREAST

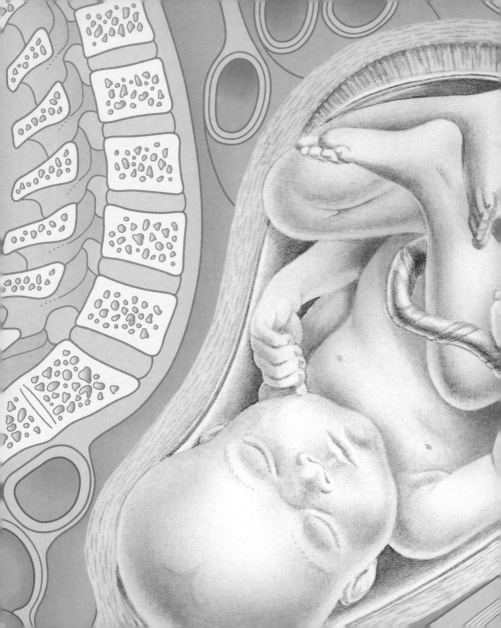

Human life cycle

From the moment an egg cell is fertilized by
a sperm, an embryo begins to form and in a
few weeks becomes a fetus. Before birth
and in a child's early years, growth and
development are rapid; when the process
of puberty is complete, adulthood
is reached. In a healthy person,
further changes are small until
the aging process causes body
systems to decline. Many of
an individual's lifelong traits
are influenced by the genes
inherited from both parents.

Pregnancy
The unborn child, or fetus,
develops inside a fluid-
filled sac within the uterus.

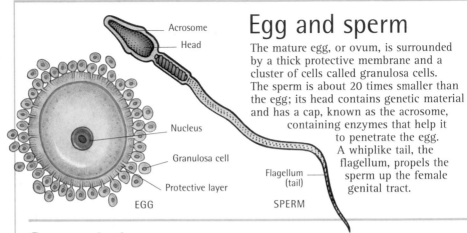

Egg and sperm

The mature egg, or ovum, is surrounded by a thick protective membrane and a cluster of cells called granulosa cells. The sperm is about 20 times smaller than the egg; its head contains genetic material and has a cap, known as the acrosome, containing enzymes that help it to penetrate the egg. A whiplike tail, the flagellum, propels the sperm up the female genital tract.

Acrosome

Head

Nucleus

Granulosa cell

Protective layer

EGG

Flagellum (tail)

SPERM

Sperm's journey

Each ejaculation releases about 400 million sperm into the vagina. As they propel themselves up through the cervix and into the uterus, many sperm die, some killed by acidic vaginal secretions.

Only a few thousand sperm will reach the far end of the fallopian tube; many will try to penetrate the protective layer around the egg but only one will succeed in entering and fertilizing the egg.

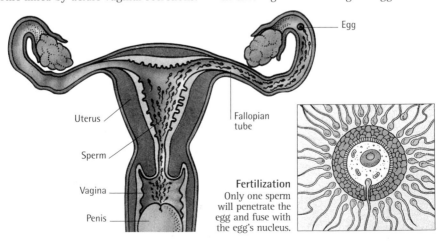

Egg

Uterus

Sperm

Vagina

Penis

Fallopian tube

Fertilization
Only one sperm will penetrate the egg and fuse with the egg's nucleus.

Division of fertilized egg

After fertilization in the fallopian tube, the nuclei of the sperm and egg fuse to form a new cell. This cell, the zygote, contains 46 chromosomes, 23 from each of the parent cells. As it travels to the uterus, the zygote divides, forming a cell cluster (the morula) by about 3 days after fertilization. The morula develops a cavity and is now known as a blastocyst, which will become the embryo.

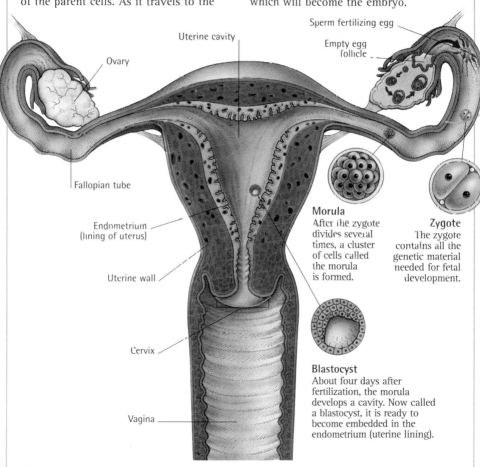

Uterine cavity

Ovary

Sperm fertilizing egg

Empty egg follicle

Fallopian tube

Endometrium (lining of uterus)

Morula
After the zygote divides several times, a cluster of cells called the morula is formed.

Zygote
The zygote contains all the genetic material needed for fetal development.

Uterine wall

Cervix

Blastocyst
About four days after fertilization, the morula develops a cavity. Now called a blastocyst, it is ready to become embedded in the endometrium (uterine lining).

Vagina

Implantation in uterus

The blastocyst, the cell cluster that forms from a fertilized egg, floats freely within the uterine cavity for about 48 hours before attaching itself to a site in the endometrium (uterine lining). By about 10 days after fertilization, the blastocyst is completely embedded. Some of its outer cells, called trophoblast cells, burrow deeply into the endometrium, eventually forming the placenta. Within the cell cluster in the blastocyst's cavity, an embryonic disk forms. This separates the cell cluster into the amniotic cavity, which will become a fluid-filled sac covering the embryo, and the yolk sac. The disk develops three germ layers from which all fetal structures derive. The main illustration shows progressive stages in the blastocyst's development.

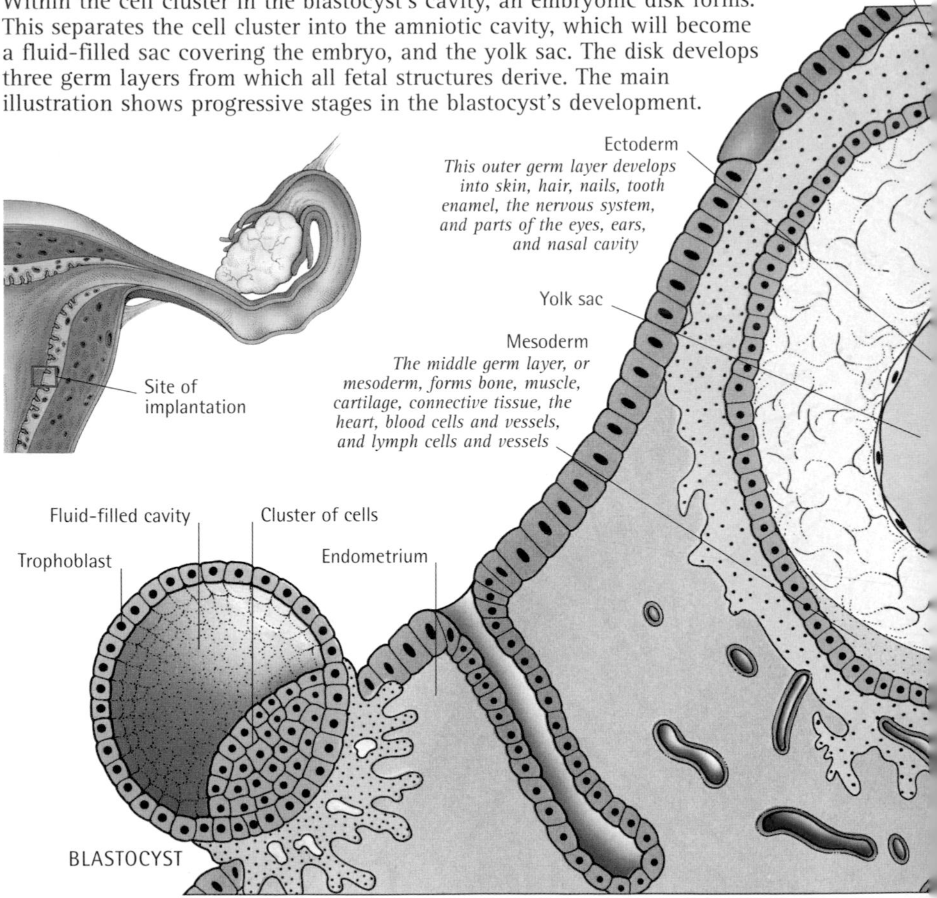

Embryonic disk

Site of implantation

Ectoderm
This outer germ layer develops into skin, hair, nails, tooth enamel, the nervous system, and parts of the eyes, ears, and nasal cavity

Yolk sac

Mesoderm
The middle germ layer, or mesoderm, forms bone, muscle, cartilage, connective tissue, the heart, blood cells and vessels, and lymph cells and vessels

Fluid-filled cavity

Cluster of cells

Trophoblast

Endometrium

BLASTOCYST

Endoderm

This inner germ layer forms the linings of many organs, such as the bladder and the digestive and respiratory organs; it also forms the lining of several glands and the tissues of the liver and pancreas

Amniotic cavity

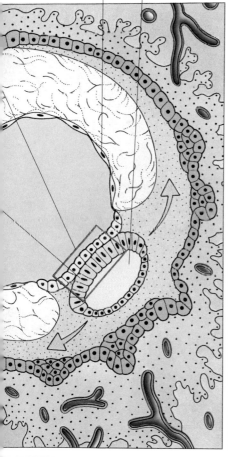

Growing embryo

After implantation, the development of the embryo is rapid and progresses through identifiable stages. At eight weeks, all the major body organs have formed and the embryo, now called a fetus, begins to move.

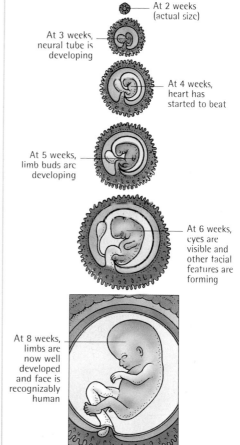

At 2 weeks (actual size)

At 3 weeks, neural tube is developing

At 4 weeks, heart has started to beat

At 5 weeks, limb buds are developing

At 6 weeks, eyes are visible and other facial features are forming

At 8 weeks, limbs are now well developed and face is recognizably human

Fetal development

Within the uterus, the fetus develops inside a sac filled with a clear fluid, called amniotic fluid, that provides protection against injury. The fluid is swallowed by the fetus and absorbed into its bloodstream; excess fluid is excreted as urine, and wastes are removed via the placenta. Oxygen and nutrients needed by the fetus are supplied from maternal blood through the placenta. By about week 32 of pregnancy, the fetus is likely to be in a head-down position.

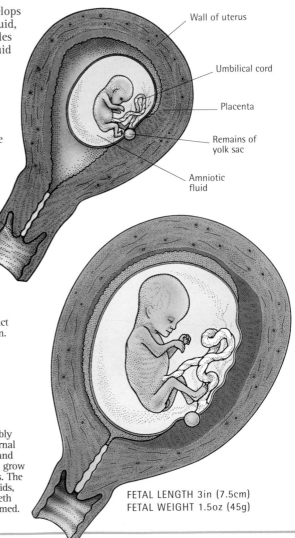

Wall of uterus

Umbilical cord

Placenta

Remains of yolk sac

Amniotic fluid

Cervix

FETAL LENGTH 1in (2.5cm)
FETAL WEIGHT 0.07oz (2g)

8 weeks
The arms, legs, and major joints of the fetus are forming and it begins to move, although these movements will not be felt by the mother at this early stage. Toes and fingers are distinct but may still be joined by webs of skin. The fetal blood cells circulate within immature blood vessels.

12 weeks
The fetus is recognizably human. Its major internal organs have formed, and nails are beginning to grow on its fingers and toes. The external ears, the eyelids, and the permanent teeth buds have usually formed.

FETAL LENGTH 3in (7.5cm)
FETAL WEIGHT 1.5oz (45g)

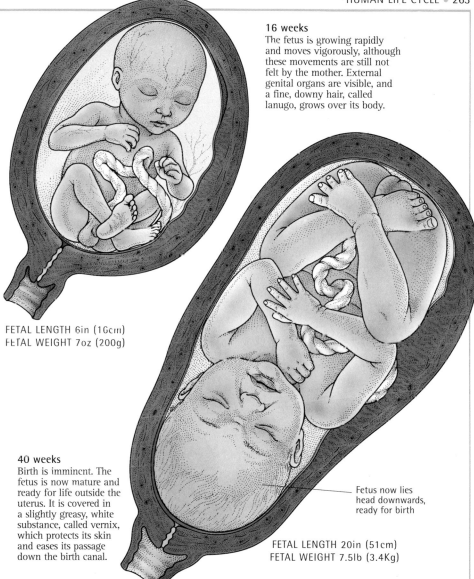

16 weeks
The fetus is growing rapidly and moves vigorously, although these movements are still not felt by the mother. External genital organs are visible, and a fine, downy hair, called lanugo, grows over its body.

FETAL LENGTH 6in (16cm)
FETAL WEIGHT 7oz (200g)

40 weeks
Birth is imminent. The fetus is now mature and ready for life outside the uterus. It is covered in a slightly greasy, white substance, called vernix, which protects its skin and eases its passage down the birth canal.

Fetus now lies head downwards, ready for birth

FETAL LENGTH 20in (51cm)
FETAL WEIGHT 7.5lb (3.4Kg)

Developing placenta

The placenta is a special organ that supplies the fetus with nutrients and oxygen, absorbs fetal waste products, and acts as a barrier against harmful substances. It derives from the trophoblast, the outer cell layer of the blastocyst that implants in the endometrium; the placenta is well established by the end of the 4th week after fertilization. Placental hormones help to maintain the endometrium so that the pregnancy continues.

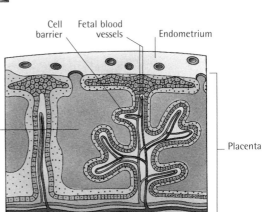

Rapid development

Specialized cells of the embedded trophoblast rapidly penetrate nearby uterine blood vessels. Blood from the mother flows from these blood vessels into spaces within the trophoblast.

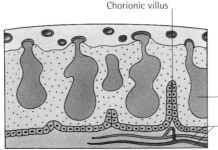

Chorionic villi

Other trophoblast cells extend finger-like projections, called chorionic villi, into the endometrium. These are surrounded by the spaces filled with maternal blood. Fetal blood vessels grow into the chorionic villi.

Oxygen transfer

Maternal and fetal blood do not make direct contact in the placenta, but are separated by a barrier of cells. Oxygen, as well as nutrients and protective antibodies, passes across the barrier to the fetus and waste products pass back to the placenta.

Umbilical cord
The placenta is attached to the center of the fetus's abdomen by the umbilical cord. The cord, a jellylike structure containing blood vessels, is about 12–35in (30–90cm) long.

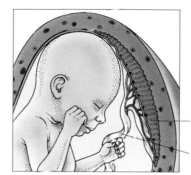

Placenta

Umbilical cord

Continued growth
The placenta continues to develop as the fetus grows, so that by the end of the pregnancy it is about 8in (20cm) wide and 1in (2.5cm) thick. It is expelled from the uterus shortly after the baby is born.

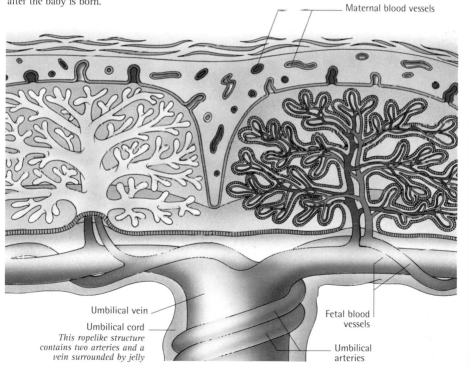

Maternal blood vessels

Umbilical vein

Umbilical cord
This ropelike structure contains two arteries and a vein surrounded by jelly

Fetal blood vessels

Umbilical arteries

Trimesters of pregnancy

Pregnancy typically lasts 40 weeks from the first day of a woman's last menstrual period. By convention, the duration of the pregnancy is divided into stages known as trimesters, each lasting about three months. During this time, a woman's body undergoes many changes to support the growing fetus and prepare for childbirth.

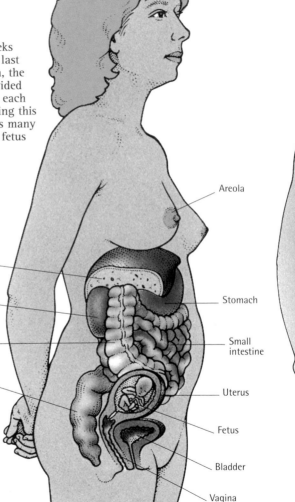

Areola

Liver

Kidney

Colon

Rectum

Stomach

Small intestine

Uterus

Fetus

Bladder

Vagina

First trimester (0–12 weeks)
The breasts become tender and begin to enlarge; the areola surrounding the nipple darkens. Vaginal discharge sometimes increases, as does the need to urinate. Vomiting and nausea are common in the early weeks. Weight begins to increase.

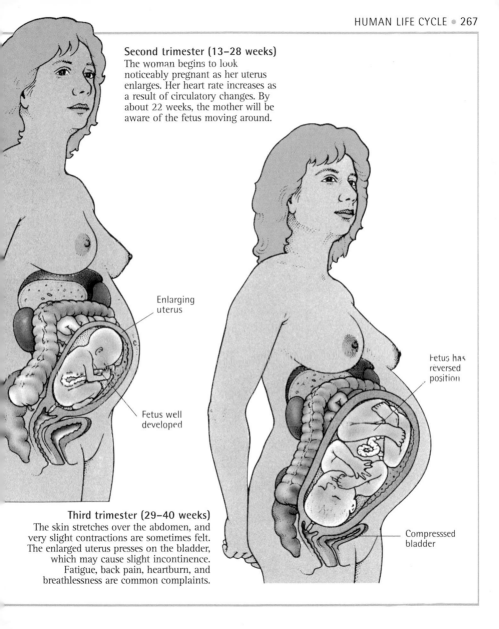

Second trimester (13–28 weeks)
The woman begins to look noticeably pregnant as her uterus enlarges. Her heart rate increases as a result of circulatory changes. By about 22 weeks, the mother will be aware of the fetus moving around.

Enlarging uterus

Fetus well developed

Fetus has reversed position

Compresssed bladder

Third trimester (29–40 weeks)
The skin stretches over the abdomen, and very slight contractions are sometimes felt. The enlarged uterus presses on the bladder, which may cause slight incontinence. Fatigue, back pain, heartburn, and breathlessness are common complaints.

Onset of labor

With the approach of childbirth, the head of the fetus drops lower into the pelvis, easing pressure on the expectant mother's lungs and abdomen. When labor begins, the mucus plug sealing off the cervix is expelled as a bloodstained discharge. Uterine contractions become stronger and more regular. The membranous sac around the amniotic fluid ruptures and the fluid leaks out through the vagina.

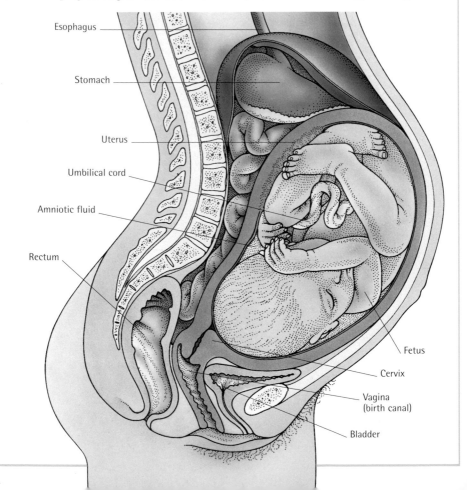

Esophagus

Stomach

Uterus

Umbilical cord

Amniotic fluid

Rectum

Fetus

Cervix

Vagina
(birth canal)

Bladder

Fetal positions

Towards the end of pregnancy, before about 30 weeks, the fetus tends to turn in the uterus. After this time, the most usual fetal position is head downwards, facing towards the woman's side, with the neck flexed forwards. Twins lie in various combinations of positions, but are commonly both head downwards. About 3 percent of full-term deliveries are breech, in which the baby's buttocks are delivered before the head. The incidence of breech delivery is much higher among premature babies.

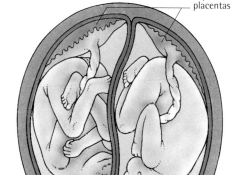

Separate placentas

Separate amniotic sacs

Vagina (birth canal)

Twins
Most commonly, both twins are head downwards in the uterus; one with head down and one with buttocks down is also a fairly common position. Other combinations can occur but are rarer.

Placenta

Umbilical cord

Frank breech
In this presentation, known as a "frank" or "incomplete" breech, the baby's hips are flexed and the legs extend alongside the body. The feet lie beside the head.

Placenta

Umbilical cord

Complete breech
In a complete breech presentation, the baby's legs are flexed at the knees and at the hips. This presentation occurs less commonly than frank breech.

Pelvic size and shape

The size and shape of a woman's pelvis are very important in determining the ease of childbirth. If the pelvis is narrow, or the baby is particularly large, a normal delivery may be prevented. Any mismatch between the dimensions of the mother's pelvis and the baby's head is termed "disproportion".

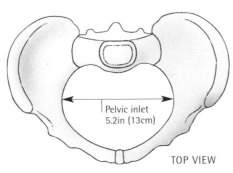

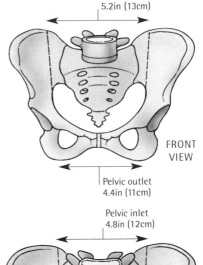

Round pelvis
The inlet of a woman's pelvis is most commonly rounded. The outlet of the pelvis is roughly diamond-shaped.

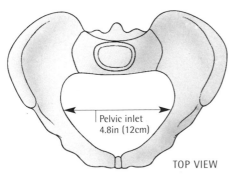

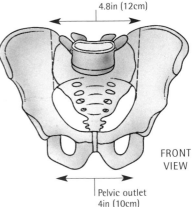

Constricted pelvis
In some women, the channel through the pelvis is narrower than usual. Childbirth through a pelvis of this shape is less easy than through a rounded one.

Changes in the cervix

The cervix is a firm band of muscle and connective tissue that forms the lower end of the uterus. In late pregnancy, the cervix softens and shortens in readiness for childbirth. Painless contractions, called Braxton-Hicks contractions, which occur in the upper segment of the uterus, pull against the lower segment and help to draw the cervix upwards. The cervix remains closed until labor begins.

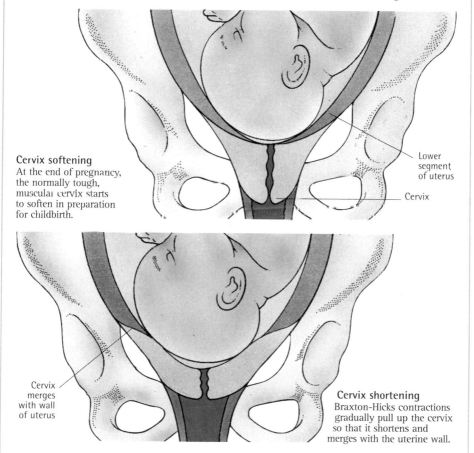

Cervix softening
At the end of pregnancy, the normally tough, muscular cervix starts to soften in preparation for childbirth.

Lower segment of uterus

Cervix

Cervix merges with wall of uterus

Cervix shortening
Braxton-Hicks contractions gradually pull up the cervix so that it shortens and merges with the uterine wall.

Engagement

During the last weeks of pregnancy, the baby's head descends into the cavity of the pelvis, a process called engagement. When this happens, many women feel the load "lightening" and pressure taken off the diaphragm, making breathing easier. Engagement usually takes place at around 36 weeks during a first pregnancy, but may not happen until the onset of labor during subsequent pregnancies.

Pelvis

Before engagement
The widest section of the baby's head has not yet passed down through the inlet of the pelvis into the pelvic cavity.

Pelvis

Entering the pelvis
The baby's head descends through the inlet of the pelvis and becomes engaged within the pelvic cavity. The height of the uterus drops and the baby's head rests against the cervix.

First stage of labor

The first stage of labor begins with the onset of regular, painful contractions of the uterus that cause the cervix to dilate (widen) progressively. The cervix is fully dilated when its opening measures around 4in (10cm) in diameter, marking the onset of the second stage of labor. For a first baby, the cervix dilates at a rate of about 0.5in (1cm) per hour but more rapidly in subsequent pregnancies.

Dilation starts
When labor begins, the cervix starts to dilate, opening up a little wider with the force of each contraction.

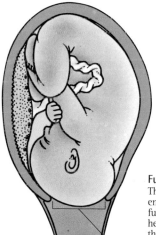

Cervix ³/₄in (2cm) dilated

Half dilated
As the contractions become stronger, the baby's head is pushed lower down; the increasing pressure of the head causes the cervix to widen further.

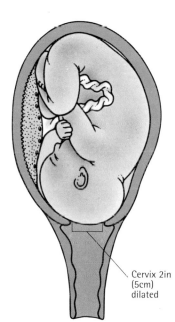

Cervix 2in (5cm) dilated

Fully dilated
The first stage of labor is ended when the cervix is fully dilated; the baby's head is able to pass into the birth canal.

Cervix 4in (10cm) dilated

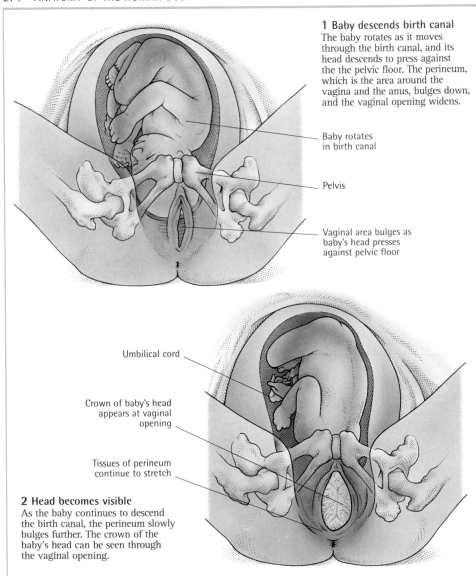

1 Baby descends birth canal

The baby rotates as it moves through the birth canal, and its head descends to press against the the pelvic floor. The perineum, which is the area around the vagina and the anus, bulges down, and the vaginal opening widens.

Baby rotates
in birth canal

Pelvis

Vaginal area bulges as
baby's head presses
against pelvic floor

Umbilical cord

Crown of baby's head
appears at vaginal
opening

Tissues of perineum
continue to stretch

2 Head becomes visible

As the baby continues to descend the birth canal, the perineum slowly bulges further. The crown of the baby's head can be seen through the vaginal opening.

Second stage of labor

In the second stage of labor, which begins after full dilation of the cervix, the contractions become stronger, and the woman feels a strong urge to push. As the baby moves out of the uterus and down the birth canal, it turns towards the mother's back. When the head emerges from the vagina, it turns sideways, and the rest of the body slides out easily. This stage of labor usually lasts about 50 minutes for first babies and about 20 minutes for subsequent babies.

3 Head emerges

The baby's head usually emerges with the face towards the mother's anus. As the shoulders move down into the pelvis, the head then turns sideways so that the alignment of the baby's body is restored.

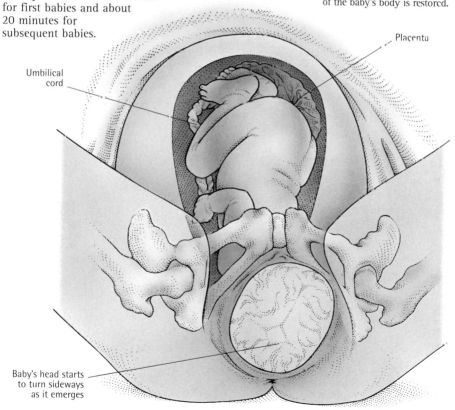

Umbilical cord

Placenta

Baby's head starts to turn sideways as it emerges

Delivery of the baby

Once the whole of the baby's head has emerged, delivery is completed rapidly with the next few contractions. First one shoulder is pushed out, then the other, after which the rest of the body emerges smoothly. At this stage, the placenta is normally still attached to the uterine wall, and will be delivered a few minutes later. The umbilical cord is clamped, to prevent blood loss, and cut.

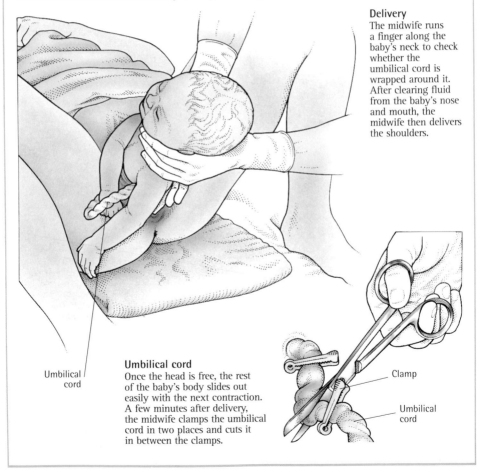

Delivery
The midwife runs a finger along the baby's neck to check whether the umbilical cord is wrapped around it. After clearing fluid from the baby's nose and mouth, the midwife then delivers the shoulders.

Umbilical cord

Umbilical cord
Once the head is free, the rest of the baby's body slides out easily with the next contraction. A few minutes after delivery, the midwife clamps the umbilical cord in two places and cuts it in between the clamps.

Clamp

Umbilical cord

Assisted delivery

An assisted delivery may be necessary if labor is not progressing satisfactorily. Forceps are useful for delivering a baby quickly – especially if the baby is in distress or the mother is exhausted or has bled excessively. An alternative procedure to forceps delivery is vacuum extraction. These methods can be used only if the cervix is fully dilated and the baby's head has started to descend.

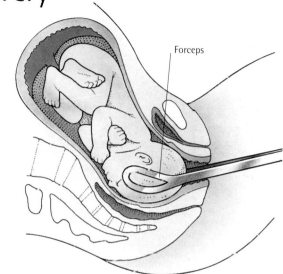

Forceps

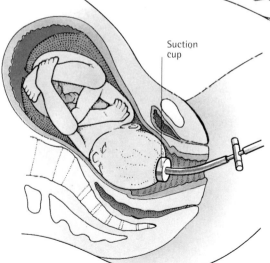

Suction cup

Forceps delivery

Obstetrical forceps consist of two curved metal blades designed to fit around the baby's head. A gentle pull on the forceps guides the head down into the vagina. Once the head emerges, the forceps are usually removed and delivery can continue normally.

Vacuum extraction

A vacuum extractor is a pump attached to a suction cup made of metal, rubber, or plastic. The cup is placed on the baby's head and with each contraction the baby is gently pulled down the birth canal.

Fetal monitoring

During labor, the condition of the fetus is monitored by measuring fetal heart rate (normally 120–160 beats per minute, or BPM). The recordings are made by strapping a metal plate to the mother's abdomen or attaching electrodes to the baby's head. Heart rate can decrease with each contraction but should quickly return to normal. Prolonged deceleration may indicate a potential problem.

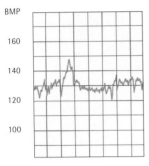

NORMAL FETAL HEART RATE

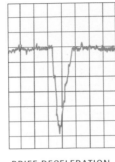

BRIEF DECELERATION

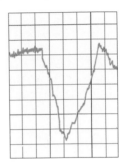

PROLONGED DECELERATION

Delivery of twins

Twins must be monitored closely during labor, and both fetal heart rates must be recorded (see above). The first baby is usually positioned head downwards. If the second twin is head down or in a breech position, it can be delivered normally; if the baby is lying transversely, it may be possible to turn it. When twins share a placenta, the umbilical cord of the first-born is clamped after delivery to prevent the second baby bleeding through the cord.

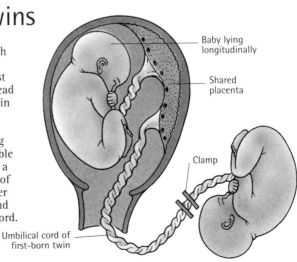

Baby lying longitudinally

Shared placenta

Clamp

Umbilical cord of first-born twin

Third stage of labor

Expulsion of the placenta from the uterus is known as the third stage of labor, and is usually completed within about 15 minutes of the birth of the baby. The umbilical cord is cut before the placenta is delivered. The placenta detaches from the uterus and is expelled; the process is sometimes assisted by giving the mother a hormone injection to increase uterine contractions.

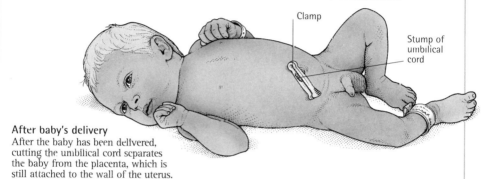

Clamp

Stump of umbilical cord

After baby's delivery
After the baby has been delivered, cutting the umbilical cord separates the baby from the placenta, which is still attached to the wall of the uterus.

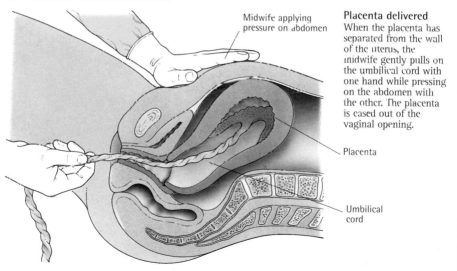

Midwife applying pressure on abdomen

Placenta delivered
When the placenta has separated from the wall of the uterus, the midwife gently pulls on the umbilical cord with one hand while pressing on the abdomen with the other. The placenta is eased out of the vaginal opening.

Placenta

Umbilical cord

Newborn baby

At birth, some aspects of the baby's head and body may appear to be different from those of an older baby. These differences are normal in the newborn and will disappear within a few days or weeks. A full-term, newborn baby weighs on average 7.7lb (3.5kg) and measures 20in (51cm) in length. During the first few days, the baby loses up to 10 percent of its birth weight, but regains this by about the tenth day. The baby is usually covered with a greasy, whitish substance called vernix, which protects its skin within the uterus and eases delivery. Premature babies are sometimes covered with a fine, downy hair, known as lanugo.

Breast
The baby's breasts are commonly swollen and may secrete a little milk. This swelling will decrease in a few weeks

Lungs
The baby's first breath expands the lungs and triggers changes in the circulation

Liver
Immaturity of the liver enzymes that break down the pigment bilirubin can cause temporary yellowing (jaundice)

Genitals
The external genitals of newborn boys and girls appear relatively large. Girls commonly have a slight vaginal discharge

Skin
Slight skin peeling can occur in the first week. Minor rashes and skin blemishes are also common during the baby's early months

Intestines
The first fecal material excreted by the baby is a thick, sticky, greenish-black substance called meconium

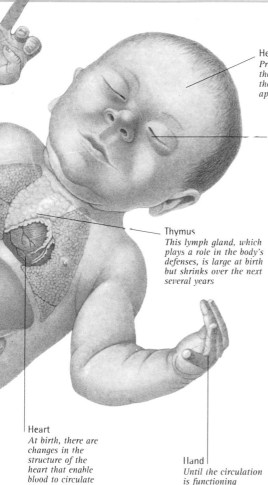

Head
Pressure during the birth can cause the skull bones to overlap, giving the baby's head an abnormal appearance for a few days

Eyes
Newborn babies can see but tend to keep their eyes shut. The eyes are often grayish-blue at first, but change color over the next few months

Thymus
This lymph gland, which plays a role in the body's defenses, is large at birth but shrinks over the next several years

Heart
At birth, there are changes in the structure of the heart that enable blood to circulate through the lungs

Hand
Until the circulation is functioning efficiently, the hands, and also the feet, may look bluish

Fontanelles

The bones of a baby's skull are not fused together but are separated by soft gaps called fontanelles. These gaps allow the skull bones to overlap during birth. By about 18 months, the fontanelles have closed.

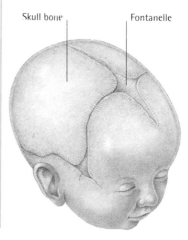

Skull bone   Fontanelle

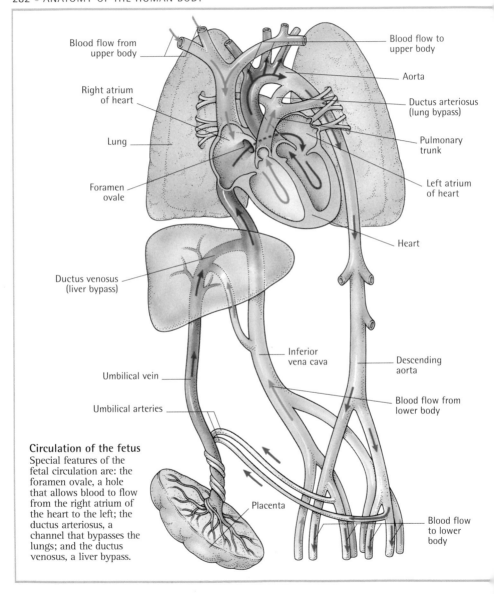

Blood flow from upper body

Right atrium of heart

Lung

Foramen ovale

Blood flow to upper body

Aorta

Ductus arteriosus (lung bypass)

Pulmonary trunk

Left atrium of heart

Heart

Ductus venosus (liver bypass)

Inferior vena cava

Descending aorta

Umbilical vein

Blood flow from lower body

Umbilical arteries

Placenta

Blood flow to lower body

Circulation of the fetus
Special features of the fetal circulation are: the foramen ovale, a hole that allows blood to flow from the right atrium of the heart to the left; the ductus arteriosus, a channel that bypasses the lungs; and the ductus venosus, a liver bypass.

Circulation before and after birth

Because exchange of oxygen occurs in the placenta, the circulation of a fetus (see left) differs from that of a baby at birth (see below). Most blood bypasses the fetal lungs through two holes: the foramen ovale in the heart and the ductus arteriosus between the pulmonary artery and the aorta. At birth, the baby's first breath triggers changes; the holes close and all blood passes through the lungs.

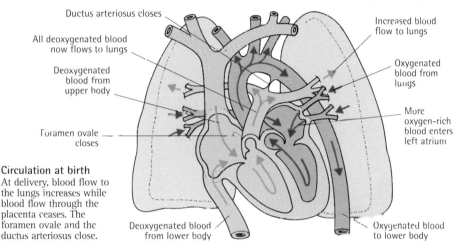

Ductus arteriosus closes

All deoxygenated blood now flows to lungs

Deoxygenated blood from upper body

Foramen ovale closes

Increased blood flow to lungs

Oxygenated blood from lungs

More oxygen-rich blood enters left atrium

Circulation at birth
At delivery, blood flow to the lungs increases while blood flow through the placenta ceases. The foramen ovale and the ductus arteriosus close.

Deoxygenated blood from lower body

Oxygenated blood to lower body

Apgar score

A system called the Apgar score is used to assess the condition of newborn babies. Factors such as heart rate are given a score at 1 and 5 minutes after birth. Healthy color in nonwhite babies is assessed by examining the mouth lining, whites of the eyes, palms of the hands, and soles of the feet.

SIGN	SCORE: 0	SCORE: 1	SCORE: 2
Heart rate	None	Below 100	Over 100
Breathing	None	Slow or irregular; weak cry	Regular; strong cry
Muscle tone	Limp	Some bending of limbs	Active movements
Response to stimulation	None	Grimace or whimpering	Cry, sneeze, or cough
Color	Pale; blue	Blue extremities	Pink

Puerperium

The period following childbirth, when a woman's body gradually returns to its pre-pregnant state, is known as the puerperium. This normally lasts about six weeks. As the placental site heals, tissue debris from the uterus is expelled in the form of a vaginal discharge called lochia. At first, the lochia is bloodstained but then becomes paler. Uterine contractions continue during the puerperium as the uterus shrinks to almost its pre-pregnant size. The cervix closes and the vagina returns to normal.

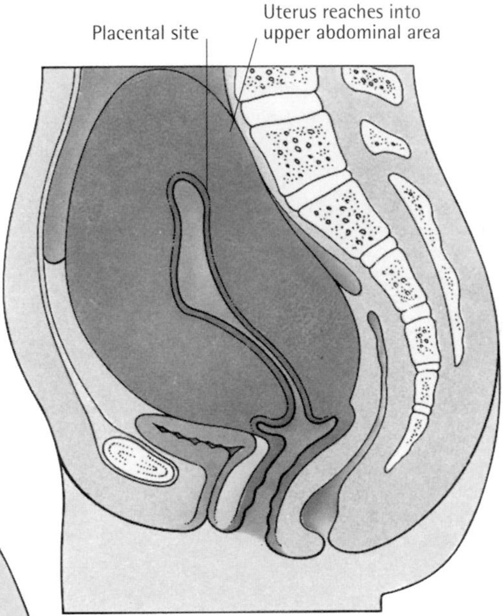

Placental site

Uterus reaches into upper abdominal area

UTERUS IMMEDIATELY AFTER CHILDBIRTH

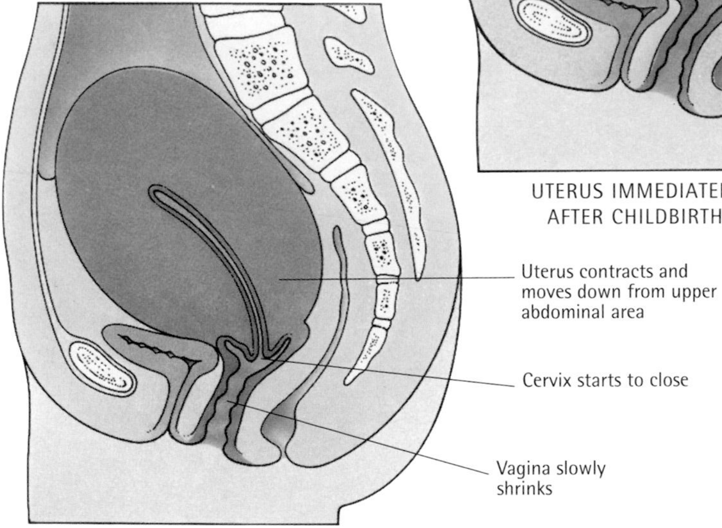

Uterus contracts and moves down from upper abdominal area

Cervix starts to close

Vagina slowly shrinks

UTERUS 1 WEEK AFTER CHILDBIRTH

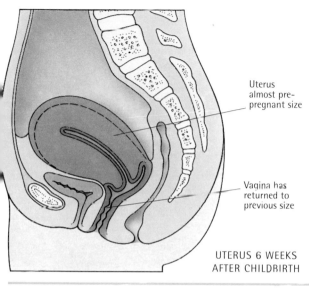

Uterus almost pre-pregnant size

Vagina has returned to previous size

UTERUS 6 WEEKS AFTER CHILDBIRTH

Cervix

In nulliparous women (those who have never given birth) the cervical opening is nearly circular. Childbirth stretches the cervix; the opening closes again but does not regain its original appearance.

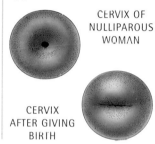

CERVIX OF NULLIPAROUS WOMAN

CERVIX AFTER GIVING BIRTH

Lactation

During pregnancy, milk-producing glands in the breast enlarge and increase in number. Lactation (the production of milk) begins in the first few days after childbirth. Immediately after delivery, the breasts produce a thick yellow fluid called colostrum, which provides the baby's first food. Sucking at the breast stimulates the release of the pituitary hormone oxytocin, which promotes the flow of milk.

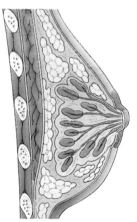

BEFORE PREGNANCY

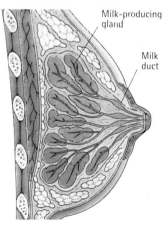

Milk-producing gland

Milk duct

DURING PREGNANCY AND LACTATION

Growth in childhood

During childhood, the development of hard bone (ossification) and the growth of bones is a continuous process. Ossification starts before birth at zones in the bone shafts (diaphyses) called primary ossification centers. In a newborn baby, only the bone shafts are ossified; the bone ends (epiphyses) are made of cartilage. As the child develops, secondary ossification centers form in the bone ends. By about the age of 18, the cartilage has all been replaced by bone and growth is complete.

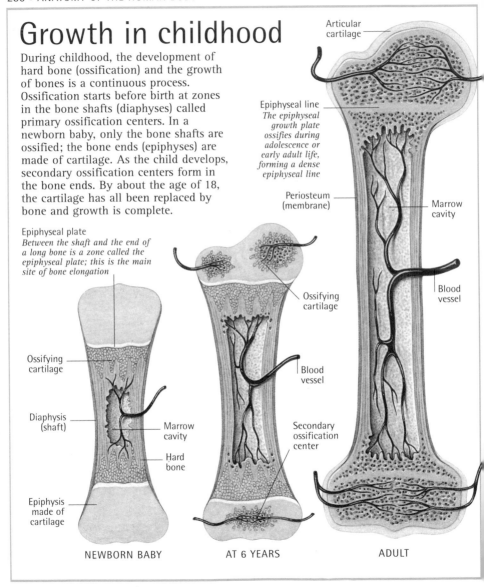

Epiphyseal plate
Between the shaft and the end of a long bone is a zone called the epiphyseal plate; this is the main site of bone elongation

Articular cartilage

Epiphyseal line
The epiphyseal growth plate ossifies during adolescence or early adult life, forming a dense epiphyseal line

Periosteum (membrane)

Marrow cavity

Blood vessel

Ossifying cartilage

Blood vessel

Ossifying cartilage

Diaphysis (shaft)

Marrow cavity

Hard bone

Secondary ossification center

Epiphysis made of cartilage

NEWBORN BABY

AT 6 YEARS

ADULT

Body proportions

In children, dramatic changes in body proportions take place. In a newborn infant, the head represents about one-quarter of the baby's total length. As the child grows, the relative sizes of the head and trunk decrease and the limbs lengthen. At final adult height, the head is only about one-eighth of body length.

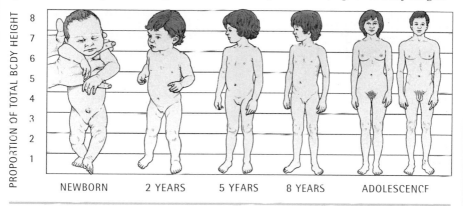

PROPORTION OF TOTAL BODY HEIGHT

NEWBORN   2 YEARS   5 YEARS   8 YEARS   ADOLESCENCE

Dental development

The first, or primary, teeth appear in a specific order (as indicated right) between the ages of 8 months and 2½ years. From about the age of 6, these teeth are gradually replaced by the 32 permanent teeth (approximate ages at which these erupt are shown far right). In some people, the third molars, also known as the wisdom teeth, never develop.

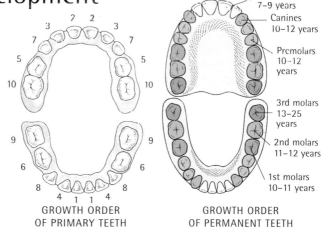

GROWTH ORDER
OF PRIMARY TEETH

Incisors
7–9 years

Canines
10–12 years

Premolars
10–12 years

3rd molars
13–25 years

2nd molars
11–12 years

1st molars
10–11 years

GROWTH ORDER
OF PERMANENT TEETH

Child development

Babies are born able to see and hear. They can also perform certain reflex movements, such as grasping an object put in the hand and turning the head towards a touch on the face, which enables the baby to find the nipple. In early childhood, basic skills of body movement, manipulation, and social behavior are acquired, and language develops. Development takes place in well-recognized steps, called developmental milestones, which occur, in most children, by predictable ages.

Developmental milestones
Each child progresses at a slightly different rate when mastering basic skills. However, the attainment of certain developmental milestones follows a predictable pattern.

AGE	MOVEMENT
1 MONTH	Lies with head to one side. Sleeps most of the time when not being fed or handled.
6 MONTHS	Sits with support. Holds head and back straight. Turns head to look around.
9 MONTHS	Attempts to crawl on all fours. Stands holding on to support for a few moments.
12 MONTHS	Walks with one or both hands held. Walks around furniture stepping sideways.
18 MONTHS	Can get up and down stairs with a helping hand or holding rail. Throws a ball.
2 YEARS	Runs around with ease. Can open doors. Kicks a ball without overbalancing.
3 YEARS	Can ride a tricycle and walk on tiptoe. Uses alternating feet to walk upstairs.
4 YEARS	Able to hop on one foot, and to run on tiptoe. Can climb up trees and ladders.
5 YEARS	Able to skip on alternate feet. Runs lightly on toes. Can dance well to music.

MANIPULATION

 Hands are normally closed at rest but grasp on to finger when palm is touched.

 Uses whole hand to grasp objects in palm. Passes objects between hands.

 Grasps between thumb and index finger. Pokes at small objects with index finger.

 Deliberately drops toys, one by one, to the ground and watches them fall.

 Can build a tower of three or four cubes. Scribbles on paper with pencil or crayon.

 Turns pages of book one at a time. Can build a tower six or seven cubes high.

 Can copy lines and circles. Able to copy a bridge made from three cubes.

 Copies some letters, such as X, V, H, T, and O. Can draw a man and a house.

 Copies squares, triangles, many letters. Writes a few letters without prompting.

SOCIAL BEHAVIOR

 Watches mother's nearby face intently. Starts smiling at about 5 or 6 weeks.

 Takes everything to mouth. Turns quickly to sound of familiar voice across room.

 Holds bottle or cup. Holds and chews solids. Babbles. Shouts to attract attention.

 Holds out arms and feet to be dressed. Understands some simple commands.

 Uses spoon well. Indicates need for toilet. Uses some words; understands many.

 Puts on shoes and socks. Makes simple sentences. Asks for food and drink.

 Understands the idea of sharing. Plays with others. Tries to tidy up. Uses fork.

 Can dress and undress. Speech grammatical and completely intelligible.

 Washes and dries face. Uses knife. Knows birthday. Can act out stories in detail.

Puberty in boys

Puberty starts at about 12 or 13 years of age in boys. Hormonal changes stimulate an increase in growth rate, alterations in behavior, enlargement of the genitals, and the appearance of secondary sexual characteristics such as facial hair. Because boys begin their final growth spurt later than girls, they have a longer period of steady growth and usually attain a greater adult height.

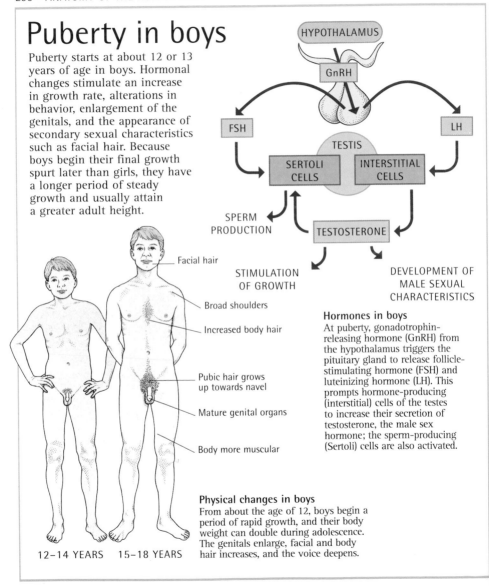

Facial hair

Broad shoulders

Increased body hair

Pubic hair grows up towards navel

Mature genital organs

Body more muscular

12–14 YEARS   15–18 YEARS

Hormones in boys

At puberty, gonadotrophin-releasing hormone (GnRH) from the hypothalamus triggers the pituitary gland to release follicle-stimulating hormone (FSH) and luteinizing hormone (LH). This prompts hormone-producing (interstitial) cells of the testes to increase their secretion of testosterone, the male sex hormone; the sperm-producing (Sertoli) cells are also activated.

Physical changes in boys

From about the age of 12, boys begin a period of rapid growth, and their body weight can double during adolescence. The genitals enlarge, facial and body hair increases, and the voice deepens.

Puberty in girls

Girls start puberty at about the age of 10 or 11. Their growth spurt starts earlier than that of boys, and girls have usually attained their adult height by about the age of 16. Menstruation does not begin until towards the end of the growth period, when the girl has reached an optimum body weight. Ovulation, when reproduction can occur, starts when a regular menstrual cycle is established.

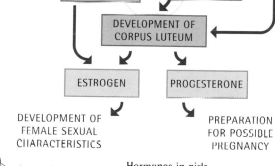

10–12 YEARS   15–16 YEARS

Mature breasts

Broader hips

Pubic hair

Hormones in girls

Triggered by gonadotrophin-releasing hormone (GnRH) from the hypothalamus, the pituary gland releases follicle-stimulating hormone (FSH) and luteinizing hormone (LH). These, in turn, stimulate the ovaries to release eggs and to produce the female sex hormones estrogen and progesterone.

Physical changes in girls

Budding of the breasts is often the first sign of puberty. This is followed by the growth of pubic and underarm hair, and by menstruation (which may be irregular at first). Fat is deposited around the hips.

Menstrual cycle

Menstruation, the periodic shedding of the endometrium (uterine lining), indicates that a woman is capable of reproduction. The menstrual cycle is regulated by hormones secreted by the pituitary gland and the ovaries. During each cycle, one or other of the two ovaries releases an egg, or ovum. If fertilization does not take place, the egg is shed, together with the endometrium, about 2 weeks later.

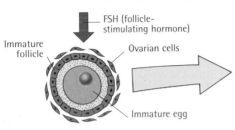

FSH (follicle-stimulating hormone)

Immature follicle

Ovarian cells

Immature egg

1 Follicle development starts
Within each ovary there are many immature follicles, each containing an immature egg cell. FSH stimulates the growth of a number of these follicles; usually only one egg in each menstrual cycle reaches maturity.

Egg

Primary follicle

Layers of follicular cells

2 Egg cell enlarges
The egg enlarges and the follicular cells multiply around it, forming layers. The follicle is now called the primary follicle.

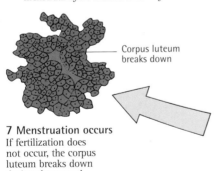

Corpus luteum breaks down

7 Menstruation occurs
If fertilization does not occur, the corpus luteum breaks down during the second week after ovulation, progesterone levels drop, and menstruation occurs.

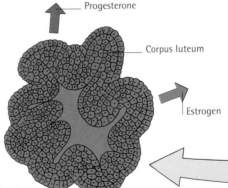

Progesterone

Corpus luteum

Estrogen

6 Corpus luteum develops
After ovulation, the ruptured follicle develops into a structure called the corpus luteum, which secretes progesterone and estrogen.

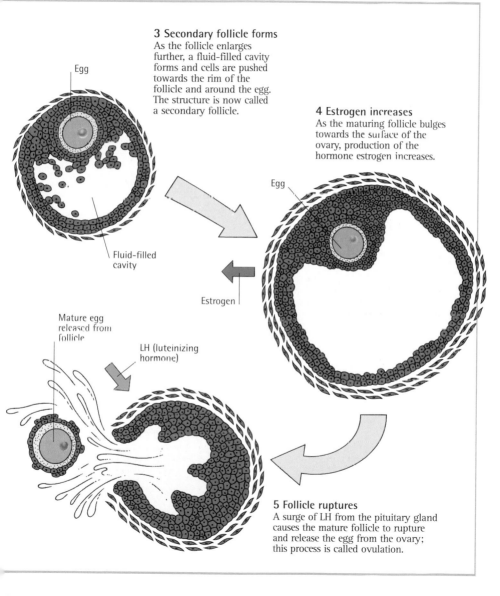

3 Secondary follicle forms
As the follicle enlarges further, a fluid-filled cavity forms and cells are pushed towards the rim of the follicle and around the egg. The structure is now called a secondary follicle.

Egg

Fluid-filled cavity

4 Estrogen increases
As the maturing follicle bulges towards the surface of the ovary, production of the hormone estrogen increases.

Egg

Estrogen

Mature egg released from follicle

LH (luteinizing hormone)

5 Follicle ruptures
A surge of LH from the pituitary gland causes the mature follicle to rupture and release the egg from the ovary; this process is called ovulation.

Changes in the endometrium

At the start of each menstrual cycle, the endometrium (uterine lining) is shed together with an unfertilized egg. After each period of bleeding, the endometrium thickens to prepare the uterus for nurturing a fertilized egg and subsequent pregnancy. If fertilization does not occur, the endometrium breaks down again and is shed, together with the unfertilized egg, as the cycle repeats itself.

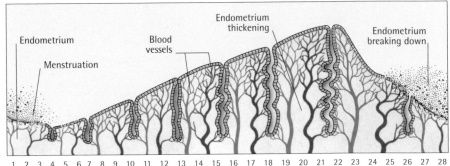

28-day cycle

The menstrual cycle typically lasts 28 days, but it can range from 23–35 days. Bleeding lasts an average of 5 days, although this can vary from 1–8 days. In the standard 28-day cycle, ovulation occurs around day 14. The volume of blood lost varies from cycle to cycle, and from woman to woman.

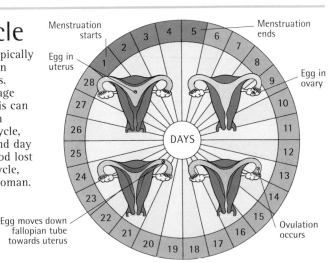

Sperm production

The production of sperm, which begins during puberty, takes place within the seminiferous tubules of the testes. Sperm develop through a complex series of events during which cells called spermatogonia change first into spermatocytes and then into spermatids. The spermatids mature into sperm, and move into the central cavity of the seminiferous tubule. From here, the sperm move to the epididymis, where they are stored.

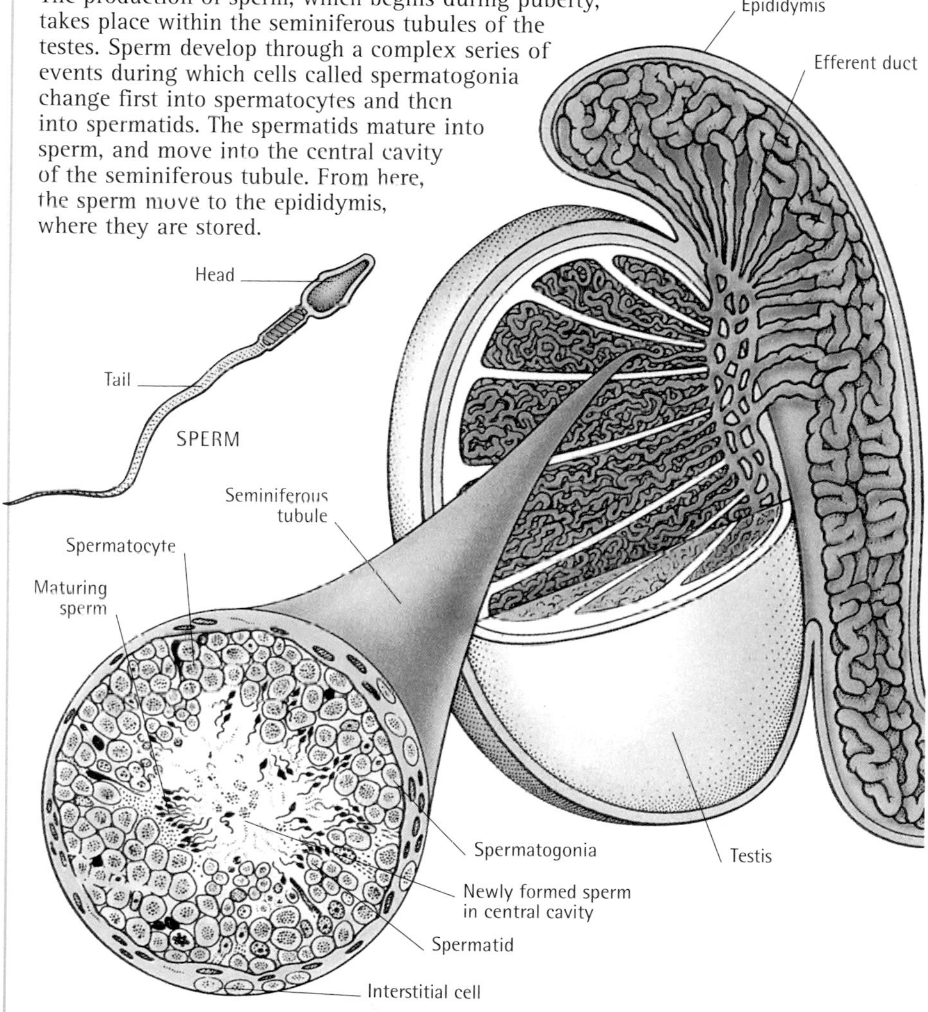

Epididymis

Efferent duct

Head

Tail

SPERM

Seminiferous tubule

Spermatocyte

Maturing sperm

Spermatogonia

Testis

Newly formed sperm in central cavity

Spermatid

Interstitial cell

Aging: skin

In an older person, the deep layers of skin tissue have fewer cells, as a result of which they lose flexibility and have less resilience. The skin becomes thinner and more fragile, and starts to wrinkle; the blood vessels in the skin are also less elastic, so that even minor injuries can cause bruising.

Aging: bone

From middle age onwards, bones start to lose density and become more brittle. The rate at which bone tissue is replaced in the normal process of renewal is exceeded by the rate at which it breaks down. By the time a person has reached the age of 70, skeletal density has often been reduced by about a third.

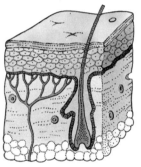

Young skin
A thick top layer and many elastic fibers in the deeper layers help to maintain the smoothness of young skin.

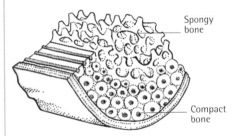

Spongy bone

Compact bone

Young bone
In a young person, bone has a thick, strong outer layer of dense compact bone and an inner core of soft, spongy bone rich in blood vessels.

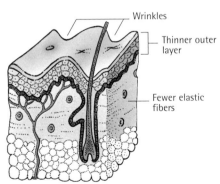

Wrinkles

Thinner outer layer

Fewer elastic fibers

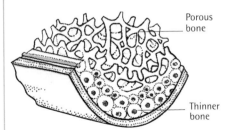

Porous bone

Thinner bone

Older bone
Bone in older people has a thinner outer layer that lacks strength. The inner, spongy bone is more porous, and contains fewer blood vessels and less calcium.

Older skin
A thinner outer layer and fewer elastic fibers in the deeper layers cause the skin of an older person to look loose, with creases and wrinkles.

Effect of age on circulation

In many older people, the internal lining of the arteries thickens with fatty deposits called plaques. These plaques may cause blood clots to form, restricting blood flow. Narrowing of the arteries forces the heart to work harder; and, like all muscles, the heart becomes weaker and less efficient with age.

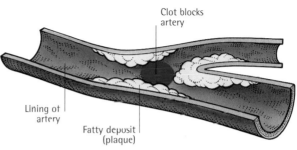

Clot blocks artery

Lining of artery

Fatty deposit (plaque)

Decline of body organs

In young people, organs such as the liver and the kidneys have a greater functional capacity than the body needs; they can sometimes compensate for damage due to disease. With age, these organs become less efficient and even a minor illness may cause them to fail.

Liver
Some decline in liver function is normal with increasing age. Deterioration may be greatly accelerated by damage from alcohol or chronic infection.

Right lobe

Left lobe

Portal vein

Hepatic duct

Hepatic artery

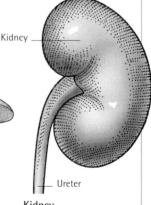

Kidney

Ureter

Kidney
By the age of 70, kidney function has usually declined to about 50 percent of its efficiency at the age of 40. The decline continues, and accelerates in old age.

Age-related hearing loss

Aging usually causes a loss of sensitivity to sounds, which may become duller or distorted so that speech becomes difficult to follow. High-pitched sounds are the first that are difficult to detect; eventually, all frequencies are affected. Hearing loss in older people may be due to degeneration of the cochlea; repeated or prolonged exposure to loud noises hastens this deterioration.

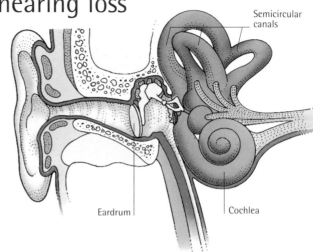

Semicircular canals

Eardrum

Cochlea

Impaired vision

With age, vision may become impaired by structural changes that affect the ability of the eyes to focus on nearby objects. Loss of tissue elasticity can stiffen the lens of the eye so that it is unable to change shape and create a clear image on the retina. Vision is also sometimes affected by degeneration of the macula, the central area of the retina, or by formation of a cataract (clouding of the lens).

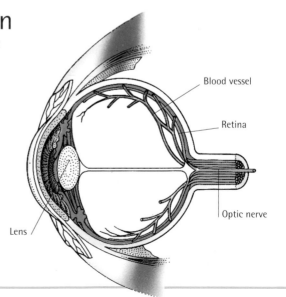

Blood vessel

Retina

Optic nerve

Lens

Menopause

Menopause, when menstruation ceases and the reproductive phase of a woman's life ends, usually occurs some time between the ages of 45 and 55. The ovaries no longer respond to follicle-stimulating hormone (FSH) from the pituitary gland, and stop producing eggs (ova). Levels of the sex hormone estrogen then drop, which can lead to physical and sometimes psychological symptoms. Hot flashes and night sweats are common symptoms; thinning and dryness of the vagina can also occur.

NORMAL   HOT FLASH

Hot flashes
These thermal images show an increase in skin temperature during a hot flash. The woman will experience a sensation of intense heat and her skin may redden.

KEY

▢ Before menopause

▢ At menopause

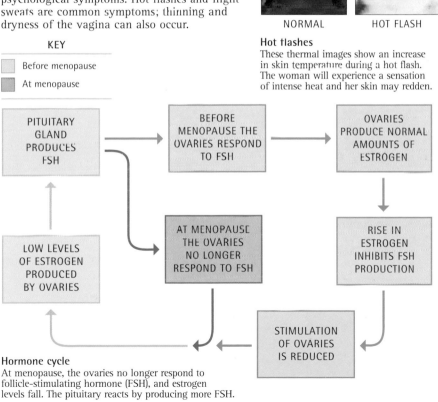

| PITUITARY GLAND PRODUCES FSH | → | BEFORE MENOPAUSE THE OVARIES RESPOND TO FSH | → | OVARIES PRODUCE NORMAL AMOUNTS OF ESTROGEN |

AT MENOPAUSE THE OVARIES NO LONGER RESPOND TO FSH

LOW LEVELS OF ESTROGEN PRODUCED BY OVARIES

RISE IN ESTROGEN INHIBITS FSH PRODUCTION

STIMULATION OF OVARIES IS REDUCED

Hormone cycle
At menopause, the ovaries no longer respond to follicle-stimulating hormone (FSH), and estrogen levels fall. The pituitary reacts by producing more FSH.

Genetic exchange

Sex cells have a particular genetic mix, produced by a form of cell division called meiosis. This process produces new cells that each contain a single set of 23 chromosomes – half the amount of genetic material contained in all other body cells (which divide by a process known as mitosis, see p.53). When sperm and egg fuse, the resulting embryo has two sets of 23 chromosomes, 46 chromosomes in all.

Centriole

Stage 1
Meiosis starts with the doubling-up of each member of the 23 pairs of chromosomes (four "doubled-up" pairs are seen here). Each of these doubled-up, X-shaped chromosomes lines up with its partner.

Stage 2
As pairs of chromosomes entwine, they exchange a random selection of DNA. This process (known as crossing over) combines genes in a way that may never be repeated.

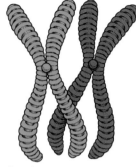

Crossing over
During this process, pairs of homologous chromosomes exchange corresponding genes (those located at the same point on each chromosome).

Stage 3
The matching (homologous) pairs of chromosomes line up in the middle of the cell. Threads form a structure called the spindle between the centrioles at the opposite poles of the cell.

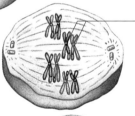

Homologous chromosomes
The chromosomes in a pair are similar but not identical, and are called homologous

Spindle thread

Stage 4
The threads of the spindle pull each of the doubled-up chromosomes in a pair to opposite sides. The cell begins to divide into two separate cells.

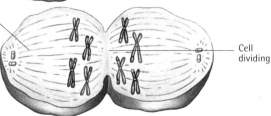

Cell dividing

New cell

Stage 5
Each new cell, complete with a new nuclear membrane, now has a doubled-up chromosome from each of the 23 pairs.

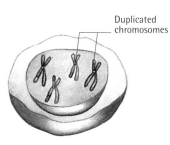

Duplicated chromosomes

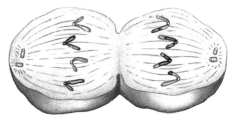

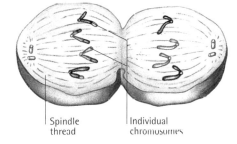

Stage 6
Spindle threads form and the chromosomes line up at the center of the cell. They separate to form individual chromosomes, which are pulled to opposite sides. The cells start to divide again.

Spindle thread

Individual chromosomes

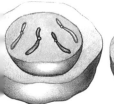

Stage 7
After division, each of the four new cells contains a unique set of 23 chromosomes that contain DNA from the original cell's 46 chromosomes.

Each new cell has a unique set of 23 chromosomes

KEY

Father's gene

Mother's gene

Transmission of genes

The mix of genes that a person receives helps to determine his or her particular characteristics, such as hair color and height. Each individual inherits half of his or her genes from the mother and half from the father. The diagram below demonstrates the transmission of genes through two generations. Only eight genes are shown here, but each body cell contains up to 100,000 genes.

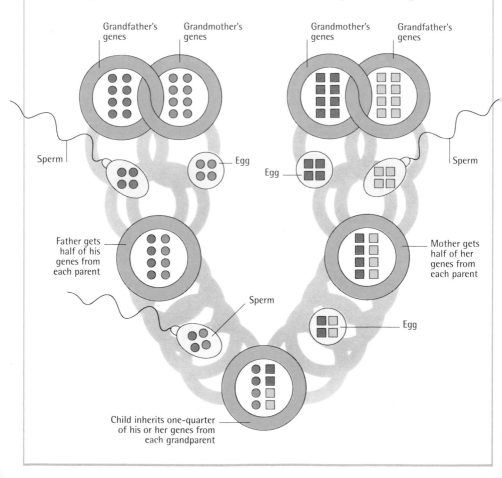

Grandfather's genes

Grandmother's genes

Grandmother's genes

Grandfather's genes

Sperm

Egg

Egg

Sperm

Father gets half of his genes from each parent

Mother gets half of her genes from each parent

Sperm

Egg

Child inherits one-quarter of his or her genes from each grandparent

Sex determination

A person's sex depends on the 23rd pair of chromosomes. The other 22 pairs carry genes for most other characteristics. In women, the sex chromosomes are both of the type known as X chromosomes. Men have two different types: an X and a Y chromosome; the Y chromosome is much smaller than the X. A sperm (which may contain either an X or a Y chromosome) determines the sex of an embryo when fusing with an egg (which is always X) at fertilization.

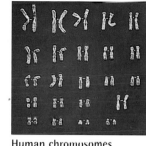

Human chromosomes
Each cell in the human body contains 46 chromosomes, arranged in 23 pairs. In the image above, the sex chromosomes (here XY) are the bottom right-hand pair.

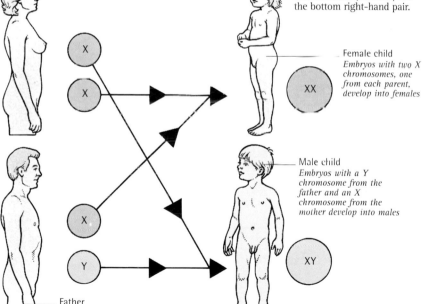

Mother
All females have two X chromosomes

Female child
Embryos with two X chromosomes, one from each parent, develop into females

Male child
Embryos with a Y chromosome from the father and an X chromosome from the mother develop into males

Father
All males have two different sex chromosomes: one X and one Y

Dominant and recessive genes

People inherit two sets of genes, one from the mother and the other from the father. Often, only one of the two genes for a particular characteristic will have an effect on a person. For example, a child who inherits both a gene for brown (dominant) eye color and a gene for blue (recessive) will have brown eyes. For a recessive trait to appear, a person must have two copies of the recessive gene.

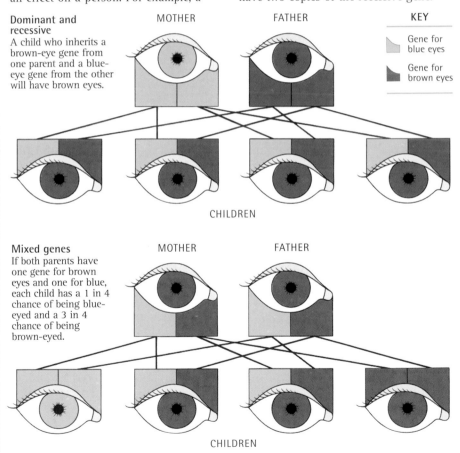

Dominant and recessive
A child who inherits a brown-eye gene from one parent and a blue-eye gene from the other will have brown eyes.

MOTHER

FATHER

KEY

Gene for blue eyes

Gene for brown eyes

CHILDREN

Mixed genes
If both parents have one gene for brown eyes and one for blue, each child has a 1 in 4 chance of being blue-eyed and a 3 in 4 chance of being brown-eyed.

MOTHER

FATHER

CHILDREN

Sex-linked inheritance

Color-blindness, and various diseases such as the blood disorder hemophilia, are due to defective recessive genes on the X (female) chromosome. A woman who inherits one normal and one abnormal gene carries the disease but is healthy because the abnormal recessive gene is "masked" by the normal dominant gene. It is rare for women to be affected by X-linked disorders. Because men have only one X chromosome, a man with the abnormal gene on his X chromosome will be affected. If he marries a woman with normal genes, all their sons will have normal genes and all their daughters will be carriers.

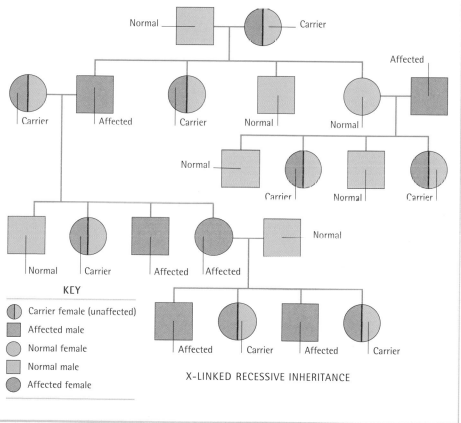

KEY

- Carrier female (unaffected)
- Affected male
- Normal female
- Normal male
- Affected female

X-LINKED RECESSIVE INHERITANCE

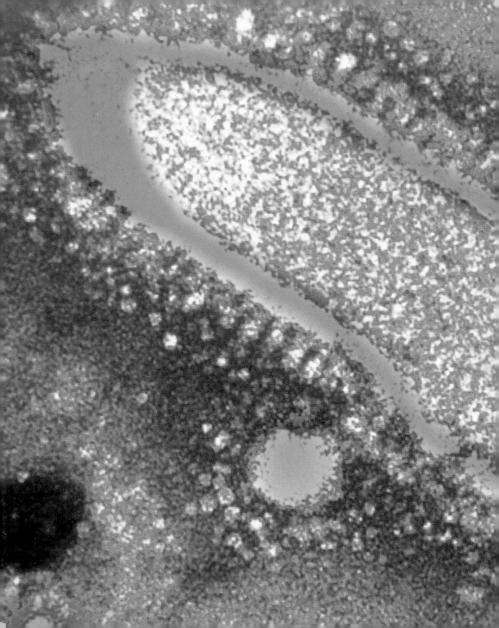

DISEASES
and
DISORDERS

Skin disorders 308

Musculoskeletal disorders 314

Nervous system disorders 330

Cardiovascular disorders 344

Infections and immune disorders 362

Respiratory disorders 376

Digestive disorders 392

Urinary disorders 406

Reproductive disorders 410

Cancer 424

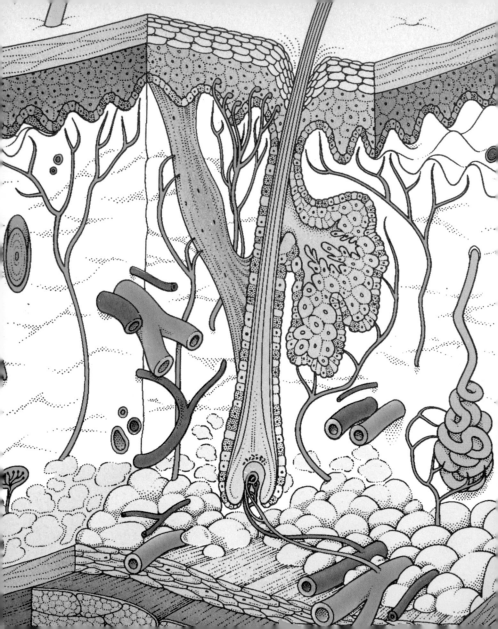

Skin disorders

The skin can be affected by a wide range of disorders. It is susceptible to physical injury and to infection by bacteria, viruses, and other microorganisms, such as fungi; overexposure to sunlight can cause cancers, in which skin cells grow abnormally. Rashes may be caused by allergic reactions, autoimmune diseases, or infections.

Skin conditions
The skin is one of the most vulnerable parts of the body. Diseases of the skin may be localized, affecting only a small site, or widespread over the body.

Wounds

Skin damage is divided into various categories. Puncture wounds, for example from a nail, have a small entry site but can penetrate deeply. Cuts are incised wounds with cleanly severed edges. Abrasions are superficial injuries where the outer skin (epidermis) is scraped off, exposing the inner layer (dermis).

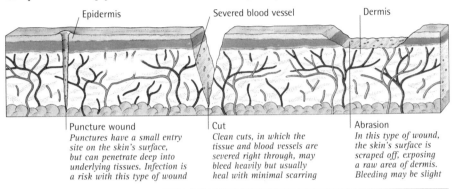

Puncture wound
Punctures have a small entry site on the skin's surface, but can penetrate deep into underlying tissues. Infection is a risk with this type of wound

Cut
Clean cuts, in which the tissue and blood vessels are severed right through, may bleed heavily but usually heal with minimal scarring

Abrasion
In this type of wound, the skin's surface is scraped off, exposing a raw area of dermis. Bleeding may be slight

Acne vulgaris

In acne vulgaris, the sebaceous (oil-producing) glands draining into the hair follicles produce excessive amounts of their secretion, called sebum. Small plugs of hardened sebum known as blackheads block the follicle openings; bacteria multiply in the trapped sebum, causing inflammation of the surrounding area.

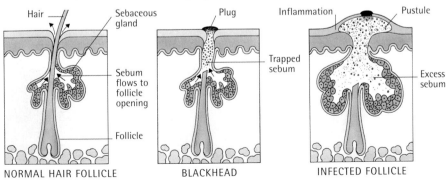

NORMAL HAIR FOLLICLE

BLACKHEAD

INFECTED FOLLICLE

Boils

Usually the result of infection by staphylococcal bacteria, a boil is a collection of pus within a hair follicle. The infected area becomes inflamed and painful. Boils most often occur in moist areas such as the armpit or groin, or at the back of the neck, where there is friction from clothing.

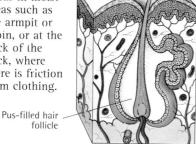

Pus-filled hair follicle

Epidermal cyst

Cysts are saclike structures containing fluid or semisolid matter; some types (epidermal cysts) develop under the skin. The most common of these are epidermoid cysts, which are caused by inflammation of a hair follicle. These contain a substance called keratin, produced by skin cells.

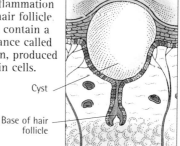

Cyst

Base of hair follicle

Warts

Warts are skin growths caused by infection with the human papillomavirus. The virus invades the prickle cells and squamous cells that make up the outer layer of skin (epidermis), causing them to overmultiply. As the excess cells are pushed upwards, a visible lump forms on the surface of the skin. The tiny black dots that are sometimes seen on warts are small blood vessels.

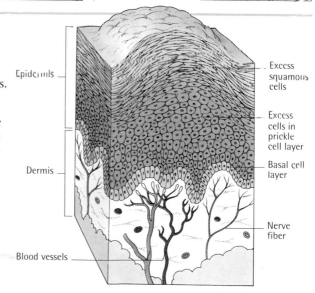

Epidermis

Dermis

Excess squamous cells

Excess cells in prickle cell layer

Basal cell layer

Nerve fiber

Blood vessels

Moles

A mole, or nevus, is caused by an overgrowth of pigment-producing skin cells known as melanocytes. Moles vary in size, color, and texture. They may be raised (as illustrated right) or flat; light or dark brown; and rough or smooth surfaced. They rarely become malignant, but changes in appearance should be reported to a doctor.

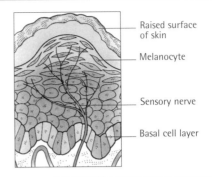

Raised surface of skin

Melanocyte

Sensory nerve

Basal cell layer

Skin cancer

Skin cancers are related to excessive exposure to sunlight. The type that occurs most frequently is basal cell carcinoma, which spreads only locally. Other types, such as squamous cell carcinoma and the rarer malignant melanoma, are more dangerous and can spread throughout the body.

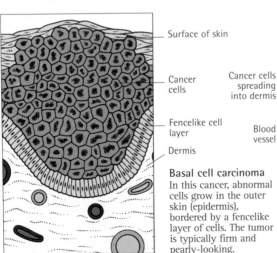

Surface of skin

Cancer cells

Fencelike cell layer

Dermis

Basal cell carcinoma
In this cancer, abnormal cells grow in the outer skin (epidermis), bordered by a fencelike layer of cells. The tumor is typically firm and pearly-looking.

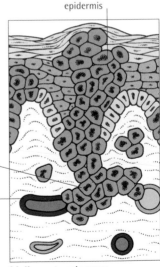

Cancer cells in epidermis

Cancer cells spreading into dermis

Blood vessel

Malignant melanoma
This skin cancer develops in pigment cells (melanocytes). The tumor cells break through layers of skin and can spread rapidly.

Rashes and discoloration

Rashes are areas of skin inflammation that may affect small patches or cover a wide area of the body. The main causes are infections and allergic reactions. However, it is not known why psoriasis and some types of eczema develop, although these rashes are very common. Vitiligo, in which the skin loses pigment but no other symptoms occur, is believed to be an autoimmune disorder.

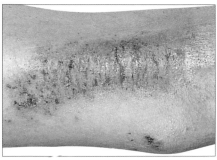

Eczema
There are several different types of this skin inflammation. Characteristic symptoms are itching, red patches and small blisters that burst, leaving the skin surface moist and crusty.

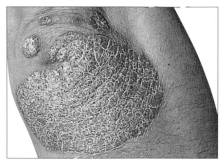

Psoriasis
This noninfectious skin disease features inflamed areas of red or pink, dry plaques with silvery, scaly surfaces. The elbows, knees, shins, scalp, and lower back are most often affected.

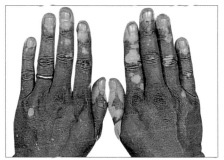

Vitiligo
In vitiligo, patches of skin lose their color. The depigmentation, which is more noticeable on dark skins, commonly affects the hands and face. Occasionally, normal skin color is regained.

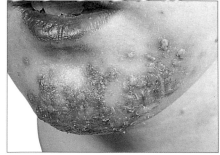

Impetigo
A highly contagious skin disorder, impetigo is caused by bacteria infecting broken skin. Tiny, fluid-filled blisters appear, usually on the face, and burst to form a yellow, itchy crust.

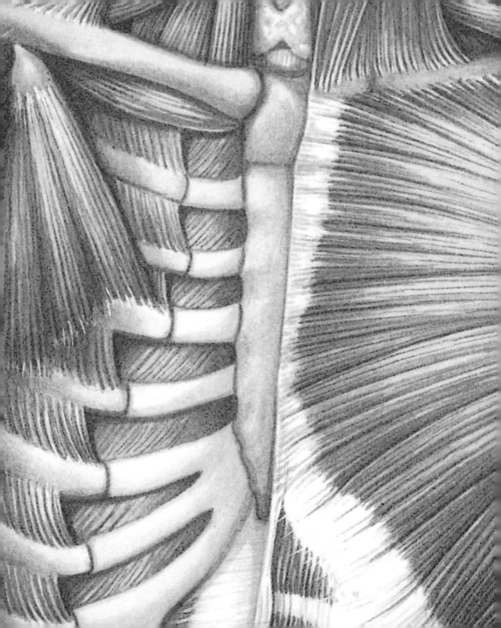

Musculoskeletal disorders

The bones, and the skeletal muscles, tendons, and ligaments that interact with them to produce the body's movement, are subjected to constant wear and tear. In general, the musculoskeletal system is more likely to be affected by injury than by disease. However, bone strength and structure may be weakened by hormonal and other disorders, particularly in older people; some inflammatory joint disorders can occur at any age.

Joint problems
The joints can be affected by disease and by injuries such as bone fractures and torn muscles or connective tissue.

Types of fracture

How a bone breaks depends on the direction and degree of force to which it is subjected. Fractures range from minor surface cracks, known as fissures, to complete breaks through the bone. If a broken bone remains beneath the skin the fracture is described as closed (simple). If the ends of broken bones project through the skin, the fracture is described as open (compound).

Transverse fracture
A powerful, direct or angled force may cause a break straight across the width of a bone.

Spiral fracture
A sharp sudden twist may break a bone diagonally across the shaft, and sometimes leave jagged edges.

Greenstick fracture
Strong force on a long bone may cause it to bend and crack obliquely on one side only. This type of fracture occurs in children, whose bones are not hardened completely.

Comminuted fracture
Direct impact can shatter a bone into several fragments.

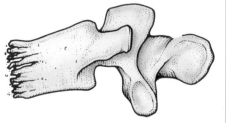

Crush fracture
Spongy bone, such as that in the vertebrae, can be crushed under compression.

Common fracture sites

Some parts of the skeleton are more vulnerable to fracture than others, although the risk of breaking a particular bone can be affected by a person's age and lifestyle. In older people, the bones are more fragile and more likely to break; a common site of fracture in an older person is the femur (thigh bone). Fractures of the ankle or collar bone often occur in young people who take part in active sports.

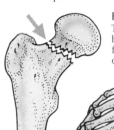

Fractured femur
The neck of the femur (thigh bone) is a common site of fracture in older people, often following a fall.

Fractured rib
Fracture of a rib is usually caused by a blow, but can be the result of stress on the ribcage through coughing or laughing. Broken ribs often heal without treatment.

Colles' fracture
This fracture of the radius, one of the long bones of the lower arm, occurs just above the wrist. It is usually the result of flexing a hand to cushion a fall.

Fractured ankle
Violent twisting of the ankle, especially during sports, can cause fracture at the lower end of either the fibula or tibia (the long bones of the leg).

Fractured scaphoid
Falling on to an outstretched hand can cause fracture of the scaphoid bone. This bone forms part of the row of wrist bones, and is just below the thumb.

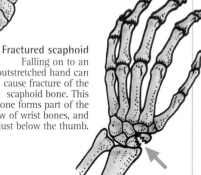

Fractured clavicle
Fracture of the clavicle (collar bone) is a common sports injury. It mostly occurs as the result of a fall on the shoulder or outstretched arm.

Osteoporosis

From middle age onwards, bones become notably thinner and more porous as levels of sex hormones, essential for bone replacement, start to decline. With advancing age, the bone-weakening disorder called osteoporosis affects nearly everyone to some degree. Women are particularly susceptible to osteoporosis when their estrogen levels fall rapidly after the menopause.

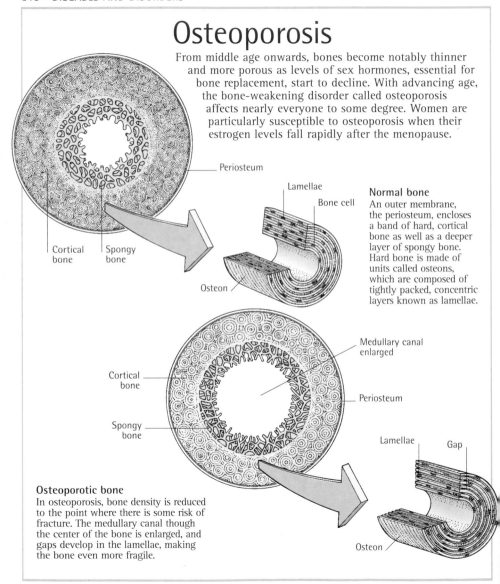

Periosteum

Cortical bone

Spongy bone

Lamellae

Bone cell

Osteon

Normal bone

An outer membrane, the periosteum, encloses a band of hard, cortical bone as well as a deeper layer of spongy bone. Hard bone is made of units called osteons, which are composed of tightly packed, concentric layers known as lamellae.

Cortical bone

Spongy bone

Medullary canal enlarged

Periosteum

Lamellae

Gap

Osteon

Osteoporotic bone

In osteoporosis, bone density is reduced to the point where there is some risk of fracture. The medullary canal though the center of the bone is enlarged, and gaps develop in the lamellae, making the bone even more fragile.

Why osteoporosis occurs

Bones are continually being broken down and then rebuilt to facilitate growth and repair. In young people, the rate of bone formation exceeds the rate at which bone cells (osteocytes) are reabsorbed by the body. This process begins to change in early adulthood, with the rate of reabsorption becoming greater than the rate of cell formation. The skeleton gradually becomes weaker and lighter.

Bone formation

Bone is built up by the deposition of minerals (mainly calcium salts) in an organic matrix of collagen fibers. Bone cells, or osteocytes, form the collagen and also aid in the laying down of calcium. Canals in the bone permit calcium to move into or out of the blood in response to hormones that regulate the body's needs.

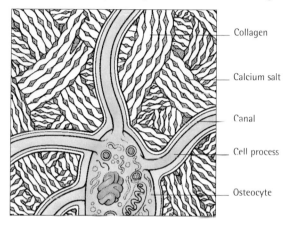

Collagen

Calcium salt

Canal

Cell process

Osteocyte

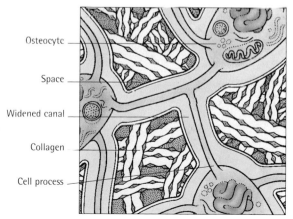

Osteocyte

Space

Widened canal

Collagen

Cell process

Bone reabsorption

As osteoporosis develops with age, both the collagen framework and deposited minerals are broken down much faster than they are formed. The canals that connect the osteocytes become wider and new spaces appear in the collagen. These changes weaken the bone.

Spinal fractures

Most spinal injuries occur as a result of severe forces of compression, or of rotation or bending beyond the spine's normal range of movement. If a spinal fracture is unstable (likely to shift), there is a risk that the spinal cord or nerves may be damaged, with loss of bodily sensation and function, or even paralysis. Bone diseases, such as osteoporosis (see pp. 318–319), can weaken the spine and may increase the likelihood of fractures.

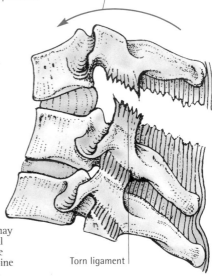

Fractures of transverse processes

Flexion force

Stable fracture of vertebrae
Fracture of a transverse process – one of the bony knobs that extend from each vertebra – is usually a minor injury because the vertebra does not shift from its normal position. This type of fracture, often the result of a direct blow, most commonly affects the lumbar vertebrae; it rarely results in nerve damage.

Unstable fracture and dislocation
If ligaments in the vertebral column tear during extreme flexion or rotation, vertebrae may slip or be pushed out of normal alignment. This type of fracture threatens the stability of the spine and endangers the spinal cord.

Torn ligament

Disc prolapse

The shock-absorbing discs that separate adjacent vertebrae have a hard outer covering and a jellylike center. If the outer layer is ruptured, through wear and tear or excessive pressure, the center protrudes and presses on a spinal root nerve or the spinal cord. The condition, which is known as a prolapsed disc, mostly occurs in the lower spine; it causes severe pain.

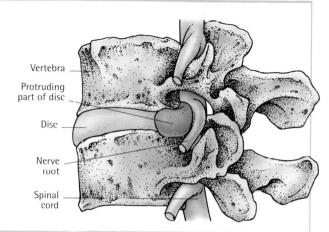

Vertebra

Protruding part of disc

Disc

Nerve root

Spinal cord

Whiplash injury

Injury to the cervical (neck) section of the spine can occur if the neck is suddenly and violently forced either forwards or backwards and then rebounds in the opposite direction. This so-called whiplash injury, often the result of a car accident, stretches the neck and may sprain or tear the ligaments and/or partially dislocate a cervical joint. Occasionally, a vertebra may fracture.

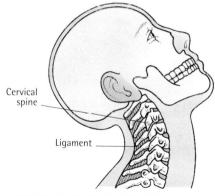

Cervical spine

Ligament

BACKWARD FORCE (HYPEREXTENSION)

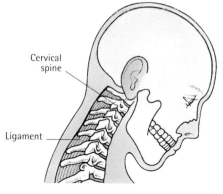

Cervical spine

Ligament

FORWARD FORCE (FLEXION)

Ligament injury

Ligaments are strong bands of fibrous tissue that link bone ends together. If the bones within a joint are pulled too far apart because of a forceful or sudden movement, some of the fibers in the ligament may be overstretched (sprained) or torn. This commonly results in swelling, pain, or muscle spasm; if the injury is severe, joint instability or dislocation may result. The ankle is one of the most common sites of ligament injury.

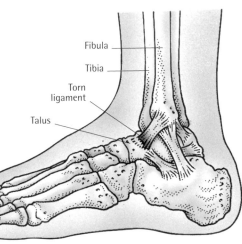

Fibula

Tibia

Torn ligament

Talus

Torn cartilage

One type of cartilage found in the body consists of firm, flexible, slightly elastic connective tissue. In the knee joint (shown left) two discs called menisci made of this cartilage act as shock absorbers, cushioning the femur (thigh bone) and the tibia (shin bone) against excessive force. Tearing of a meniscus by a sudden twisting movement is a common injury in sports such as football.

Femur (thigh bone)

Menisci (cartilage pads)

Torn meniscus

Tibia (shin bone)

Rheumatoid arthritis

Arthritis is a general term used to describe several different disorders that cause painful, swollen joints. Rheumatoid arthritis is thought to be an autoimmune disorder that develops when the immune system produces antibodies that attack body tissues. The joints become inflamed, swollen, stiff, and deformed. The disorder can occur in any joint in the body, but the fingers are among those most commonly affected.

Metacarpal

Phalanx

Inflamed synovial membrane

EARLY STAGE   LATE STAGE

Inflamed joint

Eroded articular cartilage

Synovial membrane spreading

Rheumatoid arthritis in the hand
The joints of the hand are common sites for rheumatoid arthritis. In severe cases, the spaces between the joints disappear and the hand and fingers become distorted.

Stages of the disease
The first symptom of rheumatoid arthritis is inflammation of the synovial membrane lining an affected joint. This membrane thickens and spreads across the joint. The articular cartilage and the bone ends are roughened and eroded.

Osteoarthritis

Osteoarthritis is the degeneration of the articular cartilage that covers the ends of bones in a joint. The disorder is more likely to affect older people, as part of the aging process, although it can be hastened by localized wear and tear. Symptoms may occur in only one joint. In the early stages of osteoarthritis, the cartilage begins to break down, becoming thinner and roughened. Eventually, the bone surfaces are exposed and rub against one another, causing pain. Affected joints may become inflamed from time to time.

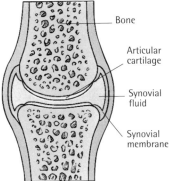

Bone

Articular cartilage

Synovial fluid

Synovial membrane

Normal joint structure
In a healthy joint, the articular cartilage that covers the bone ends is lubricated by synovial fluid. This fluid, which is secreted by the synovial membrane lining the inner surface of the joint, facilitates movement.

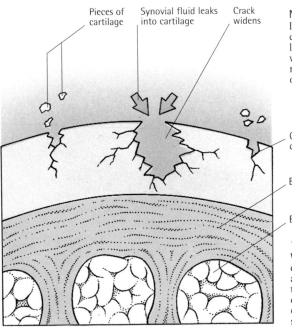

Pieces of cartilage

Synovial fluid leaks into cartilage

Crack widens

Cracked articular cartilage

Bone

Bone marrow

1 Cartilage weakens
When cartilage cells (articular chondrocytes) die, surface cracks appear; this allows synovial fluid to leak in, causing greater cartilage degeneration. Pieces of weakened cartilage break off, irritating the synovial membrane.

2 Gap develops

Eventually, a gap develops in the cartilage and spreads down to the underlying bone. Blood vessels begin to grow into the affected area, and a plug made of tissue called fibrocartilage develops to fill the gap.

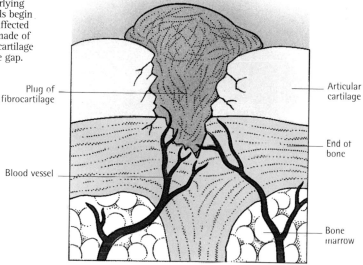

Plug of fibrocartilage

Articular cartilage

End of bone

Blood vessel

Bone marrow

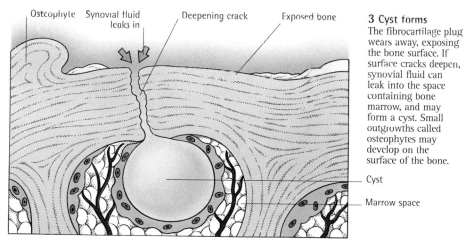

Osteophyte

Synovial fluid leaks in

Deepening crack

Exposed bone

Cyst

Marrow space

3 Cyst forms

The fibrocartilage plug wears away, exposing the bone surface. If surface cracks deepen, synovial fluid can leak into the space containing bone marrow, and may form a cyst. Small outgrowths called osteophytes may develop on the surface of the bone.

Muscle strains and tears

Overexertion or sudden pulling or twisting movements, such as those occurring during sports, can cause damage to muscle fibers. For example, vigorous shoulder action may tear a deltoid or pectoral muscle where it attaches to the humerus (upper arm bone). Moderate injury is referred to as a strain; severe damage is called a muscle tear. Bleeding inside the injured muscle can cause severe pain and swelling; visible bruising around the area may also occur.

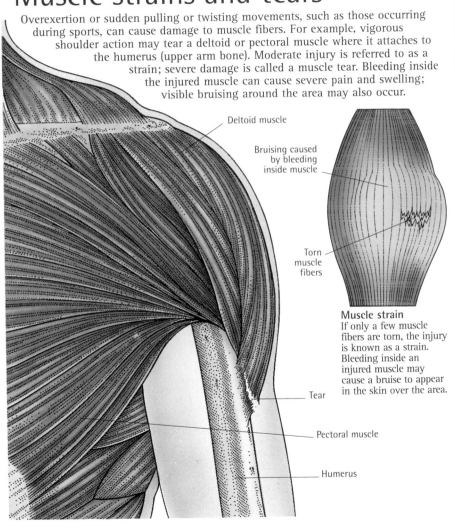

Deltoid muscle

Bruising caused by bleeding inside muscle

Torn muscle fibers

Muscle strain
If only a few muscle fibers are torn, the injury is known as a strain. Bleeding inside an injured muscle may cause a bruise to appear in the skin over the area.

Tear

Pectoral muscle

Humerus

Repetitive strain injury

Several conditions caused by the constant repetition of particular movements are referred to as repetitive strain injury (RSI). Irritation of the flexor and extensor tendons in the wrist and hand is a common injury that often affects keyboard operators, causing pain when the fingers are moved. RSI can also lead to another disorder called carpal tunnel syndrome; this is due to pressure on the median nerve as it passes through a gap under a ligament at the front of the wrist.

RSI
The symptoms of RSI include pain, aching, and tingling in the injured area. Sometimes, movement may be weakened or restricted.

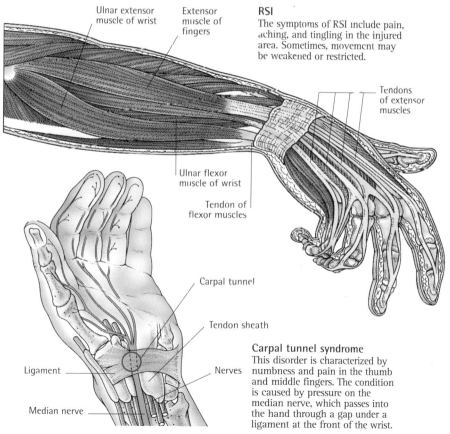

Ulnar extensor muscle of wrist

Extensor muscle of fingers

Tendons of extensor muscles

Ulnar flexor muscle of wrist

Tendon of flexor muscles

Carpal tunnel

Tendon sheath

Ligament

Nerves

Median nerve

Carpal tunnel syndrome
This disorder is characterized by numbness and pain in the thumb and middle fingers. The condition is caused by pressure on the median nerve, which passes into the hand through a gap under a ligament at the front of the wrist.

Tenosynovitis and tendinitis

Injuries involving tendons may affect the inner lining
of the fibrous sheaths that enclose some tendons
(tenosynovitis) or the tendon itself (tendinitis).
Tenosynovitis may be the result of overstretching
or repetitive movements. Tendinitis can occur
when movement creates excessive friction
between a tendon and an adjacent bone.

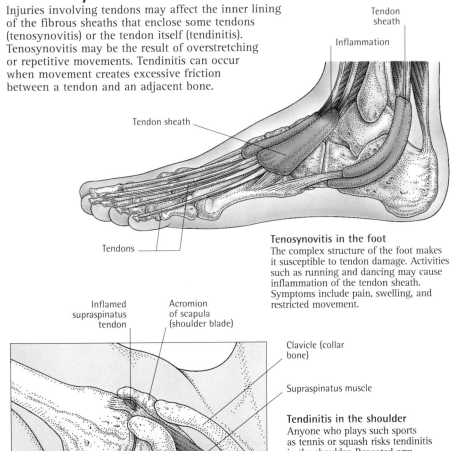

Tendon sheath

Inflammation

Tendon sheath

Tendons

Tenosynovitis in the foot

The complex structure of the foot makes
it susceptible to tendon damage. Activities
such as running and dancing may cause
inflammation of the tendon sheath.
Symptoms include pain, swelling, and
restricted movement.

Inflamed
supraspinatus
tendon

Acromion
of scapula
(shoulder blade)

Clavicle (collar
bone)

Supraspinatus muscle

Tendinitis in the shoulder

Anyone who plays such sports
as tennis or squash risks tendinitis
in the shoulder. Repeated arm
lifting causes friction between
the supraspinatus tendon in the
shoulder and a bony projection
on the scapula (shoulder blade)
called the acromion.

Tendon tears

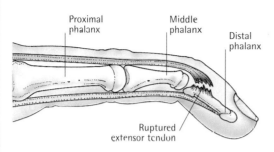

Proximal phalanx

Middle phalanx

Distal phalanx

Ruptured extensor tendon

A sudden, powerful movement can severely damage a tendon and possibly tear it away from the bone. For example, a forceful blow on the tip of a finger can bend the finger forwards, tearing the extensor tendon from its attachment on the small bone known as the distal phalanx.

Soft-tissue inflammation

Inflammation of the tissues surrounding bones and joints is caused by naturally occurring chemicals in the body released in response to tissue damage. These chemicals stimulate nerve endings, producing pain. They also cause blood vessels to dilate and leak fluid; white blood cells are attracted to the damaged area. The result is localized redness, heat, and swelling. Inflammation is often treated with anti-inflammatory drugs, which reduce symptoms by blocking the production of chemicals.

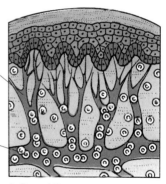

Dilated blood vessel

Increased white blood cells

Inflamed tissue
Blood vessels in the injured area are dilated and abundant white blood cells accumulate, causing heat, redness, pain, and swelling.

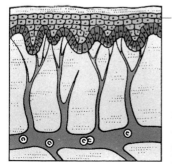

Reduced swelling

Swelling reduced
Drugs inhibit the production of chemicals that cause inflammation. Blood vessels return to normal, numbers of white blood cells decrease, and swelling, redness, and pain are reduced.

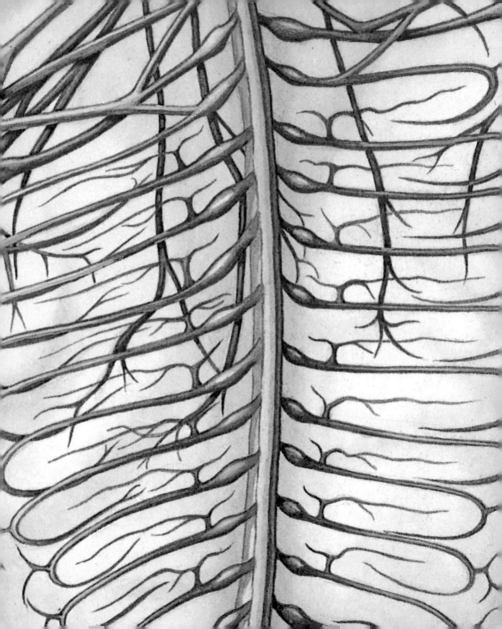

Nervous system disorders

Disruption of electrical impulses in the brain and spinal cord, or the nerves leading to and from them, may impair both physical and mental functions. The nervous system can be affected by infections, injury, tumors, and degenerative conditions. One of the most common nervous system disorders is stroke, caused by blockage or leakage of blood vessels supplying the brain and sometimes resulting in permanent disability.

Spinal cord injury
Damage to the spinal cord can cause loss of sensation or even paralysis in various parts of the body.

Epilepsy

Epilepsy is a result of abnormal electrical activity in the brain; the disorder causes uncontrollable periodic seizures. It may affect the whole brain (generalized seizures), causing loss of consciousness, or just part of the brain (partial seizures), which may or may not cause loss of consciousness. There are two main types of generalized seizure: tonic–clonic seizures, which cause unconsciousness and twitching of the whole body; and petit mal (absence) seizures, which mostly occur in children, and cause only brief unconsciousness and little abnormal movement. An electroencephalogram (EEG), which records brain impulses, may help to diagnose epilepsy.

NORMAL EEG

EEG DURING A SIMPLE PARTIAL SEIZURE

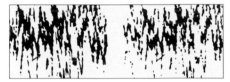

EEG DURING A GENERALIZED SEIZURE

Multiple sclerosis

Multiple sclerosis (MS) is a disorder that causes a range of disabilities affecting movement, vision, speech, and other functions. MS may be caused by the body's immune system attacking its own tissues, damaging the myelin sheaths that protect nerve fibers. Scavenger cells called macrophages remove damaged myelin, leaving the fibers exposed and interfering with conduction of impulses.

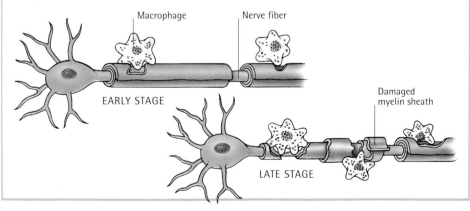

Macrophage

Nerve fiber

EARLY STAGE

Damaged myelin sheath

LATE STAGE

Parkinson's disease

Parkinson's disease is a degenerative condition that affects the basal ganglia (nerve cell clusters) in the brain. The basal ganglia help to control body movement; to function, they need dopamine, a neurotransmitter released by a brain structure called the substantia nigra. In Parkinson's disease, the substantia nigra is damaged, resulting in lack of dopamine. The disease causes stiffness and trembling of the muscles, and interferes with walking, speech, and facial expression.

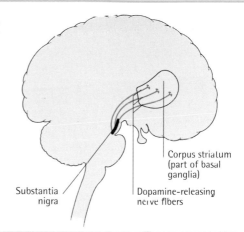

Substantia
nigra

Corpus striatum
(part of basal
ganglia)

Dopamine-releasing
nerve fibers

Dementia

Dementia is the deterioration of mental ability due to brain disease. It is one of the most common features of Alzheimer's disease, in which brain cells degenerate and a protein called amyloid builds up in the brain. Blockage of blood vessels in the brain, obstructing blood flow and resulting in small strokes, can also cause brain damage and dementia.

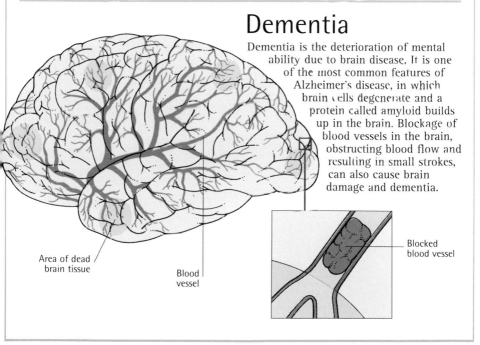

Area of dead
brain tissue

Blood
vessel

Blocked
blood vessel

Stroke

Stroke is damage to part of the brain, due to interruption of its blood supply, commonly caused by blockage of an artery. Starved of oxygen, affected brain cells cease to function; movement, vision, and speech may be impaired. Bleeding within brain tissue (see p.335) is another cause of strokes.

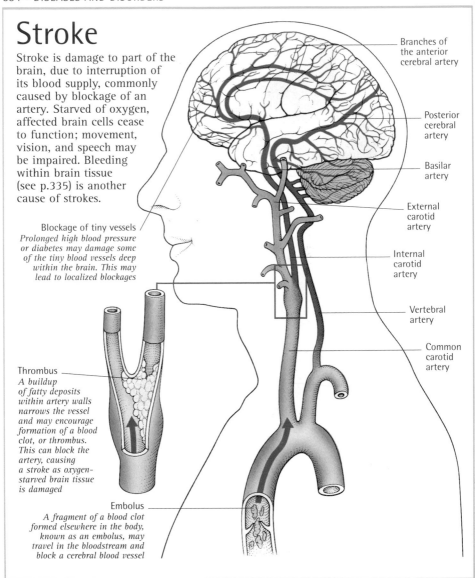

Branches of the anterior cerebral artery

Posterior cerebral artery

Basilar artery

External carotid artery

Internal carotid artery

Vertebral artery

Common carotid artery

Blockage of tiny vessels
Prolonged high blood pressure or diabetes may damage some of the tiny blood vessels deep within the brain. This may lead to localized blockages

Thrombus
A buildup of fatty deposits within artery walls narrows the vessel and may encourage formation of a blood clot, or thrombus. This can block the artery, causing a stroke as oxygen-starved brain tissue is damaged

Embolus
A fragment of a blood clot formed elsewhere in the body, known as an embolus, may travel in the bloodstream and block a cerebral blood vessel

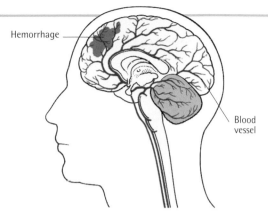

Hemorrhage

Blood vessel

Bleeding within brain tissue

Bleeding within the brain, known as intracerebral hemorrhage, is a major cause of stroke in older people who have abnormally high blood pressure (hypertension). The extra strain put on small arteries in the brain by constant high blood pressure can cause these vessels to balloon out and rupture.

Subarachnoid hemorrhage

Subarachnoid hemorrhage occurs when blood leaks into the subarachnoid space, an area between the membranes covering the brain (the meninges). In young people, strokes may be caused by brain hemorrhage due to congenital arterial defects. Common defects are swellings on arteries called berry aneurysms, which may rupture, and malformed connections between cerebral blood vessels.

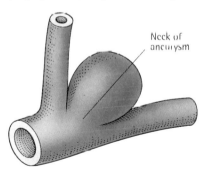

Neck of aneurysm

Berry aneurysm
A berry aneurysm usually forms at an arterial junction near the base of the brain. These berrylike swellings develop from weak points that are probably present at birth, and can rupture spontaneously.

Venule

Arteriole

Capillaries

NORMAL CAPILLARIES

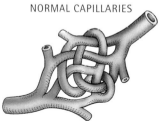

ABNORMAL CAPILLARIES

Arteriovenous malformation
In this congenital defect, there are poorer capillary connections than normal between arterioles and venules. This causes increased pressure and leakage of blood.

Transient ischemic attack

A transient ischemic attack (TIA) is a temporary interruption of the blood supply to the brain, resulting in strokelike symptoms that usually last from 2–30 minutes, but not more than 24 hours. It is usually due to blockage of a cerebral artery by an embolus, a tiny blood clot or fragment of fat from elsewhere in the body.

Skull

Blocked area of carotid artery

Spine

Blockage
An embolus lodging in a cerebral artery temporarily deprives part of the brain of oxygenated blood, resulting in symptoms similar to those of a stroke.

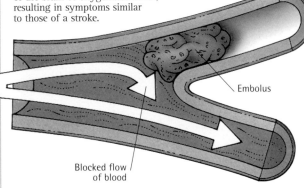

Embolus

Blocked flow of blood

Blocked artery
This color-enhanced X-ray shows how the carotid artery leading to the brain has narrowed, probably because of blockage with fatty deposits. Transient ischemic attacks are a common outcome of such blockages.

Dispersed particles

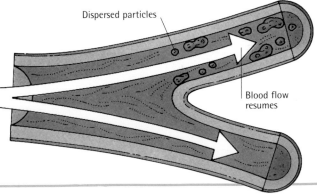

Blood flow resumes

Dispersal
As normal blood flow breaks up and disperses the blood clot, oxygenated blood again reaches the starved section of the brain and symptoms of TIA disappear.

Migraine

The cause of migraine headaches is unknown, but the symptoms are linked with excess release of serotonin (a neurotransmitter) in the brain that changes the diameter of the blood vessels. Migraine attacks take several forms and may include pain, dizziness, and visual disturbances, often accompanied by nausea and sometimes vomiting.

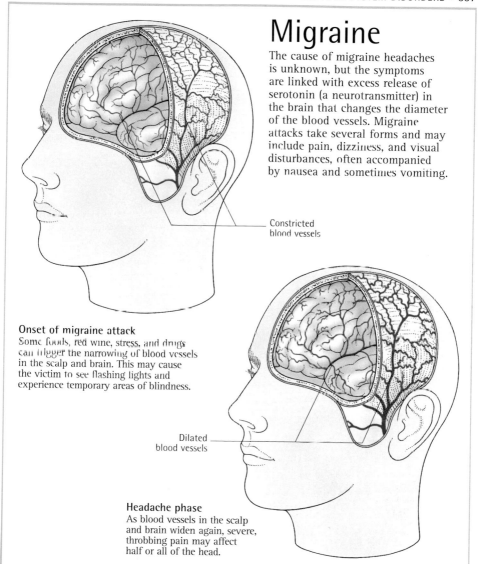

Constricted
blood vessels

Onset of migraine attack
Some foods, red wine, stress, and drugs can trigger the narrowing of blood vessels in the scalp and brain. This may cause the victim to see flashing lights and experience temporary areas of blindness.

Dilated
blood vessels

Headache phase
As blood vessels in the scalp and brain widen again, severe, throbbing pain may affect half or all of the head.

Brain infections

A wide variety of viruses, bacteria, and parasites can infect the brain. Some viral and parasitic brain infections are due to mosquito or other insect bites, while others develop from general infections such as mumps and measles. Infectious organisms can affect the brain itself and the membranes (meninges) that surround the brain. Infections usually reach the brain through the bloodstream, but can also spread from an ear or nose infection or a wound that penetrates the skull.

Brain tissue
Infection of brain tissue, known as encephalitis, is a serious disorder. It starts with a headache and fever and can result in permanent mental impairment or death

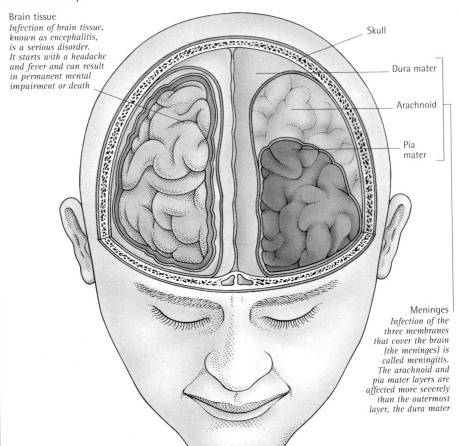

Skull

Dura mater

Arachnoid

Pia mater

Meninges
Infection of the three membranes that cover the brain (the meninges) is called meningitis. The arachnoid and pia mater layers are affected more severely than the outermost layer, the dura mater

Meningitis

Meningitis, inflammation of the membranes (meninges) covering the brain, is caused by various types of bacterial or viral infection. The disease produces flulike symptoms. Bacterial meningitis can be fatal; one type, meningococcal meningitis, is characterized by a red rash. Viral meningitis, which is milder, tends to occur in winter epidemics.

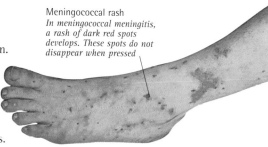

Meningococcal rash
In meningococcal meningitis, a rash of dark red spots develops. These spots do not disappear when pressed

Brain abscesses and tumors

Abscesses and tumors can develop either on or within the brain. Abscesses are pus-filled swellings, usually caused by bacterial infection. Brain tumors are abnormal growths that may be either malignant or benign. Both disorders cause increased pressure inside the skull; symptoms include headaches, vomiting, and muscle weakness.

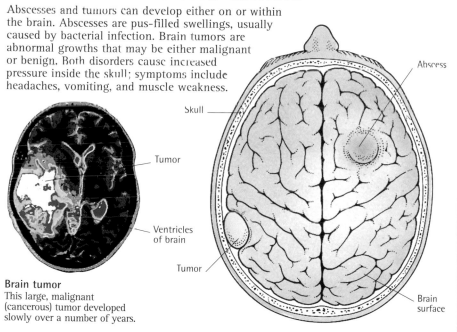

Tumor

Ventricles of brain

Tumor

Brain tumor
This large, malignant (cancerous) tumor developed slowly over a number of years.

Abscess

Skull

Tumor

Tumor

Brain surface

Bleeding within the skull

A head injury that does not penetrate the skull, known as a closed injury, can cause internal bleeding. There may be no immediate symptoms, although brief loss of consciousness may occur, followed by impaired brain function for a few minutes or for several hours. A closed head injury can be fatal if bleeding is untreated. If blood collects, symptoms such as drowsiness, headache, confusion, and change in personality may appear, followed by coma or even death.

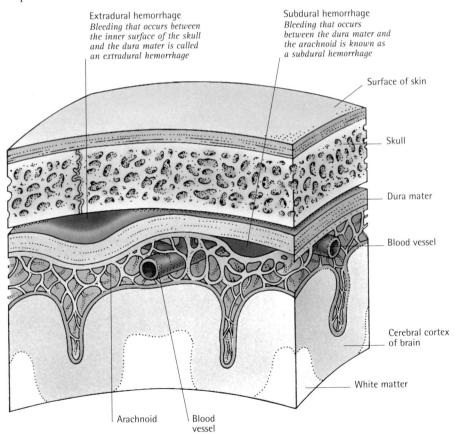

Extradural hemorrhage
Bleeding that occurs between the inner surface of the skull and the dura mater is called an extradural hemorrhage

Subdural hemorrhage
Bleeding that occurs between the dura mater and the arachnoid is known as a subdural hemorrhage

Surface of skin

Skull

Dura mater

Blood vessel

Cerebral cortex of brain

White matter

Arachnoid

Blood vessel

Paralysis

Damage to motor areas of the brain or neural pathways of the spinal cord can cause paralysis of various areas of the body. Voluntary muscle activity, including automatic functions such as breathing, may be affected, and there may be a loss of sensation. Consciousness or intellectual function are not usually affected by paralysis.

KEY

Area of body affected

Site of damage

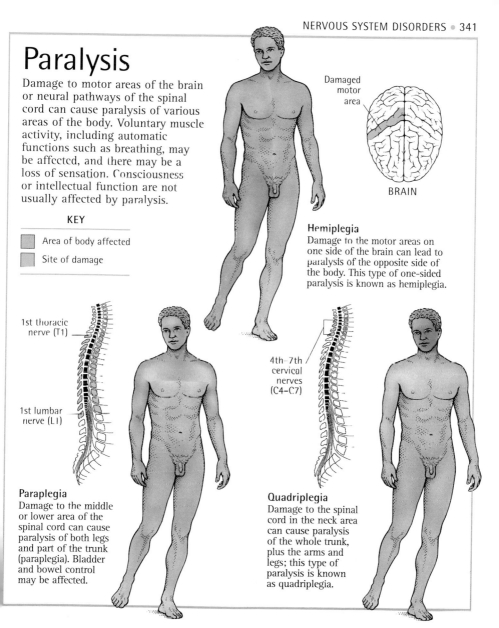

Damaged motor area

BRAIN

Hemiplegia

Damage to the motor areas on one side of the brain can lead to paralysis of the opposite side of the body. This type of one-sided paralysis is known as hemiplegia.

1st thoracic nerve (T1)

1st lumbar nerve (L1)

4th–7th cervical nerves (C4–C7)

Paraplegia

Damage to the middle or lower area of the spinal cord can cause paralysis of both legs and part of the trunk (paraplegia). Bladder and bowel control may be affected.

Quadriplegia

Damage to the spinal cord in the neck area can cause paralysis of the whole trunk, plus the arms and legs; this type of paralysis is known as quadriplegia.

Loss of hearing

A type of hearing loss, known as sensorineural deafness, occurs when nerve impulses between the ear and the brain are poorly generated or transmitted as a result of damage to inner ear structures or to the acoustic nerve pathways. This may be due to congenital defects; other causes include damage by some drugs, prolonged exposure to loud noise, or deterioration of ear structures with age. Other types of hearing loss result from impaired transmission of sound waves through the outer or middle ear (conductive deafness); a common cause is a middle ear infection, otitis media, which can lead to sticky secretions in the middle ear.

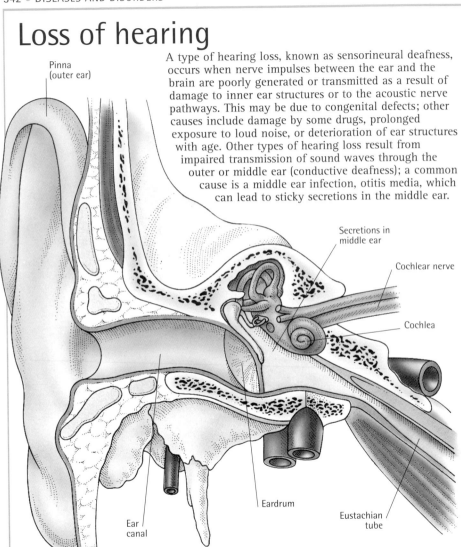

Pinna
(outer ear)

Secretions in
middle ear

Cochlear nerve

Cochlea

Eardrum

Ear
canal

Eustachian
tube

Visual problems

Inability to focus clearly can be caused by variations in the shape of the eye. In nearsightedness (myopia), the eyeball is too long; distant objects focus in front of the retinal surface instead of on it. Farsightedness (hypermetropia) occurs when the eyeball is too short, causing near objects to focus behind the retina.

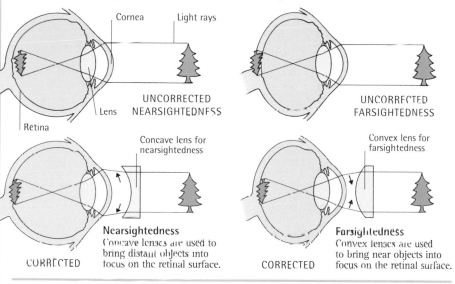

Nearsightedness
Concave lenses are used to bring distant objects into focus on the retinal surface.

Farsightedness
Convex lenses are used to bring near objects into focus on the retinal surface.

Glaucoma

In this disorder there is a build-up in the front of the eye of aqueous humor, a fluid that is usually secreted into the eye and drained out at the same rate. If outflow is blocked by a fault in the drainage channel between the back of the cornea and the iris, fluid pressure rises. This can damage the optic nerve, leading to permanent loss of vision.

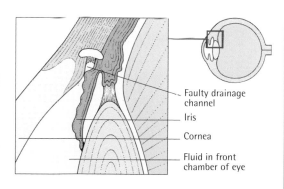

Faulty drainage channel

Iris

Cornea

Fluid in front chamber of eye

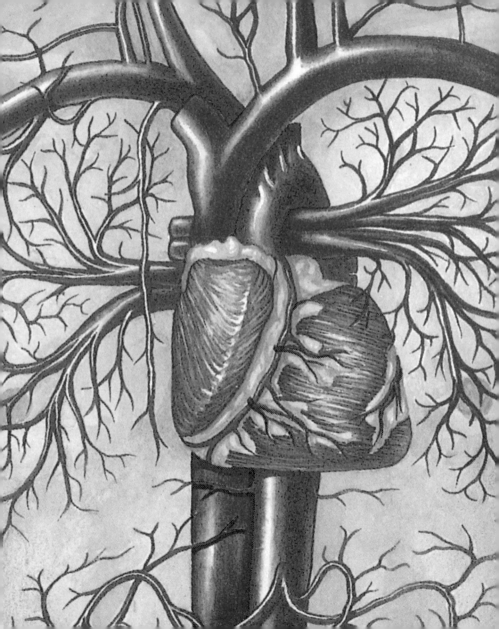

Cardiovascular disorders

Disorders of the heart and circulatory system are among the leading causes of ill-health in the developed world. Common heart diseases include structural defects; damage due to restricted blood supply caused by narrowed arteries; disorders of the heart muscle; and viral infections. Blood circulation may be hindered by obstructions such as blood clots or fatty deposits, or by weakened arteries or malfunctioning valves.

Heart problems
Hypertension, a congenital defect, or coronary artery disease are the most frequent causes of heart problems.

Atherosclerosis

In this disorder, the arteries become narrowed because of a buildup of fatty deposits. Atherosclerosis occurring in the arteries supplying blood to the heart is a major cause of heart damage. The disorder begins with the accumulation of excess fats and cholesterol in the blood. These substances infiltrate the lining of arteries, gradually forming deposits called atheroma, or plaques. Over time, plaques can build up to form masses that impede blood flow.

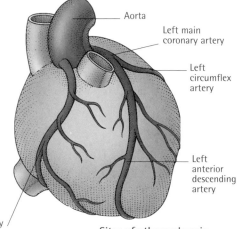

Aorta

Left main coronary artery

Left circumflex artery

Left anterior descending artery

Right coronary artery

Sites of atherosclerosis
In the heart, atherosclerosis can occur anywhere in the main coronary arteries or in smaller branches. Plaques usually build up at stress points in the artery, such as branch junctions.

Narrowed arterial channel

Fatty core of plaque

Fibrous cap

Plaques
The fatty deposits known as plaques consist of a fatty core topped by a fibrous cap. If blood turbulence roughens the surface of the plaques, platelets and blood cells can collect, creating a blood clot that may block the artery completely. This may lead to embolism, in which a fragment of the clot breaks off and travels in the bloodstream to another site.

Angina

Chest pain that comes on with exertion, called angina, is a sign that the heart muscle is not receiving sufficient oxygenated blood. Narrowing of the coronary arteries due to atherosclerosis (see p.346) is the most common cause. The typically constrictive or cramping pain, which is caused by the buildup of toxic wastes in the heart, begins behind the breast bone and may radiate into the neck, jaw, and arms.

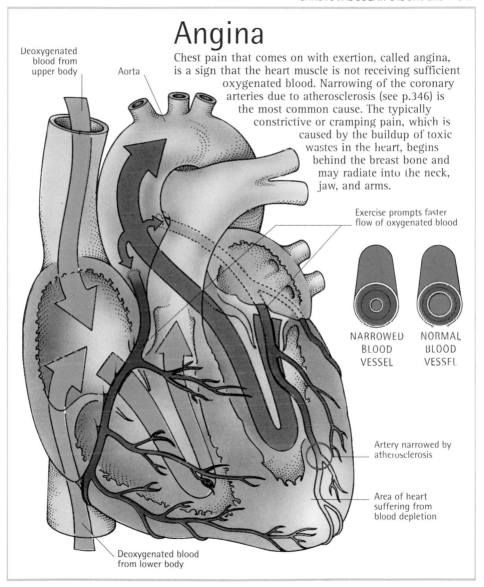

Deoxygenated blood from upper body

Aorta

Exercise prompts faster flow of oxygenated blood

NARROWED BLOOD VESSEL

NORMAL BLOOD VESSEL

Artery narrowed by atherosclerosis

Area of heart suffering from blood depletion

Deoxygenated blood from lower body

Heart attack

A heart attack, known medically as a myocardial infarction, is the death of part of the heart muscle. This is caused by blockage of a coronary artery, which deprives an area of the heart of its blood supply. The symptoms usually include severe chest pain that is not relieved by rest, sweating, nausea, and shortness of breath. If the attack leads to complete stoppage of the heart, known as cardiac arrest, death may follow.

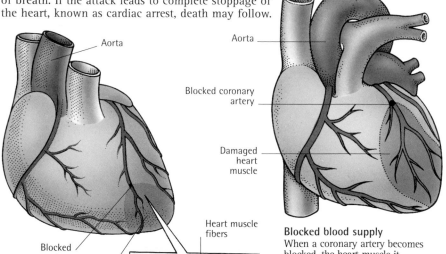

Aorta

Aorta

Blocked coronary artery

Damaged heart muscle

Blocked coronary artery

Damaged area

Heart muscle fibers

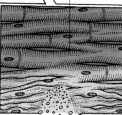

Blocked blood supply
When a coronary artery becomes blocked, the heart muscle it supplies dies. The severity of a heart attack depends on the extent of the blockage and the level of activity of the heart tissue.

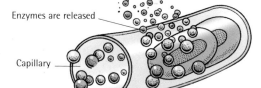

Enzymes are released

Capillary

Enzyme release
Tissue damage during a heart attack results in the release of certain enzymes from heart muscle fibers. The extent of the damage can be revealed by measuring the level of these enzymes in the blood.

Arrhythmias

If the rhythm of heartbeat is erratic, or the heart rate is unusually slow or fast, the condition is known as an arrhythmia. These abnormal heart patterns are often caused by blocked coronary arteries.

Arrhythmias, various types of which are shown below, are classified by number of heartbeats per minute, type of rhythm, and the area of the heart from which the irregular beat originates.

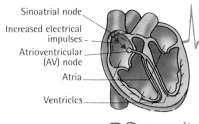

Sinoatrial node
Increased electrical impulses
Atrioventricular (AV) node
Atria
Ventricles

Sinus tachycardia
This regular but rapid heartbeat (over 100 beats per minute) is due to increased electrical impulses from the sinoatrial node, the heart's natural pacemaker. Sinus tachycardia can be caused by stress or exercise, or stimulants such as caffeine.

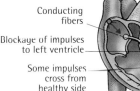

Conducting fibers
Blockage of impulses to left ventricle
Some impulses cross from healthy side

Bundle-branch block
Damage to a branch of the bundle of conducting fibers in the heart impedes the electrical impulses to a ventricle. In complete bundle-branch block, both branches are blocked and the heart rate slows.

Irregular impulses through atria
Variable blockage at AV node

Atrial fibrillation
Abnormal electrical activity in the atria causes a random and rapid heartbeat (300–500 beats per minute). Because a proportion of these beats pass through the atrioventricular node, the ventricles also beat irregularly (120–180 beats per minute).

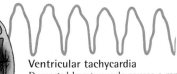

Damaged muscle
Slowed conduction through damaged area causes circular impulses

Ventricular tachycardia
Damaged heart muscle causes a rapid and ineffective heartbeat (120–220 beats per minute), which is initiated by abnormal, circular, impulses from a ventricle instead of the sinoatrial node.

Heart valve disorders

Effective pumping by the heart depends on all four heart valves operating properly. There are two main types of disorder that may affect one or more valves. Stenosis, in which a valve outlet is too narrow, may be congenital (present at birth) or due to disease or aging. Incompetence, or insufficiency, in which the valve cusps do not close properly, may be due to coronary artery disease.

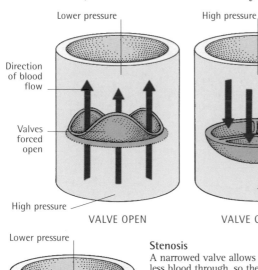

Lower pressure

Direction of blood flow

Valves forced open

High pressure

VALVE OPEN

High pressure

Pressure of blood flow keeps valve cusps closed

Lower pressure

VALVE CLOSED

Normal aortic valve
As the ventricles contract, high pressure forces the valve open, allowing blood through (far left). When the ventricles relax and fill with blood, the pressure is higher on the other side of the valve (left); this closes the valve tightly and prevents backflow of blood.

Lower pressure

Narrowed valve restricts blood flow

Higher pressure

Stenosis
A narrowed valve allows less blood through, so the heart must pump harder to maintain blood flow.

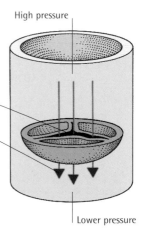

High pressure

Cusps not closed fully

Leakage of blood

Lower pressure

Incompetence
The leakage of blood back into the ventricles can occur when the cusps of a valve fail to close completely.

Heart murmurs

Normally, blood flow in the heart cannot be heard. Abnormal noises, known as heart murmurs, are caused by turbulent blood flow due to structural defects within the heart. Blood rushing around and through the cusps of a narrowed (stenosed) valve, or leaking back through an incompetent valve and colliding with onrushing blood, are both conditions that produce murmurs.

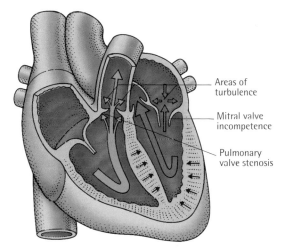

Areas of turbulence

Mitral valve incompetence

Pulmonary valve stenosis

Replacement valves

Defective heart valves that cannot be rectified by surgery may need to be replaced. Artificial valves, which are usually made from metal and plastic, are of two main types: those using a caged-ball mechanism and those with a tilting disc mechanism. Both are efficient and long-lasting but tend to cause blood clots, which may necessitate drug treatment. Replacement valves are also made from animal or human tissues. These are less durable but do not cause clots.

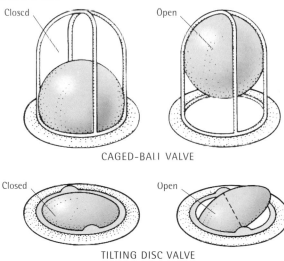

Closed

Open

CAGED-BALL VALVE

Closed

Open

TILTING DISC VALVE

Congenital heart defects

Sometimes, the fetal heart fails to develop normally, leading to congenital defects (those that are present at birth). In many cases, the cause is unknown. However, abnormalities in fetal heart structure can occur if a woman contracts a viral infection (particularly rubella) during early pregnancy, or if a pregnant woman has diabetes that is not well controlled. Defects are also associated with children who have Down's syndrome (caused by a chromosomal abnormality). Most congenital heart defects affect only one part of the heart; multiple defects occur more rarely.

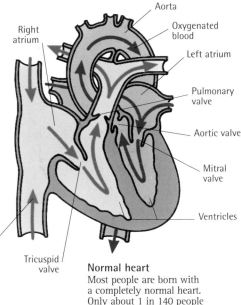

Aorta

Oxygenated blood

Right atrium

Left atrium

Pulmonary valve

Aortic valve

Mitral valve

Ventricles

Deoxygenated blood

Tricuspid valve

Normal heart
Most people are born with a completely normal heart. Only about 1 in 140 people has a congenital heart defect.

Region of aortic narrowing

Coarctation of the aorta
In this defect, a short section of the aorta is narrowed, which results in reduced blood flow to the lower body. An infant may be pale and find it difficult to breathe or eat. Corrective surgery is usually needed.

Reduced blood flow

Atrial septal defect

A hole in the septum, the wall that separates the atria and ventricles (heart chambers), allows too much blood to flow to the lungs. Often occurring in children with Down's syndrome, atrial septal defect generally needs surgery to avoid problems later in life.

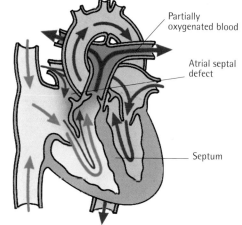

Partially oxygenated blood

Atrial septal defect

Septum

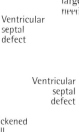

Displaced aorta

Narrowed pulmonary valve

Ventricular septal defect

Ventricular septal defect

A hole in the septum separating the ventricles means that blood from the left ventricle enters the right, and too much blood is pumped into the lungs. Small holes often close as a child grows; large holes may need surgery.

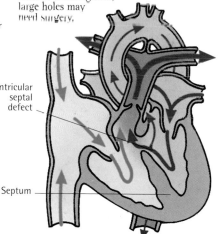

Ventricular septal defect

Septum

Thickened wall

Tetralogy of Fallot

This is a combination of four structural defects: a hole in the septum between the ventricles; a thickened right ventricular wall; a displaced aorta; and a narrowed pulmonary valve. Symptoms are a bluish tinge to the skin and breathlessness. Corrective surgery is necessary.

Heart muscle disease

The most common disorders of the heart muscle are inflammation, known as myocarditis, and various types of noninflammatory disease called cardiomyopathy. Myocarditis is usually caused by viral infection, but may also result from rheumatic fever, or exposure to radiation or drugs. Cardiomyopathy may be due to a genetic disorder, a vitamin or mineral deficiency, or excessive alcohol; three types are described below.

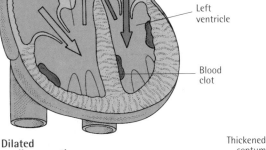

Right ventricle

Left ventricle

Blood clot

Thickened wall of left ventricle

Thickened septum

Dilated cardiomyopathy

In this disorder, the muscular walls of the heart dilate, reducing the force of heart contractions. As a result, not enough blood is ejected with each heartbeat and less oxygen reaches the body tissues. In many cases, blood clots may form on the inner walls of the heart.

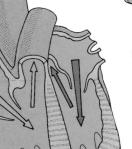

Rigid ventricular wall

Hypertrophic cardiomyopathy

This condition is usually inherited; the cause is not known. Overgrowth of heart muscle fibers causes thickened tissues, especially in the left ventricle and the septum, preventing normal filling with blood.

Restrictive cardiomyopathy

This disease is caused by scar tissue or deposits of iron or protein in the heart. The walls of the ventricles become rigid, which prevents them from filling normally with blood.

Pericarditis

A membranous bag, the pericardium, surrounds the heart. Inflammation of this membrane, known as pericarditis, is sometimes caused by a viral infection or a heart attack. The disorder may also be a complication of a bacterial infection such as rheumatic fever, or due to cancer, autoimmune disease, kidney failure, or a penetrating wound of the pericardium. Symptoms include chest pain and fever.

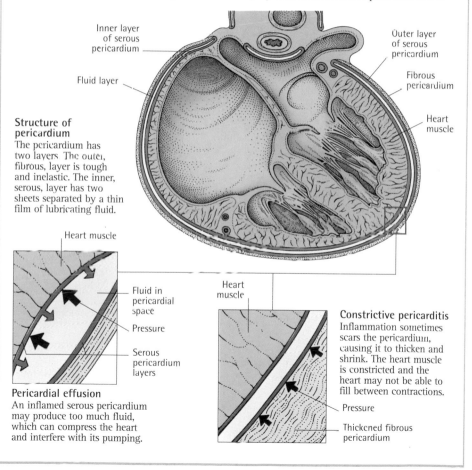

Inner layer of serous pericardium

Fluid layer

Outer layer of serous pericardium

Fibrous pericardium

Heart muscle

Structure of pericardium
The pericardium has two layers. The outer, fibrous, layer is tough and inelastic. The inner, serous, layer has two sheets separated by a thin film of lubricating fluid.

Heart muscle

Fluid in pericardial space

Pressure

Serous pericardium layers

Pericardial effusion
An inflamed serous pericardium may produce too much fluid, which can compress the heart and interfere with its pumping.

Heart muscle

Constrictive pericarditis
Inflammation sometimes scars the pericardium, causing it to thicken and shrink. The heart muscle is constricted and the heart may not be able to fill between contractions.

Pressure

Thickened fibrous pericardium

Heart failure

Heart failure, also known as ventricular failure, does not mean that the heart stops beating; it means that blood cannot be pumped effectively to the lungs and body tissues. Symptoms of the disease, which are related to which side of the heart is affected, include coughing, fatigue, edema (fluid in tissues), and shortage of breath.

RIGHT-SIDED HEART FAILURE

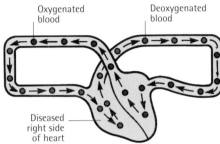

Oxygenated blood

Deoxygenated blood

Diseased right side of heart

1 Right side of heart fails
A diseased right side of the heart, possibly due to a valve defect or a respiratory disorder, is unable to pump blood to the lungs as fast as it returns from the body through the veins.

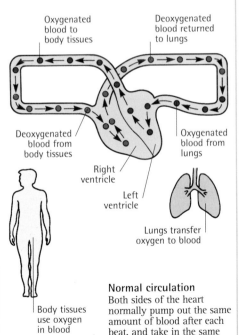

Oxygenated blood to body tissues

Deoxygenated blood returned to lungs

Deoxygenated blood from body tissues

Right ventricle

Left ventricle

Oxygenated blood from lungs

Lungs transfer oxygen to blood

Body tissues use oxygen in blood

Normal circulation
Both sides of the heart normally pump out the same amount of blood after each beat, and take in the same amount as they pump out. There is no blood congestion anywhere in the circulation.

LEFT-SIDED HEART FAILURE

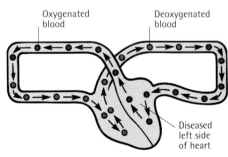

Oxygenated blood

Deoxygenated blood

Diseased left side of heart

1 Left side of heart fails
The left side of the heart may fail to pump blood out to the body as fast as it returns from the lungs through the pulmonary veins. This may be due to a heart structure defect or an arrhythmia.

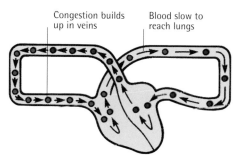

2 Veins become congested
Blood begins to pool in the right side of the heart. The veins, continuing their attempt to return blood, become congested.

3 Capillaries leak
Increased congestion causes raised pressure in the veins, which forces fluid out through the capillary walls. Tissues in the ankles and sometimes the lower back swell as fluid accumulates.

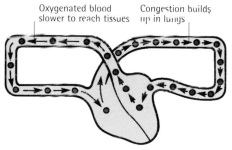

2 Lungs become congested
Blood that cannot re-enter the circulation begins to back up into the pulmonary veins and the lungs, causing congestion.

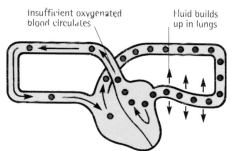

3 Fluid collects in lungs
The pressure from increasing congestion causes fluid to collect in the lungs, preventing efficient transfer of oxygen to the blood. This leads to coughing, breathlessness, and fatigue.

Thrombosis

A blood clot (thrombus) can form within a vein or artery, impeding normal circulation of blood. This condition, known as thrombosis, is especially likely to occur if an artery contains fatty deposits (plaques). If a thrombus blocks one of the arteries supplying the heart, it can be life-threatening. Thrombi in the arteries supplying the brain are a major cause of stroke (see p.334).

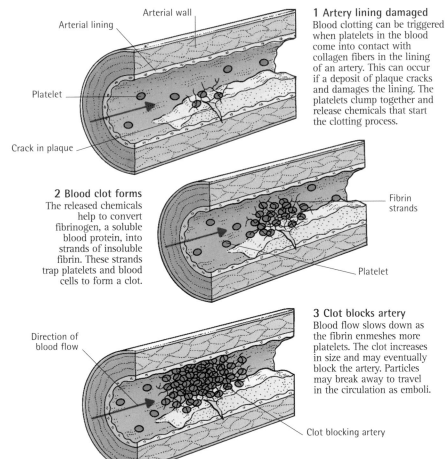

Arterial wall

Arterial lining

Platelet

Crack in plaque

1 Artery lining damaged
Blood clotting can be triggered when platelets in the blood come into contact with collagen fibers in the lining of an artery. This can occur if a deposit of plaque cracks and damages the lining. The platelets clump together and release chemicals that start the clotting process.

2 Blood clot forms
The released chemicals help to convert fibrinogen, a soluble blood protein, into strands of insoluble fibrin. These strands trap platelets and blood cells to form a clot.

Fibrin strands

Platelet

3 Clot blocks artery
Blood flow slows down as the fibrin enmeshes more platelets. The clot increases in size and may eventually block the artery. Particles may break away to travel in the circulation as emboli.

Direction of blood flow

Clot blocking artery

Embolism

An embolism occurs when particles of material travel in the bloodstream from their point of origin and lodge in another site. Such fragments (emboli) may be a part of, or the whole of, a blood clot (thrombus) that has detached from its original site. Emboli may also be composed of atheromatous debris, cholesterol, air, or fat from the marrow of fractured bones. Some embolisms, such those affecting the pulmonary artery, can be fatal.

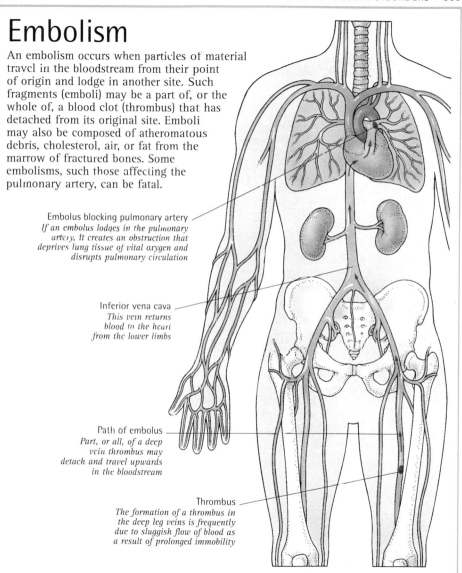

Embolus blocking pulmonary artery
If an embolus lodges in the pulmonary artery, it creates an obstruction that deprives lung tissue of vital oxygen and disrupts pulmonary circulation

Inferior vena cava
This vein returns blood to the heart from the lower limbs

Path of embolus
Part, or all, of a deep vein thrombus may detach and travel upwards in the bloodstream

Thrombus
The formation of a thrombus in the deep leg veins is frequently due to sluggish flow of blood as a result of prolonged immobility

Aneurysm

An aneurysm is an abnormal swelling in a weakened arterial wall. The defect may be due to disease or an injury, or be congenital (present at birth). Although aneurysms can occur anywhere, the most common site is the aorta, the body's main artery that arises from the heart. In older people, aneurysms develop more frequently in the abdominal aorta, at a point just below the kidneys.

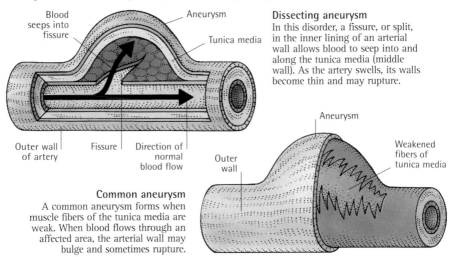

Dissecting aneurysm
In this disorder, a fissure, or split, in the inner lining of an arterial wall allows blood to seep into and along the tunica media (middle wall). As the artery swells, its walls become thin and may rupture.

Common aneurysm
A common aneurysm forms when muscle fibers of the tunica media are weak. When blood flows through an affected area, the arterial wall may bulge and sometimes rupture.

Varicose veins

Defective valves in deep, lower-leg veins can cause blood to drain backwards and pool in the superficial veins nearest the skin surface. These veins, known as varicose veins, become swollen, distorted, and painful. The condition most commonly occurs on the back and sides of the calf. In severe cases, the skin over the affected veins becomes thin, dry, and discolored, and may eventually ulcerate.

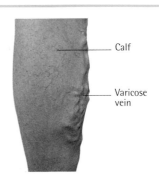

Hypertension

Hypertension is persistent, abnormally elevated blood pressure. Although this may cause no symptoms, it increases the risk of stroke, heart attack, and other circulatory diseases. Blood pressure is recorded in millimetres of mercury (mmHg). Normal blood pressure in young adults is about 110/75mmHg. The first number is the systolic pressure, taken just after the ventricles contract; the second is the diastolic pressure, taken when the ventricles relax.

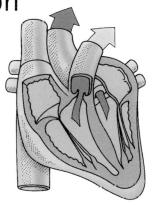

SYSTOLE: VENTRICLES CONTRACT AND FORCE BLOOD OUT

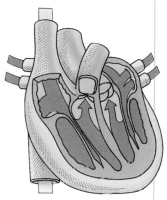

DIASTOLE: VENTRICLES RELAX AND FILL WITH BLOOD

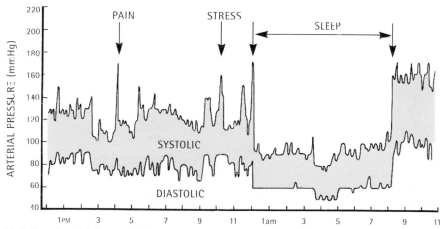

Variations in a 24-hour period
The chart above shows how blood pressure varies greatly in response to various stimuli, such as pain or stress. Variations of this kind are normal.

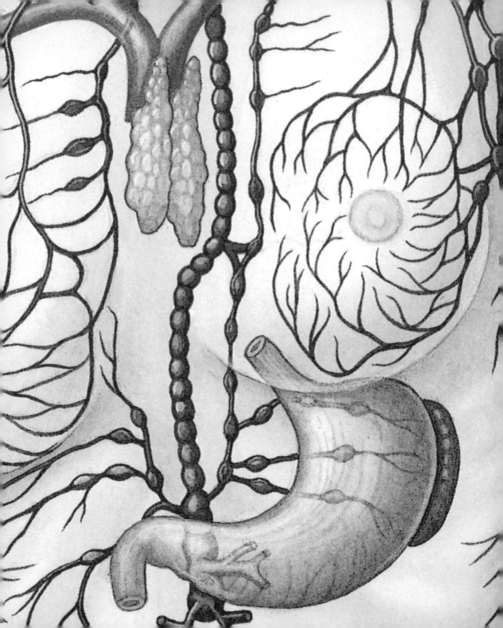

Infections and immune disorders

The human body is constantly infected by organisms: bacteria, viruses, fungi, and protozoa. The immune system produces antibodies that fight these infections and sometimes help to prevent reinfection. There are two types of immune system disorders. In allergies and autoimmune diseases, the immune system overreacts; in immunodeficiency diseases, the defense systems are too weak to deal with threats to health.

Swollen lymph nodes
Swelling of the lymph nodes is a common symptom of many infectious diseases.

Bacteria

Bacteria are single-celled microorganisms that live in soil, water, and air, and may also inhabit the body. Many bacteria are harmless, or even beneficial, but some enter internal tissues and cause disease. Their shapes vary and include oval (as illustrated), spherical, rod-like, and spiral.

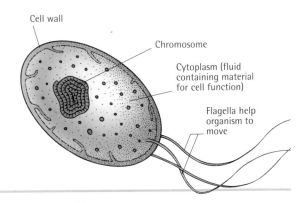

Cell wall

Chromosome

Cytoplasm (fluid containing material for cell function)

Flagella help organism to move

How bacteria cause disease

Disease-causing bacteria can enter the body in several ways: via the airways or digestive tract, during sexual contact, or through breaks in the skin. Some bacteria produce poisonous substances called toxins, which are released into the body. These toxins can destroy or alter the function of body cells. Less commonly, bacteria may directly enter cells and destroy them.

1 Release of toxins
Toxins released by disease-causing bacteria may alter certain chemical reactions in cells, so that normal cell function is disrupted or the cell dies.

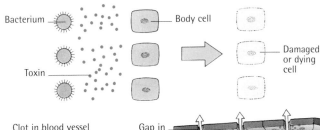

Bacterium

Toxin

Body cell

Damaged or dying cell

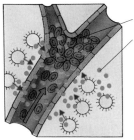

Clot in blood vessel

Toxins from bacteria

2 Blood clots
Some toxins cause blood to clot in small vessels. Areas of tissue supplied by these vessels may be deprived of blood and damaged.

Gap in wall

Fluid leaking into tissue

Toxins

Blood vessel

3 Leakage of blood
Toxins can lead to damage of blood vessel walls, causing leakage and subsequent decrease of blood pressure.

Resistance to antibiotics

Many bacteria have developed ways of resisting antibiotics. The most effective mechanism is the rapid transfer of plasmids – small packages of the cell's DNA, its genetic material – between bacterial populations. Plasmids may contain resistant genes: the bacteria that receive these plasmids inherit the resistant genes.

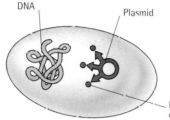

The role of plasmids
Plasmids can allow bacteria to produce enzymes that inactivate drugs. They may also enable a bacterium to alter its receptor sites (where antibiotics normally bind).

Exchange of genes
Conjugation of two rod-shaped bacteria is shown above. DNA is passed from one to the other through a tube called the pilus.

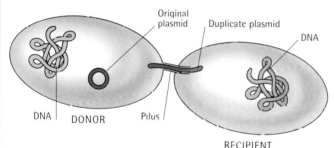

Conjugation
Plasmid transfer takes place during a process known as conjugation. The plasmid duplicates itself in a "donor" bacterium. This copy passes through a tube, the pilus, to the recipient cell.

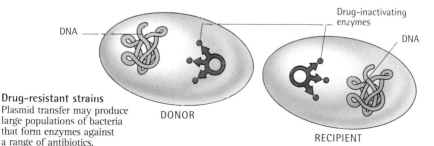

Drug-resistant strains
Plasmid transfer may produce large populations of bacteria that form enzymes against a range of antibiotics.

Viruses

Billions of viruses can cover a pinhead. These tiny organisms are parasites that can reproduce only within living cells. Each virus has a core of genetic material, composed of either DNA or RNA, and one or two protein shells. Surface proteins (antigens) stud the outer shell.

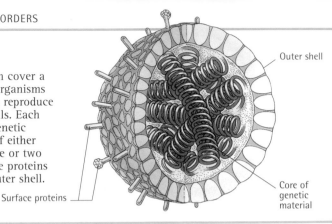

Outer shell

Core of genetic material

Surface proteins

Types of virus

Viruses can be broadly classified as RNA or DNA viruses, depending on the main constituent of their genetic material.

Further classification is governed by the size, shape, and symmetry of the virus. Some common families are listed below.

FAMILY		TYPE AND DISEASES
Adenoviruses		DNA viruses. Cause infections of the tonsils, respiratory tract, and eyes (such as conjunctivitis).
Papovaviruses		DNA viruses. Initiate benign, or noncancerous, tumors such as warts on the hands and feet. May be linked to some cancers.
Herpesviruses		DNA viruses. Cause cold sores, genital herpes, chickenpox, shingles, and mononucleosis.
Picornaviruses		RNA viruses. Cause a variety of illnesses, including myocarditis, polio, viral hepatitis, one form of meningitis, and the common cold.
Retroviruses		RNA viruses that can convert RNA into DNA. They cause AIDS and a type of leukemia.
Orthomyxo-viruses		RNA viruses. Cause influenza, symptoms of which include fever, cough, a sore throat, and aching limbs.
Paramyxo-viruses		RNA viruses. Cause mumps, measles, and respiratory infections such as croup.

How viral diseases occur

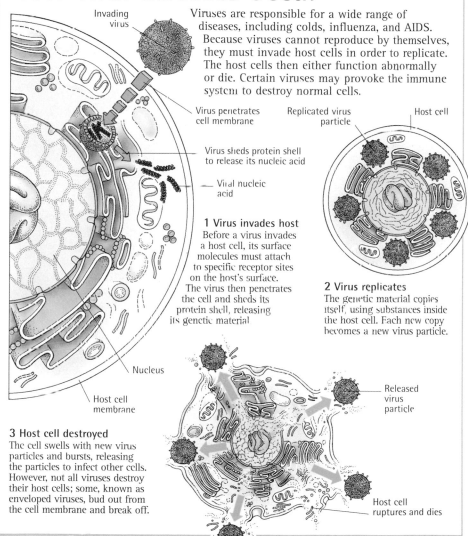

Invading virus

Viruses are responsible for a wide range of diseases, including colds, influenza, and AIDS. Because viruses cannot reproduce by themselves, they must invade host cells in order to replicate. The host cells then either function abnormally or die. Certain viruses may provoke the immune system to destroy normal cells.

Virus penetrates cell membrane

Replicated virus particle

Host cell

Virus sheds protein shell to release its nucleic acid

Viral nucleic acid

1 Virus invades host
Before a virus invades a host cell, its surface molecules must attach to specific receptor sites on the host's surface. The virus then penetrates the cell and sheds its protein shell, releasing its genetic material

2 Virus replicates
The genetic material copies itself, using substances inside the host cell. Each new copy becomes a new virus particle.

Nucleus

Host cell membrane

3 Host cell destroyed
The cell swells with new virus particles and bursts, releasing the particles to infect other cells. However, not all viruses destroy their host cells; some, known as enveloped viruses, bud out from the cell membrane and break off.

Released virus particle

Host cell ruptures and dies

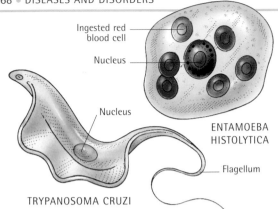

Ingested red blood cell

Nucleus

ENTAMOEBA HISTOLYTICA

Nucleus

Flagellum

TRYPANOSOMA CRUZI

Protozoa

Protozoa are primitive, single-celled animals, some of which are parasites that can cause serious disease in humans, including malaria. They have a nucleus and many, such as *Trypanosoma cruzi*, have a tail (flagellum) to aid movement. Some, such as *Entamoeba histolytica*, can ingest red blood cells and food particles.

Malaria

The protozoal infection malaria affects millions of people worldwide. It is caused by four types of plasmodia protozoan and is transmitted by the bite of the female *Anopheles* mosquito. Malaria produces chills and a high fever; the disease recurs if not treated effectively. One type, *Plasmodium falciparum*, damages vital organs such as the kidneys and brain, and can be rapidly fatal.

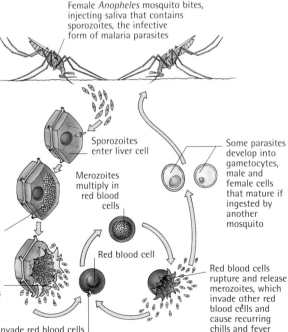

Female *Anopheles* mosquito bites, injecting saliva that contains sporozoites, the infective form of malaria parasites

Sporozoites enter liver cell

Merozoites multiply in red blood cells

Some parasites develop into gametocytes, male and female cells that mature if ingested by another mosquito

In liver, sporozoites multiply and develop into merozoites, another form of the parasite

Red blood cell

Merozoites are released from liver cells into bloodstream

Red blood cells rupture and release merozoites, which invade other red blood cells and cause recurring chills and fever

Merozoites invade red blood cells

Fungi

Fungi are simple organisms that scavenge dead or rotting tissue. Some types live harmlessly on humans without multiplying; other fungi can cause superficial diseases of the hair, nails, or mucous membranes. More rarely, fungal infections may attack vital internal organs such as the lungs, and can be fatal, particularly in people who have reduced immunity to disease.

Types of fungi

Fungi that cause infection are of two main types: filamentous and yeasts. Filamentous fungi grow in long branching threads called hyphae. The yeasts are single-celled and include the fungi that cause common infections such as thrush.

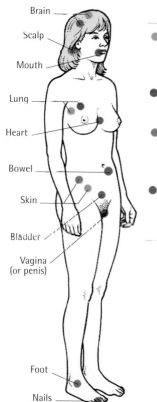

Brain
Scalp
Mouth
Lung
Heart
Bowel
Skin
Bladder
Vagina (or penis)
Foot
Nails

KEY

● Cryptococcosis
This infection causes meningitis and pneumonia and can also affect the skin and bones.

● Aspergillosis
This is a fungal infection that can affect the lungs.

● Dermatophytosis
This skin infection, also called tinea, most commonly affects the scalp, feet, or nails.

● Candidiasis
Candida fungi infect the mouth and genitals, and occur in the heart, bowel, bladder, and brain.

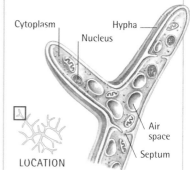

Cytoplasm
Hypha
Nucleus
Air space
Septum

LOCATION

Filamentous fungi
The tubular branches (hyphae) of filamentous fungi are sometimes divided by cross walls called septa.

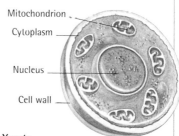

Mitochondrion
Cytoplasm
Nucleus
Cell wall

Yeasts
These single-celled fungi occur in colonies. They spread by cell division.

Immunization

There are two types of immunization against infectious disease: active and passive. Active immunization, or vaccination, uses an altered form of a specific disease organism to stimulate the immune system into making antibodies. Passive immunization uses blood extracts containing antibodies to the disease.

ACTIVE IMMUNIZATION

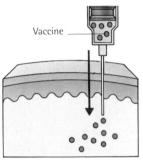

Vaccine

1 Vaccine injected
Vaccine with dead or harmless forms of an organism is injected into a healthy person.

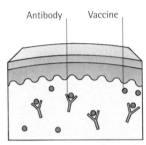

Antibody   Vaccine

2 Immune system triggered
In response, antibodies are produced. The immune system will "remember" the organism.

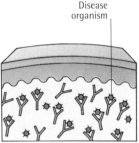

Disease organism

3 Infection halted
In any subsequent infection, the immune system will produce large numbers of antibodies.

PASSIVE IMMUNIZATION

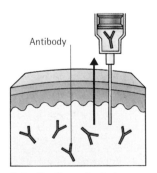

Antibody

1 Antibodies collected
Blood with antibodies is taken from humans or animals with immunity to the infection.

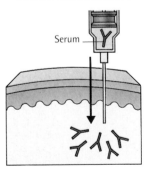

Serum

2 Serum injected
Serum extracted from this blood is injected into the person needing protection.

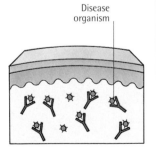

Disease organism

3 Protection given
Antibodies in the serum either attack a current infection or provide short-term protection.

Genetically engineered viruses

Genetic engineering is a term describing a technique that alters the genetic material (DNA) of an organism by inserting the genes from another organism. This technique is used to create some vaccines. The gene of the surface antigen (protein) of a virus, such as hepatitis B, is inserted into the DNA of other organisms, such as bacteria; when the bacteria multiply, large amounts of replicated viral material are produced and can be used as vaccines to stimulate an immune response.

Gene from viral DNA inserted into DNA of bacterium

Replicated bacterium

Copy of viral DNA

Replication of material

When bacteria containing altered DNA (genetic material) multiply, each new bacterium will contain a copy of the altered DNA.

Gene of surface antigen extracted from DNA of virus

Surface antigen will trigger immune response when used as vaccine

Hepatitis B virus

Used as a vaccine, the virus's surface antigen will give protection against the disease. The gene for the surface antigen is extracted from the viral DNA and inserted in the DNA of another organism to be replicated.

Allergic response

Allergy is an overreaction by the immune system to a normally harmless substance that may be inhaled, swallowed, or touched. Such substances, or allergens, trigger a response in certain cells in the skin and the lining of the airways, lungs, and stomach. The cells, known as mast cells, release an irritant called histamine that produces an allergic response such as asthma or a skin rash.

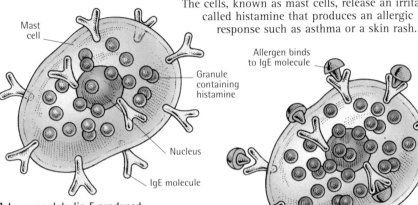

Mast cell

Granule containing histamine

Nucleus

IgE molecule

Allergen binds to IgE molecule

1 Immunoglobulin E produced
Allergens provoke the immune system to produce an antibody called immunoglobulin E (IgE), molecules of which coat the surface of mast cells.

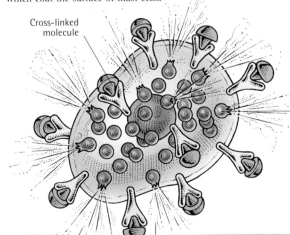

Cross-linked molecule

2 Allergens bind to IgE
In an allergic person, subsequent exposure to allergens causes them to bind to the IgE molecules. This is known as cross-linking.

Granules release histamine into tissues

3 Histamine released
Cross-linking causes granules inside the mast cell to release the inflammatory substance histamine, triggering an allergic response.

Autoimmune disorders

Sometimes the immune system forms antibodies not against invaders such as bacteria but against the body's own tissues. These inappropriate attacks, known as autoimmune disorders, may be directed against a particular organ, such as the thyroid gland, or may cause more general illness (see table below). Such diseases are more common in middle age and affect women more often than men.

DISORDER	DESCRIPTION
Addison's disease	Damaged adrenal glands result in low blood pressure and weakness, lowering the body's ability to respond to stress.
Insulin-dependent diabetes mellitus	Clusters of pancreatic cells called islets of Langerhans cannot produce sufficient insulin, which causes high blood glucose levels.
Hemolytic anemia	This autoimmune form of anemia shortens the lifespan of red blood cells, causing loss of energy, pallor, headaches, and breathlessness.
Graves' disease	The thyroid becomes overactive and may become enlarged, forming a goiter. There is loss of weight, restlessness, and tremor.
Multiple sclerosis	Damage to nerve fiber coverings causes muscle weakness, disordered sensations, and problems with speech and vision.
Myasthenia gravis	Damage to the junctions between nerves and muscles causes muscle weakness and fatigue, especially noticeable in muscles of the face.
Systemic lupus erythematosus	Damaged tissue causes progressive loss of function, particularly in the kidneys, lungs, and joints. A distinctive rash appears on the face.
Vitiligo	Loss of pigment-producing cells (melanocytes) in skin cause multiple, irregular, pale patches to appear, especially on the hands and face.

HIV infection and AIDS

Acquired immune deficiency syndrome, or AIDS, is caused by the human immunodeficiency virus (HIV). The virus (shown below) destroys one type of white blood cell, the CD$_4$ "helper" lymphocyte.

As cell numbers fall, the immune system becomes less effective and the body is more susceptible to disease. HIV is spread by infected blood and other body fluids, for example during sexual intercourse.

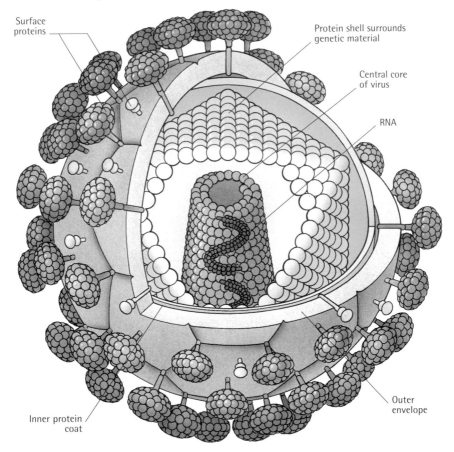

Surface proteins

Protein shell surrounds genetic material

Central core of virus

RNA

Inner protein coat

Outer envelope

Effects of AIDS

Many people infected with HIV have no symptoms for many years, and are known as asymptomatic carriers. In later stages of the infection, they lose weight and develop night sweats, fever, and diarrhea. If full-blown AIDS develops, which may be up to 14 years after infection, people become susceptible to a variety of infections and to certain cancers.

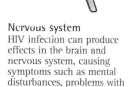

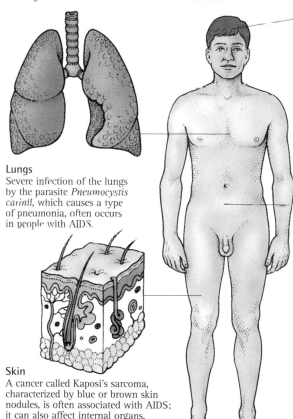

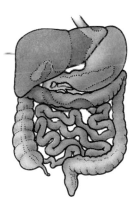

Nervous system
HIV infection can produce effects in the brain and nervous system, causing symptoms such as mental disturbances, problems with vision, and paralysis.

Lungs
Severe infection of the lungs by the parasite *Pneumocystis carinii*, which causes a type of pneumonia, often occurs in people with AIDS.

Digestive system
Persistent diarrhea, often due to protozoal infection of the gastrointestinal tract, is one of the most common illnesses associated with AIDS.

Skin
A cancer called Kaposi's sarcoma, characterized by blue or brown skin nodules, is often associated with AIDS; it can also affect internal organs.

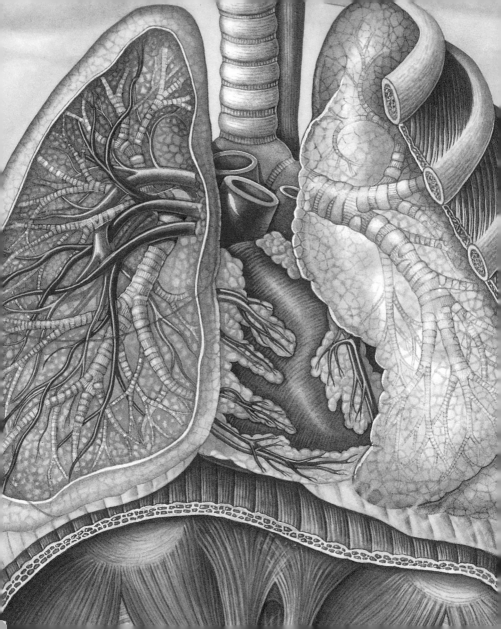

Respiratory disorders

During an intake of breath, microorganisms such as viruses and bacteria can easily be inhaled. These may cause mild illnesses affecting only the upper airways, like the common cold, or more serious diseases of the lungs, such as bronchitis or pneumonia. Other types of disorder sometimes affecting the lungs include inflammation of the tissues due to allergies, damage resulting from inhalation of irritant dust particles, and cancers.

Respiratory illness
The most common diseases of the respiratory tract, especially of the upper air passages, are caused by infections.

Upper airway infections

Infections of the upper airways, also known as upper respiratory tract infections, affect the nasal sinuses, pharynx, and larynx. Such disorders occur when droplets contaminated by viruses, or sometimes bacteria, are inhaled. Infection often results in the inflammation and swelling of mucous membranes that line the nose and throat.

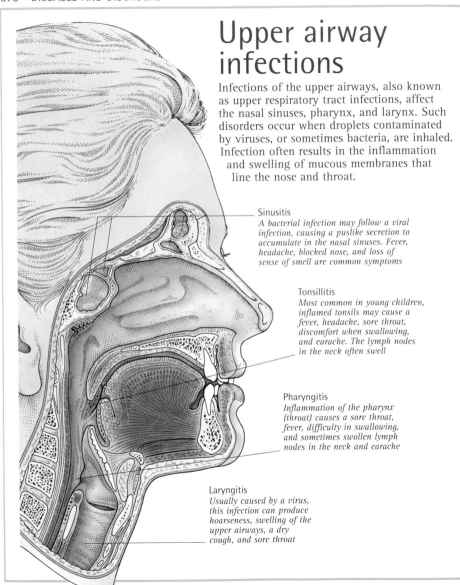

Sinusitis
A bacterial infection may follow a viral infection, causing a puslike secretion to accumulate in the nasal sinuses. Fever, headache, blocked nose, and loss of sense of smell are common symptoms

Tonsillitis
Most common in young children, inflamed tonsils may cause a fever, headache, sore throat, discomfort when swallowing, and earache. The lymph nodes in the neck often swell

Pharyngitis
Inflammation of the pharynx (throat) causes a sore throat, fever, difficulty in swallowing, and sometimes swollen lymph nodes in the neck and earache

Laryngitis
Usually caused by a virus, this infection can produce hoarseness, swelling of the upper airways, a dry cough, and sore throat

Common cold

About 200 types of virus cause the common cold. Infection is passed from person to person in droplets released by coughs and sneezes. Cold viruses can also be passed on by hand-to-hand contact or by handling contaminated objects. The symptoms include a runny nose, sore throat, headache, and a cough. Normally, the body's immune system quickly overcomes the infectious organisms.

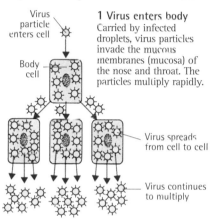

1 Virus enters body
Carried by infected droplets, virus particles invade the mucous membranes (mucosa) of the nose and throat. The particles multiply rapidly.

Virus particle enters cell

Body cell

Virus spreads from cell to cell

Virus continues to multiply

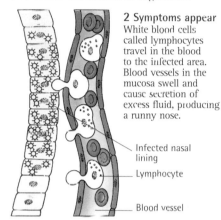

2 Symptoms appear
White blood cells called lymphocytes travel in the blood to the infected area. Blood vessels in the mucosa swell and cause secretion of excess fluid, producing a runny nose.

Infected nasal lining

Lymphocyte

Blood vessel

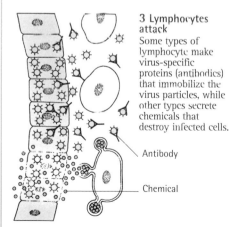

3 Lymphocytes attack
Some types of lymphocyte make virus-specific proteins (antibodies) that immobilize the virus particles, while other types secrete chemicals that destroy infected cells.

Antibody

Chemical

Phagocyte

Debris of virus particles

4 Phagocytes engulf debris
Phagocytes, another type of white blood cell, engulf the debris of immobilized virus particles and damaged cells. Symptoms of the cold soon subside.

Influenza

Commonly called "flu", influenza is a serious viral infection that causes fever, chills, muscle pain, headache, weakness, and cough. Influenza viruses, one type of which is shown here (right), are of three main strains: A, B, and C. Some viruses change their structures so that they cannot be recognized by the immune system. If a new strain emerges in this way, most people have no immunity to it and an epidemic can occur.

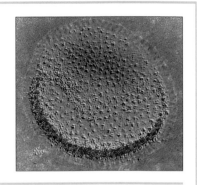

Sinusitis

In the skull bones surrounding the eyes and nose there are several air-filled cavities called sinuses. Bacterial infection of the mucous membranes lining the sinuses, which may follow a viral infection such as a cold, can cause an inflammation known as sinusitis. This disorder causes headache, pain and tenderness in the face, nasal congestion, and sometimes a nasal discharge.

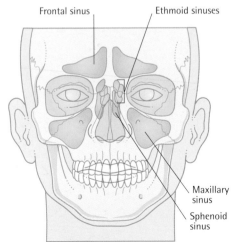

Frontal sinus

Ethmoid sinuses

Maxillary sinus

Sphenoid sinus

FRONT VIEW

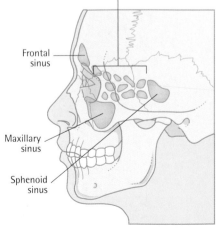

Ethmoid sinuses

Frontal sinus

Maxillary sinus

Sphenoid sinus

SIDE VIEW

Acute bronchitis

Acute bronchitis, an inflammation of the bronchi (airways) that join the trachea (windpipe) to the lungs, is usually a complication of a viral infection such as influenza. The lining or deeper tissues of the bronchi become inflamed and swollen, narrowing the lumen (inner space). Abundant amounts of mucus are secreted, causing a sputum-producing cough and sometimes wheezing.

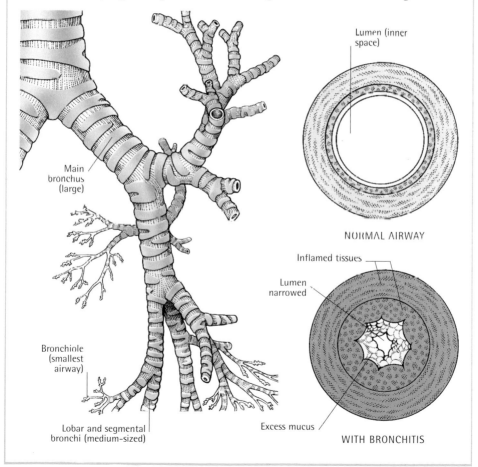

Lumen (inner space)

Main bronchus (large)

Bronchiole (smallest airway)

Lobar and segmental bronchi (medium-sized)

NORMAL AIRWAY

Inflamed tissues

Lumen narrowed

Excess mucus

WITH BRONCHITIS

Pneumonia

In pneumonia, the smallest bronchioles and areas of alveolar tissue become inflamed. There are two main types of the disorder: lobar pneumonia, which affects one lobe of the lung; and bronchopneumonia, which affects patches in one or both lungs. Pneumonia is usually a result of bacterial infection, but is sometimes caused by viruses. Symptoms include coughing, breathlessness, fever, and joint and muscle pain.

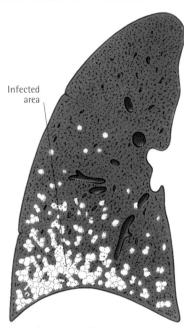

Infected area

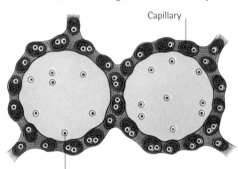

Capillary

Macrophage

Bronchopneumonia
This type of pneumonia mainly affects the chronically ill, the elderly, and the very young. The scattered white areas shown in the illustration (above) are patches of inflamed lung tissue.

Healthy alveoli
The alveoli (tiny air sacs in the lungs) contain white blood cells called macrophages, which ingest inhaled irritants but respond slowly to infective organisms.

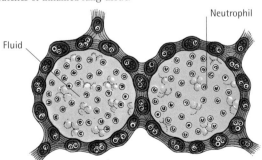

Neutrophil

Fluid

Infected alveoli
Infection triggers changes in the capillaries in the alveolar walls. This allows in another type of white blood cell, called neutrophils, which attack the invading organisms. Fluid also flows in and accumulates.

Pleural effusion

Pleural effusion is an accumulation of excess fluid between the two-layered membrane, or pleura, that lines the inside of the chest cavity and the outside of the lungs. This may be caused by inflammation of the pleura, resulting from infections such as pneumonia or tuberculosis, or sometimes from heart failure. If extensive, pleural effusion can cause difficulty with breathing.

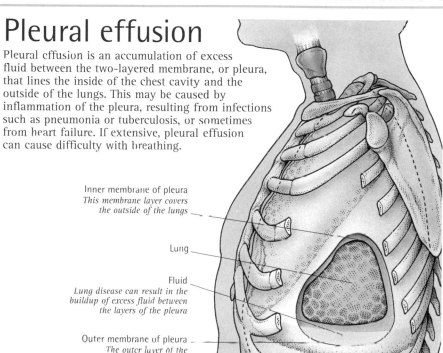

Inner membrane of pleura
This membrane layer covers the outside of the lungs

Lung

Fluid
Lung disease can result in the buildup of excess fluid between the layers of the pleura

Outer membrane of pleura
The outer layer of the pleura lines the chest cavity

Legionnaires' disease

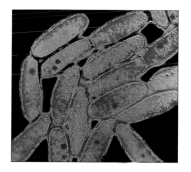

This rare infection, first identified in 1976, is a form of pneumonia (see p. 382). Legionnaires' disease is caused by the bacterium *Legionella pneumophila* (shown left), which thrives in water-cooled air conditioning systems and in plumbing systems where water stagnates. Symptoms include a high fever, coughing up of sputum, chills, muscle aches, confusion, severe headache, and abdominal pain and diarrhea. Elderly people are particularly at risk.

Pulmonary hypertension

Pulmonary hypertension is abnormally high blood pressure in the arteries taking blood to the lungs. The disorder occurs when there is resistance to blood flow through the lungs, usually because of blockage of an artery by a blood clot or because of scarring of lung tissue due to disease. The right ventricle of the heart must pump more vigorously than usual, causing thickening of muscle in the heart and artery walls. Blood pressure rises and is transmitted back through the heart.

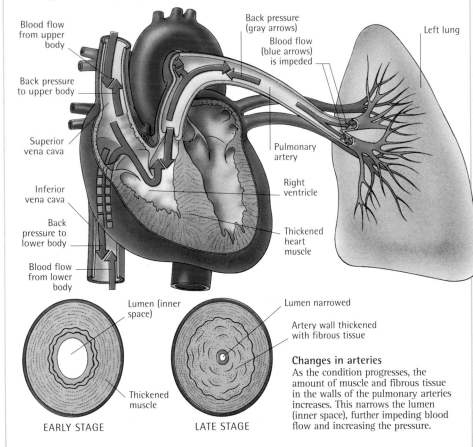

Blood flow from upper body

Back pressure (gray arrows)

Blood flow (blue arrows) is impeded

Left lung

Back pressure to upper body

Superior vena cava

Pulmonary artery

Inferior vena cava

Right ventricle

Back pressure to lower body

Thickened heart muscle

Blood flow from lower body

Lumen (inner space)

Lumen narrowed

Artery wall thickened with fibrous tissue

Thickened muscle

Changes in arteries

As the condition progresses, the amount of muscle and fibrous tissue in the walls of the pulmonary arteries increases. This narrows the lumen (inner space), further impeding blood flow and increasing the pressure.

EARLY STAGE

LATE STAGE

Pneumothorax

In a pneumothorax, air enters the space between the layers of the pleura, the double-layered membrane that covers the outside of the lungs and the inside of the chest cavity. Air can enter from the lung or from outside the body, altering the pressure, and causing the lung to collapse. The condition can occur spontaneously or as a result of injury. Chest pain and breathlessness are common symptoms.

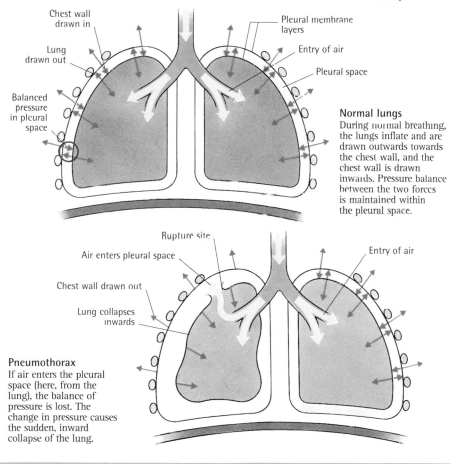

Chest wall drawn in

Lung drawn out

Balanced pressure in pleural space

Pleural membrane layers

Entry of air

Pleural space

Normal lungs
During normal breathing, the lungs inflate and are drawn outwards towards the chest wall, and the chest wall is drawn inwards. Pressure balance between the two forces is maintained within the pleural space.

Rupture site

Air enters pleural space

Chest wall drawn out

Lung collapses inwards

Entry of air

Pneumothorax
If air enters the pleural space (here, from the lung), the balance of pressure is lost. The change in pressure causes the sudden, inward collapse of the lung.

Fibrosing alveolitis

Also called idiopathic pulmonary fibrosis (IPF), fibrosing alveolitis is believed to be an autoimmune disorder of unknown cause. In some cases it occurs with other immune disorders such as rheumatoid arthritis. IPF, which usually affects either the whole or part of both lungs, causes scarring (fibrosis) and thickening of the alveoli, the air sacs of the lungs. Severe breathlessness is a common symptom.

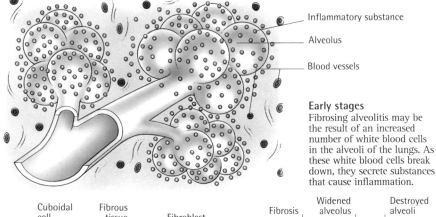

Inflammatory substance

Alveolus

Blood vessels

Early stages

Fibrosing alveolitis may be the result of an increased number of white blood cells in the alveoli of the lungs. As these white blood cells break down, they secrete substances that cause inflammation.

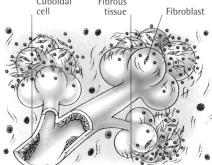

Cuboidal cell   Fibrous tissue   Fibroblast

Growth of fibrous tissue

The inflammation causes cells called fibroblasts to produce an overgrowth of fibrous tissue. Thick cells replace the usual thin cells that line the bronchi, restricting the passage of oxygen.

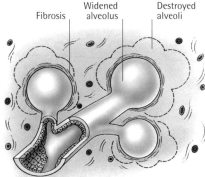

Fibrosis   Widened alveolus   Destroyed alveoli

Late stages

Formation of scar tissue (fibrosis) occurs, which gradually destroys the alveolar walls. The remaining alveoli widen. Scar tissue may contract, restricting lung expansion.

Silicosis

Silicosis is one of several occupational diseases caused by inhalation of dust particles. It is a condition in which scarring (fibrosis) of lung tissue occurs due to irritation by silica dust, usually in the form of quartz. People at risk are those such as quarry workers, stone masons, and pottery workers. Symptoms, which include breathlessness, may not become apparent for years.

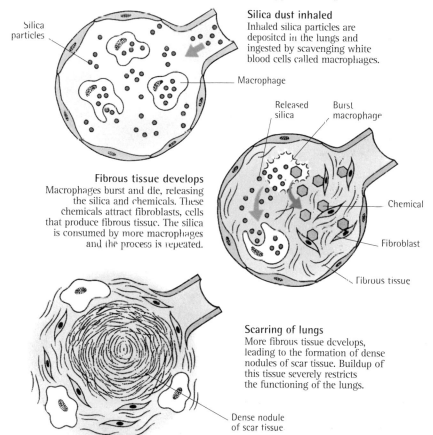

Silica dust inhaled
Inhaled silica particles are deposited in the lungs and ingested by scavenging white blood cells called macrophages.

Silica particles

Macrophage

Released silica

Burst macrophage

Chemical

Fibroblast

Fibrous tissue

Fibrous tissue develops
Macrophages burst and die, releasing the silica and chemicals. These chemicals attract fibroblasts, cells that produce fibrous tissue. The silica is consumed by more macrophages and the process is repeated.

Scarring of lungs
More fibrous tissue develops, leading to the formation of dense nodules of scar tissue. Buildup of this tissue severely restricts the functioning of the lungs.

Dense nodule of scar tissue

Chronic bronchitis

Chronic bronchitis is persistent inflammation of the bronchi (airways in the lungs). The disorder may be the result of air pollution but it is most commonly caused by smoking, which stimulates the linings of the bronchi to produce too much mucus. The main symptom is a phlegm-producing cough that gradually worsens. Hoarseness and breathlessness may also occur.

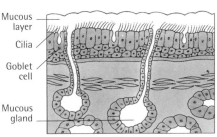

Healthy bronchi
In the lining of normal airways, mucus produced by mucous glands and specialized cells (goblet cells) is moved up into the throat by cilia (tiny hairs) to be coughed up or swallowed.

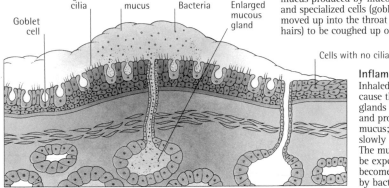

Inflamed bronchi
Inhaled irritants cause the mucous glands to enlarge and produce more mucus; the cilia are slowly destroyed. The mucus cannot be expelled and becomes infected by bacteria.

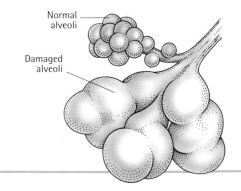

Emphysema

In emphysema, the alveoli (tiny air sacs) in the lungs become enlarged due to damage by cigarette smoke or other pollutants, which cause the alveolar walls to break down and merge. The formation of fewer and larger alveoli reduces the surface area for gas exchange in the lungs. Symptoms include breathlessness and a cough.

Asthma

Asthma is inflammation and reversible constriction of the airways in the lungs, which causes recurrent episodes of breathlessness, wheezing, and sometimes a dry cough. The disorder can be triggered by an allergic reaction to certain substances, such as house dust or animal fur, but some types of asthma have no known cause. The condition often starts in childhood.

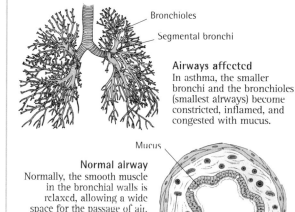

Bronchioles

Segmental bronchi

Airways affected
In asthma, the smaller bronchi and the bronchioles (smallest airways) become constricted, inflamed, and congested with mucus.

Mucus

Normal airway
Normally, the smooth muscle in the bronchial walls is relaxed, allowing a wide space for the passage of air.

Blood vessels

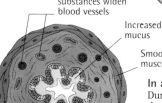

Inflammatory substances widen blood vessels

Increased mucus

Relaxed smooth muscle

Smooth muscle contracts

In an asthma attack
During an attack of asthma, the muscle walls of the airways contract; increased mucus and inflamed tissues further narrow the airways.

Inflammation and swelling

Treating asthma

The symptoms of allergic asthma are caused mainly by irritant substances such as histamine, which is released by certain cells (mast cells) in an allergic reaction. One way of treating this type of asthma is with drugs called mast-cell stabilizers that inhibit histamine production.

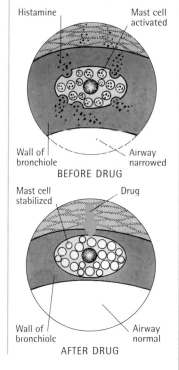

Histamine

Mast cell activated

Wall of bronchiole

Airway narrowed

BEFORE DRUG

Mast cell stabilized

Drug

Wall of bronchiole

Airway normal

AFTER DRUG

Lung cancer

Lung cancer, the growth of abnormal cells in the lungs, is usually triggered by inhaled irritants. The most common cause of the disease is cigarette smoke, which contains carcinogenic (cancer causing) substances. In most cases of lung cancer, a tumor develops in the bronchi, where it may cause coughing, pain, and obstruction of breathing.

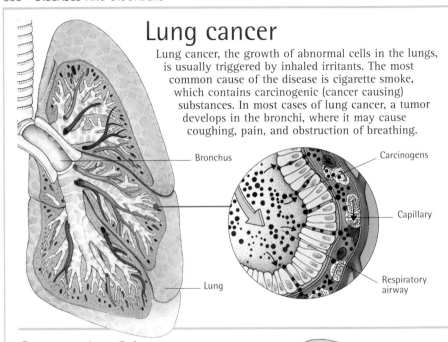

Bronchus

Carcinogens

Capillary

Lung

Respiratory airway

Spread of lung cancer

Cells from a cancerous tumor in the lungs (the primary tumor) may spread to other areas of the body. If cancerous tissue develops at a new site (see diagram), it is called a metastasis. Metastases in bones may cause pain and fractures; in the brain, paralysis, confusion, and many other symptoms; in the liver, weight loss and nausea; in the lymph nodes, impairment of the immune system.

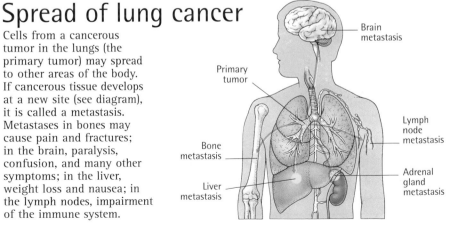

Brain metastasis

Primary tumor

Lymph node metastasis

Bone metastasis

Adrenal gland metastasis

Liver metastasis

How smoking damages the lungs

The tar formed when tobacco is burned contains many different chemicals, and is strongly carcinogenic. Over time, inhaled smoke drawn into the airways flattens the top layer of cells in the bronchial linings, and their tiny surface hairs, or cilia, are gradually destroyed. Basal cells in the layers below multiply rapidly to replace the damaged cells, and some of these new basal cells become cancerous.

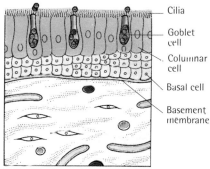

Cilia
Goblet cell
Columnar cell
Basal cell
Basement membrane

1 Healthy bronchi
Normal bronchi are lined with goblet cells, which produce lubricating mucus, and columnar cells topped by cilia. Under these are basal cells, which constantly replace damaged columnar cells.

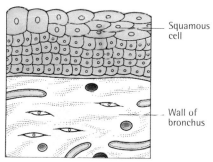

Squamous cell

Wall of bronchus

2 Columnar cells flattened
Over a number of years, columnar cells continually damaged by inhaled smoke become flattened and turn into squamous (scalelike) cells, which gradually lose their cilia.

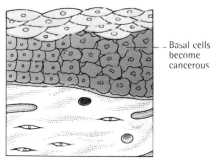

Basal cells become cancerous

3 Basal cells multiply
To replace damaged squamous cells, the basal cells start to multiply at an increased rate. Some of these new cells become cancerous.

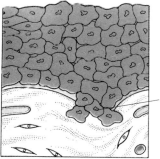

Multiplying cancer cells break through basement membrane

4 Cancer cells spread
The cancer cells multiply, replacing healthy cells. They may break through the basement membrane of the bronchial lining and spread to new sites.

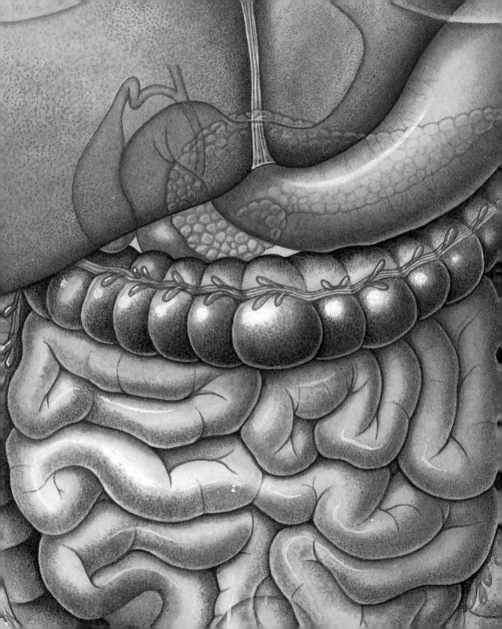

Digestive disorders

Problems with the digestive organs occur frequently and can affect people at any age. Some common disorders of the stomach and intestines are related to diet and may be aggravated by anxiety. Liver disease can be caused by viral infection and is also a common result of long-term alcohol abuse. Cancers can develop in any part of the digestive system.

Intestinal diseases
If the digestive process is disrupted by intestinal diseases, there may be symptoms such as pain and diarrhea.

Hernia

A hernia is the protrusion of an organ or tissue out of the body cavity that it normally occupies. There are various types of abdominal hernia, two examples of which – sliding hiatal hernia and paraesophageal hiatal hernia – are shown here. The disorder occurs mostly in overweight, middle-aged, or elderly people. The first symptom is usually acid reflux, and there may be some pain.

Sliding hiatal hernia
This is one of the most common types of hernia. It occurs when the lower esophagus and upper stomach slide through an opening (hiatus) in the diaphragm. The symptoms include acid reflux and heartburn.

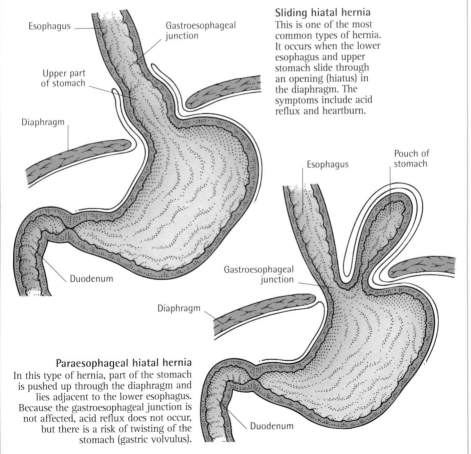

Esophagus

Gastroesophageal junction

Upper part of stomach

Diaphragm

Duodenum

Esophagus

Pouch of stomach

Gastroesophageal junction

Diaphragm

Duodenum

Paraesophageal hiatal hernia
In this type of hernia, part of the stomach is pushed up through the diaphragm and lies adjacent to the lower esophagus. Because the gastroesophageal junction is not affected, acid reflux does not occur, but there is a risk of twisting of the stomach (gastric volvulus).

Peptic ulcer

Areas of eroded tissue in the lining of the stomach or duodenum are called peptic ulcers. Normally, the linings are protected from acidic digestive juices by mucus, bicarbonate, and other secretions, but if this barrier is breached the juices can damage the lining. Peptic ulcers are often linked to infection with the bacterium *Helicobacter pylori*, which can increase production of stomach acid.

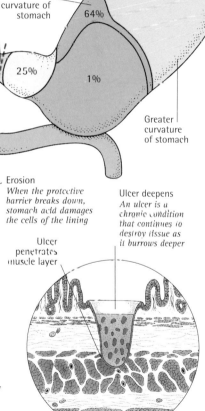

Esophagus

10%

Lesser curvature of stomach

64%

Pylorus

Duodenal bulb

90%

25%

1%

10%

Greater curvature of stomach

Sites of peptic ulcers
Peptic ulcers are found mainly in the first part of the duodenum, the duodenal bulb. The illustration shows the percentage of ulcers occurring at particular sites.

Mucosa

Submucosa

Muscle layer

Erosion
When the protective barrier breaks down, stomach acid damages the cells of the lining

Ulcer deepens
An ulcer is a chronic condition that continues to destroy tissue as it burrows deeper

Ulcer penetrates muscle layer

Development of a peptic ulcer
In the early stages of an ulcer (above) the mucous layer, or mucosa, is only partly destroyed, producing a shallow area of damage known as an erosion. Later, the ulcer penetrates the entire mucosa and the submucosal layer beneath it (right), and may eventually burrow into, or through, the muscle layer of the organ.

Alcoholic liver disease

Persistent alcohol abuse leads to liver damage and possibly irreversible scarring (cirrhosis). Initially, excess alcohol consumption causes an abnormal accumulation of fat in liver cells, a condition called fatty liver. Sometimes, alcohol damage causes inflammation, known as alcoholic hepatitis. Both conditions can lead to liver failure if alcohol abuse continues. In the early stages, alcoholic liver disease may produce no symptoms. Later, there may be nausea, discomfort, jaundice, and weight loss.

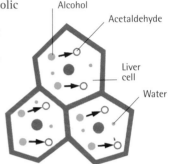

Alcohol

Acetaldehyde

Liver cell

Water

How damage occurs

Some alcohol is excreted unchanged, but most is converted by enzymes in the liver into a substance called acetaldehyde. Both alcohol and acetaldehyde are poisonous to liver cells.

Fat-laden cell

Damaged tissue

Fatty liver

The liver cells become infiltrated with globules of fat, which results in enlargement of the liver.

Alcoholic hepatitis

As a result of the production of acetaldehyde, liver cells become inflamed and damaged so that liver function is impaired.

Scar tissue

Cirrhosis

In cirrhosis, bands of scar tissue separate nodules of overgrown cells. At this stage, liver damage is irreversible.

Portal hypertension

Portal hypertension, which is raised pressure in the portal vein carrying blood to the liver, can occur if blood flow in the liver is obstructed. In most cases, the obstruction is caused by scar tissue (cirrhosis) due to alcohol damage. The back pressure causes enlargement of veins in the lower esophagus and upper stomach. The veins may rupture, resulting in sudden, heavy bleeding.

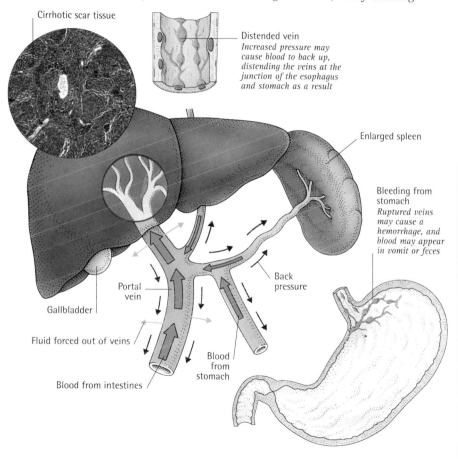

Cirrhotic scar tissue

Distended vein
Increased pressure may cause blood to back up, distending the veins at the junction of the esophagus and stomach as a result

Enlarged spleen

Bleeding from stomach
Ruptured veins may cause a hemorrhage, and blood may appear in vomit or feces

Portal vein

Back pressure

Gallbladder

Fluid forced out of veins

Blood from stomach

Blood from intestines

Hepatitis

Hepatitis is inflammation of the liver, commonly caused by infection with hepatitis A, B (illustrated right), or C viruses. Viral hepatitis is usually a short-lived illness. However, sometimes the inflammation persists, particularly if it is due to the C virus, and may eventually lead to scarring (cirrhosis). Other causes of hepatitis include adverse drug reactions, alcohol poisoning, and bacterial infection.

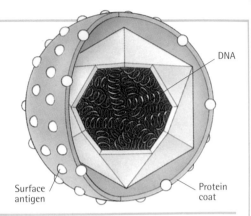

DNA

Surface antigen

Protein coat

Liver abscess

An abscess, or collection of pus, in the liver can be caused by infection with bacteria or amebae (single-celled organisms). Bacteria may spread from another infected site, such as the appendix. Amebae are transmitted through contaminated food or water, and can be carried into the large intestine, causing ulceration. From the intestine, they may spread to the liver, and cause liver enlargement, fever, pain, nausea, and weight loss.

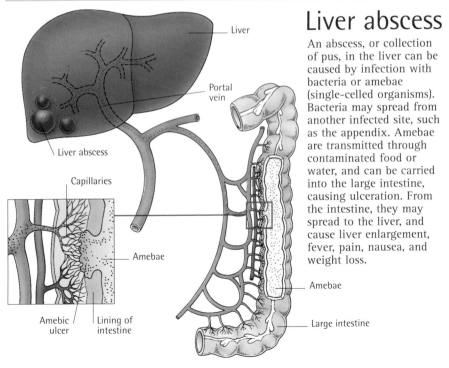

Liver

Portal vein

Liver abscess

Capillaries

Amebae

Amebic ulcer

Lining of intestine

Amebae

Large intestine

Cancer of the pancreas

In pancreatic cancer, healthy cells in the pancreatic tissue are replaced by groups of irregularly shaped malignant cells. Most of these cancers occur in the pancreatic head, especially around an area called the ampulla of Vater, where the pancreatic duct enters the duodenum. The main symptom is a dull pain in the upper abdomen penetrating to the back. Other symptoms include loss of appetite and weight, and jaundice. Pancreatic cancer is most common in elderly people.

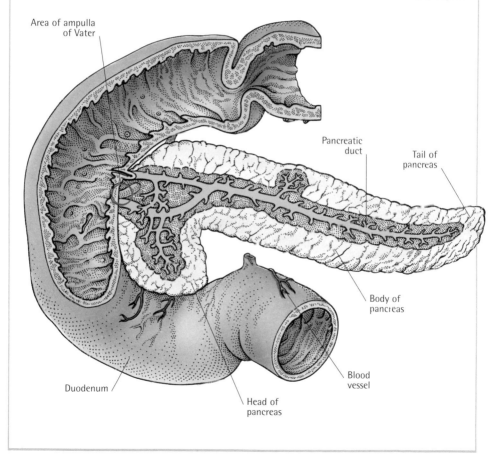

Area of ampulla of Vater

Pancreatic duct

Tail of pancreas

Body of pancreas

Duodenum

Blood vessel

Head of pancreas

Gallstones

Gallstones are solid accumulations of bile pigment and cholesterol that form in the gallbladder. They are caused by an imbalance in the chemical composition of bile. Stones that travel from the gallbladder may become impacted in either the cystic duct or the common bile duct, or pass through into the duodenum.

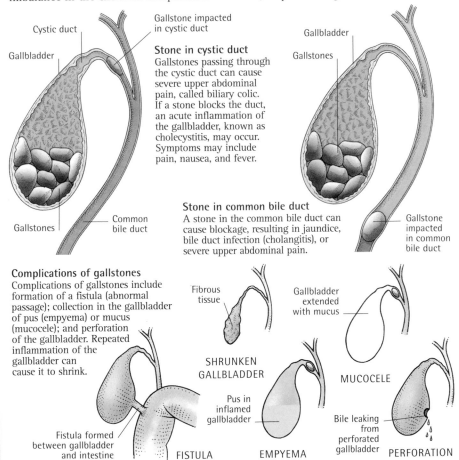

Cystic duct

Gallbladder

Gallstone impacted in cystic duct

Stone in cystic duct
Gallstones passing through the cystic duct can cause severe upper abdominal pain, called biliary colic. If a stone blocks the duct, an acute inflammation of the gallbladder, known as cholecystitis, may occur. Symptoms may include pain, nausea, and fever.

Gallbladder

Gallstones

Gallstones

Common bile duct

Stone in common bile duct
A stone in the common bile duct can cause blockage, resulting in jaundice, bile duct infection (cholangitis), or severe upper abdominal pain.

Gallstone impacted in common bile duct

Complications of gallstones
Complications of gallstones include formation of a fistula (abnormal passage); collection in the gallbladder of pus (empyema) or mucus (mucocele); and perforation of the gallbladder. Repeated inflammation of the gallbladder can cause it to shrink.

Fibrous tissue

Gallbladder extended with mucus

SHRUNKEN GALLBLADDER

MUCOCELE

Pus in inflamed gallbladder

Bile leaking from perforated gallbladder

Fistula formed between gallbladder and intestine

FISTULA

EMPYEMA

PERFORATION

Inflammatory bowel disease

The term inflammatory bowel disease includes ulcerative colitis and Crohn's disease, both of which result in chronic intestinal inflammation and ulceration. These disorders may be caused by the immune system attacking the body's own tissues. The causes are unknown but genetic predisposition is likely. Symptoms include fever, bleeding from the rectum, abdominal pain, and diarrhea.

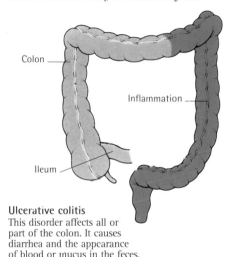

Ulcerative colitis
This disorder affects all or part of the colon. It causes diarrhea and the appearance of blood or mucus in the feces.

Crohn's disease
In this disorder, narrowing, or stricture, and patches of inflammation may occur anywhere in the digestive tract.

Irritable bowel syndrome

This chronic condition is a combination of attacks of diarrhea and intermittent constipation. It can involve disturbance of muscular movement in the large intestine. The cause of irritable bowel syndrome (IBS) is unknown but the disorder can be aggravated by stress, and may be linked to sensitivity to particular foods. Other symptoms of IBS include abdominal cramps, bloating, and passage of mucus with the feces.

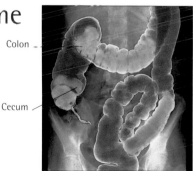

Intestinal obstruction

Partial or complete obstruction of the intestine, preventing normal digestion and movement of food, can be caused by many disorders. Among the most common are twisting of the intestine, known as volvulus, and hernia, in which part of the intestine protrudes through a weakness in the abdominal wall. In the rare disorder called mesenteric infarction, loss of blood supply causes part of the intestine to die.

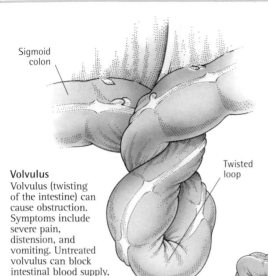

Sigmoid colon

Twisted loop

Volvulus
Volvulus (twisting of the intestine) can cause obstruction. Symptoms include severe pain, distension, and vomiting. Untreated volvulus can block intestinal blood supply, causing tissue death.

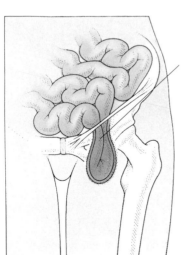

Femoral hernia

Blockage in blood vessel leading to intestine

Affected area of intestine

Femoral hernia
In a femoral hernia, the intestine protrudes through the narrow femoral canal between the abdomen and the thigh. It can become trapped, causing obstruction and severe pain.

Mesenteric infarction
In this serious condition, blockage of a blood vessel in the intestinal membrane (mesentery) cuts off blood supply to a section of intestine, resulting in tissue death.

Intussusception

This condition occurs when part of the intestine telescopes in on itself, forming a tube within a tube. Intussusception is most likely to affect the junction between the ileum and cecum. It is rare, and occurs most commonly in very young children. Symptoms include severe abdominal pain and feces that resemble redcurrant jelly. Urgent treatment is needed, because blood supply may be cut off in the affected area and tissue death may follow.

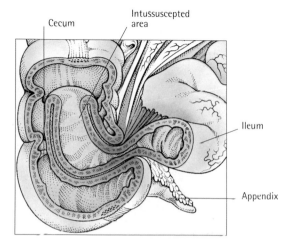

Cecum

Intussuscepted area

Ileum

Appendix

Appendicitis

Appendicitis (inflammation of the appendix) is common, especially in children. The disorder is sometimes caused by blockage of the appendix, a short, closed tube that projects from the cecum, or by ulceration of its lining. The main symptoms are acute pain and tenderness in the lower right abdomen. Other symptoms include mild fever, nausea, and vomiting. Surgery is the usual treatment. Left untreated, an inflamed appendix may rupture, leading to bacterial infection of the peritoneum (membrane lining the abdominal cavity).

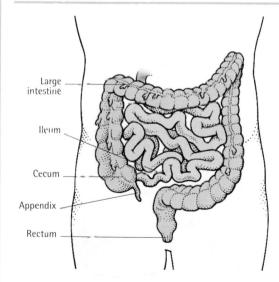

Large intestine

Ileum

Cecum

Appendix

Rectum

Diverticular disease

The term diverticular disease includes diverticulosis (the presence of small pouches, or diverticula, in the intestinal wall) and diverticulitis (inflammation of these pouches). The disorders most often affect the lower colon, and are more common in elderly people. Symptoms do not always occur but may include abdominal pain and swelling, diarrhea, constipation, intestinal gas, and rectal bleeding.

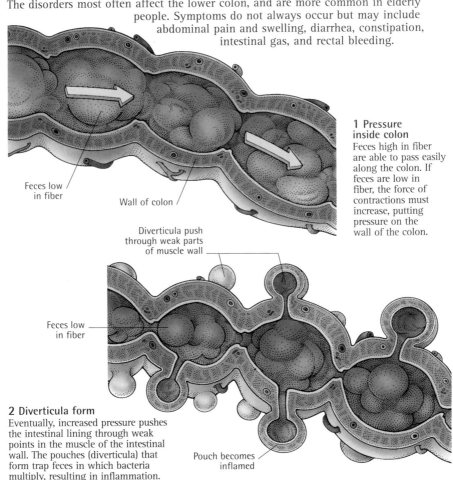

Feces low in fiber

Wall of colon

1 Pressure inside colon
Feces high in fiber are able to pass easily along the colon. If feces are low in fiber, the force of contractions must increase, putting pressure on the wall of the colon.

Diverticula push through weak parts of muscle wall

Feces low in fiber

2 Diverticula form
Eventually, increased pressure pushes the intestinal lining through weak points in the muscle of the intestinal wall. The pouches (diverticula) that form trap feces in which bacteria multiply, resulting in inflammation.

Pouch becomes inflamed

Cancer of the colon

Cancerous growths in the colon often first develop as polyps in the mucosa, or lining, of the intestinal wall. The cancer may invade the intestinal wall or spread to nearby lymph nodes, and then to more distant organs. Symptoms are blood in the feces, a change in bowel habits, and pain in the abdomen. Hereditary factors are known to increase the risk of cancer of the colon. It is also suggested that the disease may be linked to a low intake of dietary fiber and also to a high intake of animal fat.

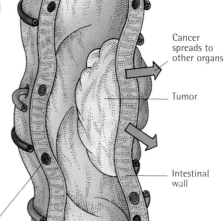

Cancer spreads to other organs

Tumor

Intestinal wall

Blood vessel

Hemorrhoids

Hemorrhoids, also called piles, are swollen (varicose) veins in the lining of the anus. The veins may protrude out of the anus, when they are known as external hemorrhoids, or occur higher up in the anal canal, when they are called internal hemorrhoids. They are usually the result of constipation and straining to pass feces. The disorder commonly causes rectal bleeding, pain on defecation, and itching around the anus.

Vein network

Rectum

Internal hemorrhoid

External hemorrhoid

Anal canal

Anus

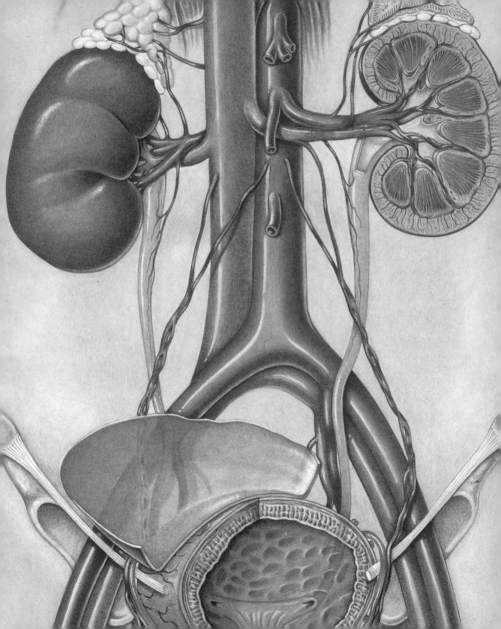

Urinary disorders

Infections can spread easily through the urinary tract, sometimes resulting in persistent, debilitating disorders. Serious illness can occur if the kidneys are damaged by disease and are unable to carry out their function of filtering wastes from the blood. One of the most common urinary disorders is incontinence, which is especially likely to occur in the elderly.

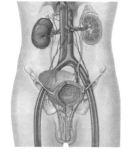

Urinary symptoms
Symptoms such as pain in the back or lower abdomen and a change in frequency of urination may indicate a urinary disorder.

Urinary tract disorders

The urinary tract is very susceptible to infection, especially in women, and is also prone to chronic disorders. Each of the urinary organs is affected by its own characteristic diseases, some of which are illustrated below. However, a disorder of any one organ can also affect other parts of the urinary system. For example, blockage of the outflow of urine may damage the kidneys as a result of back pressure.

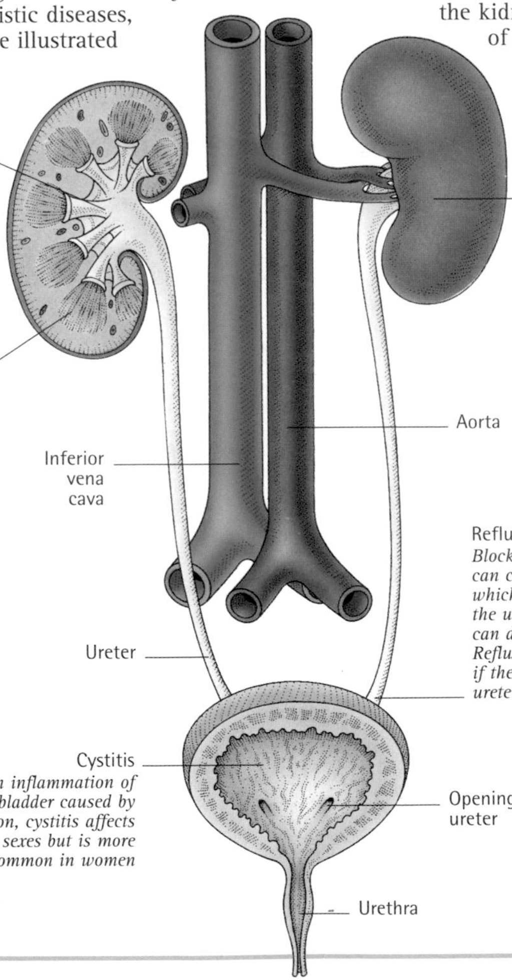

Pyelonephritis
This acute infection of the urine-collecting system of the kidney may be linked with bladder infection

Glomerulonephritis
Inflammation of the filtering units of the kidney (glomeruli) is often related to an autoimmune disorder

Diabetic nephropathy
Long-term diabetes mellitus can lead to this complication as a result of changes in the kidney's small blood vessels. It often progresses to kidney failure

Aorta

Inferior vena cava

Reflux
Blockage of the urethra can cause back pressure, which forces urine up the ureters; such reflux can damage the kidneys. Reflux can also occur if the openings of the ureters are too relaxed

Ureter

Cystitis
An inflammation of the bladder caused by infection, cystitis affects both sexes but is more common in women

Opening of ureter

Urethra

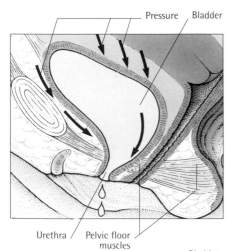

Incontinence

Urinary incontinence is the involuntary passing of urine. The disorder occurs more commonly in women than in men, mainly because women often have a weakness in the pelvic floor muscles following childbirth. Incontinence is especially common in elderly people as a result of dementia; damage to the brain or spinal cord is another possible cause.

Stress incontinence
Weak pelvic floor muscles may allow small amounts of urine to escape during strenuous exercise, or when a person laughs or coughs.

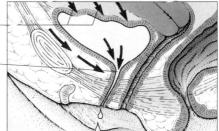

Urge incontinence
In this type of incontinence, there is an urgent desire to urinate, sometimes triggered by sudden movement. Once urination starts, the bladder contracts involuntarily until it is empty.

Kidney stones

Concentrated substances in the urine may slowly crystallize to form stones in the urine-collecting part of the kidneys. These stones, which can cause intense pain, also sometimes form in the ureters or bladder. Small stones may be excreted unnoticed in urine but larger ones can damage the ureters and block urine flow.

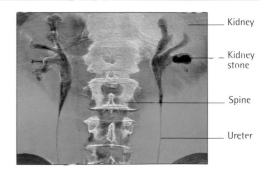

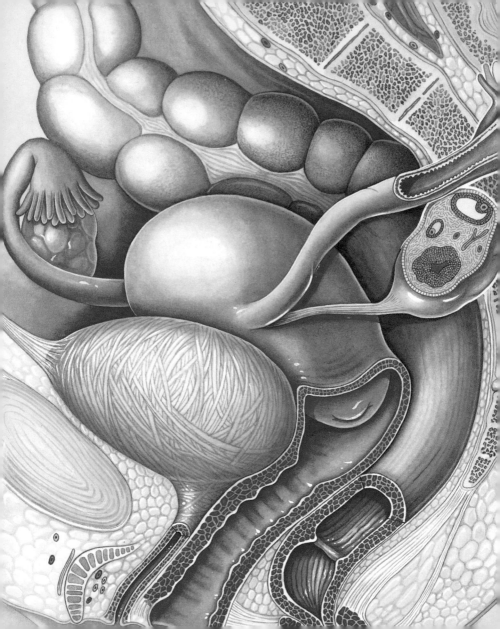

Reproductive disorders

There are many diseases that can affect the male and female reproductive tracts. These include sexually transmitted infections and cancers and nonmalignant growths of the uterus, ovaries, and testes. In both sexes, certain disorders can result in infertility. In women, hormonal changes are a common cause of minor breast disorders.

Disorders of the uterus
Apart from complications following childbirth, disorders of the uterus include fibroids, prolapse, endometriosis, and cancer.

Common breast disorders

The breasts are strongly influenced by female sex hormones, and common symptoms, such as lumpiness and pain, are often related to hormonal changes of the menstrual cycle or during pregnancy. Most types of breast lump are noncancerous and many do not need treatment, although all should be investigated by a doctor. Breast cancer, the first symptom of which is usually a painless lump, affects about one in nine women.

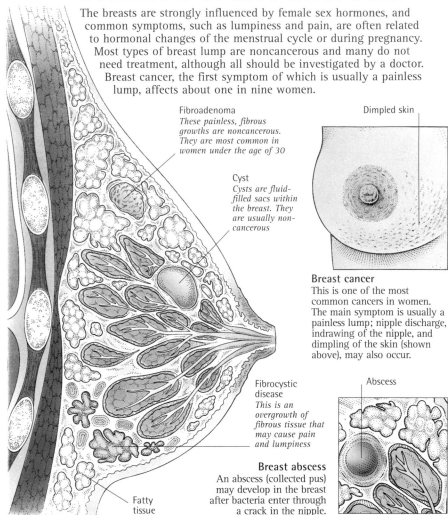

Fibroadenoma
These painless, fibrous growths are noncancerous. They are most common in women under the age of 30

Cyst
Cysts are fluid-filled sacs within the breast. They are usually noncancerous

Dimpled skin

Breast cancer
This is one of the most common cancers in women. The main symptom is usually a painless lump; nipple discharge, indrawing of the nipple, and dimpling of the skin (shown above), may also occur.

Fibrocystic disease
This is an overgrowth of fibrous tissue that may cause pain and lumpiness

Abscess

Breast abscess
An abscess (collected pus) may develop in the breast after bacteria enter through a crack in the nipple.

Fatty tissue

Endometriosis

In this disorder, fragments of endometrium (uterine lining) migrate through the fallopian tubes to other parts of the pelvic cavity, and implant in other organs. These fragments can bleed and cause pain during menstruation and sexual intercourse, and may form cysts. Some possible sites of endometriosis are shown right. Rarely, fragments of endometrium travel to more distant sites, such as the lungs.

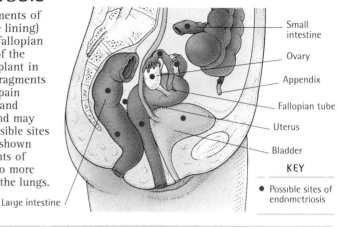

Small intestine

Ovary

Appendix

Fallopian tube

Uterus

Bladder

KEY

● Possible sites of endometriosis

Large intestine

Prolapsed uterus

A prolapsed (displaced) uterus is mostly the result of pregnancy and childbirth, which can weaken the ligaments that support the uterus. The uterus sags from its normal position, sometimes distorting the vagina and causing urinary and bowel problems. Surgery, or the insertion of a ring pessary, may be necessary.

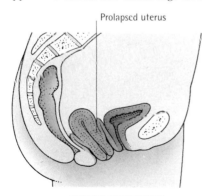

Prolapsed uterus

BEFORE TREATMENT

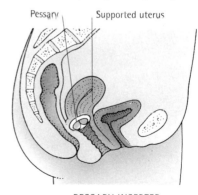

Pessary   Supported uterus

PESSARY INSERTED

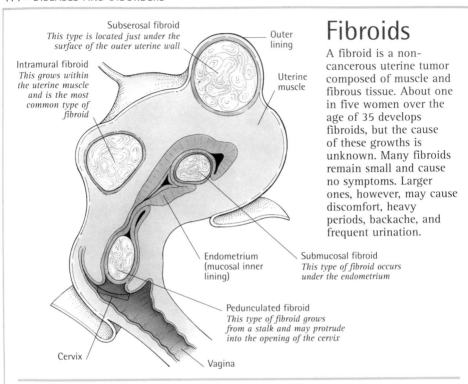

Subserosal fibroid
This type is located just under the surface of the outer uterine wall

Outer lining

Intramural fibroid
This grows within the uterine muscle and is the most common type of fibroid

Uterine muscle

Endometrium (mucosal inner lining)

Submucosal fibroid
This type of fibroid occurs under the endometrium

Pedunculated fibroid
This type of fibroid grows from a stalk and may protrude into the opening of the cervix

Cervix

Vagina

Fibroids

A fibroid is a non-cancerous uterine tumor composed of muscle and fibrous tissue. About one in five women over the age of 35 develops fibroids, but the cause of these growths is unknown. Many fibroids remain small and cause no symptoms. Larger ones, however, may cause discomfort, heavy periods, backache, and frequent urination.

Cervical cancer

Cancer of the neck of the uterus (cervix) is one of the most common cancers in women. The cause of the disease is uncertain, but changes in cervical cells may be associated with the human papillomavirus, which causes genital warts. In the early stages of cervical cancer there are often no symptoms; later, there may be a bloodstained discharge.

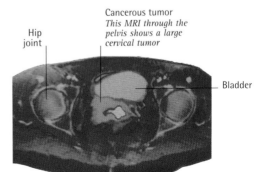

Cancerous tumor
This MRI through the pelvis shows a large cervical tumor

Hip joint

Bladder

Ovarian disorders

The most common disorder of the ovaries is the occurrence of one or more cysts. These fluid-filled swellings may develop at any age, and can occasionally grow to a very large size. Most cysts are not malignant and do not cause symptoms. Complications, such as a cyst becoming twisted, are uncommon. Multiple small cysts sometimes develop because of an imbalance of sex hormones.

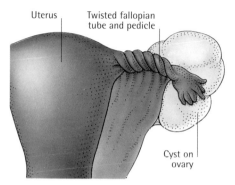

Twisted pedicle
The stalk, or pedicle, of a cyst may become twisted, cutting off the blood supply and causing sudden, severe abdominal pain.

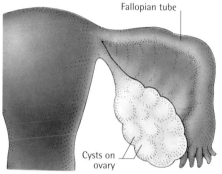

Polycystic ovaries
Many small cysts may form on the ovaries due to a hormone imbalance. This disorder is a common cause of infertility.

Ovarian cancer

Cancer of the ovary is responsible for more deaths than any other cancer of the female reproductive system. This is essentially because ovarian cancer is rarely detected in its early stages, while it may still be curable. Symptoms, which usually appear only when the cancer has spread to other organs, may include abdominal pain and discomfort, and abnormal vaginal bleeding. The illustration (right) shows possible sites of ovarian cancer, the most common site being the layer of epithelial tissue on the surface of the ovary.

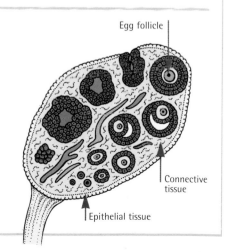

Testicular disorders

The most common symptom of testicular disorder is swelling. This may be due to injury, to an accumulation of fluid in the space around the testis (known as a hydrocele), or to an accumulation of sperm or blood. Some swellings are associated with a fever, possibly due to infection. Most are painless and harmless, but all should be checked by a doctor. Only rarely is a swelling a sign of testicular cancer (see below).

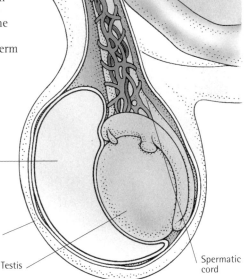

Hydrocele
This swelling, which is a collection of straw-colored fluid around the testis, very commonly affects middle-aged men

Scrotum

Testis

Spermatic cord

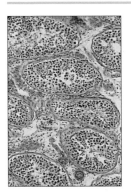

NORMAL TISSUE

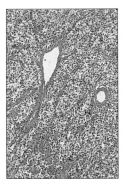

ABNORMAL TISSUE

Testicular cancer

Cancer of the testis is a rare disorder that is most likely to occur in men under the age of 40, especially those who had an undescended testis in childhood. The first symptom of the disease is usually a firm, painless swelling. In a few cases there is pain and inflammation. The photographs illustrate normal testicular tissue (far left) and the changes in tissue structure that take place when cells become cancerous (left).

Enlarged prostate

The prostate gland located at the base of the male bladder surrounds the urethra. A noncancerous condition in which this gland becomes enlarged is very common in men over the age of 50. An enlarged prostate may constrict and distort the urethra, causing a weakened flow of urine. Because the bladder cannot be emptied completely, there is a frequent need to urinate; the bladder may become distended.

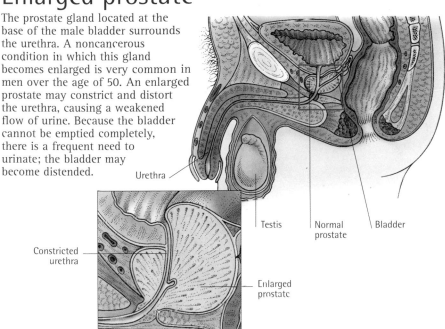

Urethra

Constricted urethra

Testis

Normal prostate

Bladder

Enlarged prostate

Prostate cancer

Prostate cancer is a malignant growth in the prostate gland. There may be no symptoms in the early stages of the disease, but as the tumor grows it can cause constriction of the urethra, resulting in difficulty urinating. Prostate cancer may spread to the bladder and ureters (the tubes running from the kidneys to the bladder), the lymph nodes, and the bones.

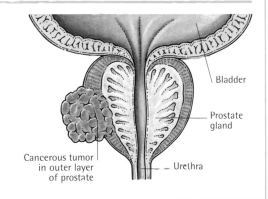

Bladder

Prostate gland

Cancerous tumor in outer layer of prostate

Urethra

Gonorrhea

Gonorrhea is one of the most common sexually transmitted infections (STIs). The disease, caused by the bacterium *Neisseria gonorrheae*, affects the urethra in both sexes; in women, it also affects the cervix, from where the infection can spread to the uterus, fallopian tubes, and ovaries. The symptoms are a discharge from the penis or vagina, and pain on urination.

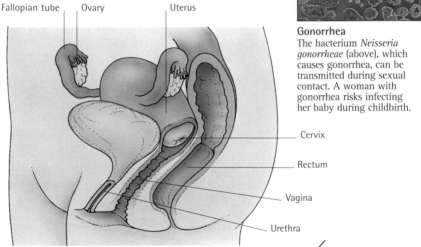

Fallopian tube   Ovary   Uterus

Cervix

Rectum

Vagina

Urethra

Gonorrhea
The bacterium *Neisseria gonorrheae* (above), which causes gonorrhea, can be transmitted during sexual contact. A woman with gonorrhea risks infecting her baby during childbirth.

Genital herpes

The common sexually transmitted infection (STI) known as genital herpes is caused by the herpes simplex virus. The disease affects both sexes and tends to be recurrent, with the first episode being the most severe. During an attack of genital herpes, crops of small blisters develop on the penis or around the vagina and on the cervix, and form painful ulcers.

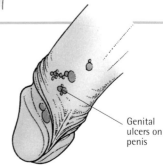

Genital ulcers on penis

Pelvic inflammatory disease

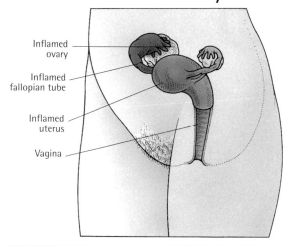

Inflamed ovary

Inflamed fallopian tube

Inflamed uterus

Vagina

Pelvic inflammatory disease (PID) is infection of the female reproductive tract. The cervix, uterus, fallopian tubes, and ovaries can all become inflamed. The disorder is usually the result of an untreated sexually transmitted infection such as gonorrhea (p.418). Symptoms include vaginal discharge, lower abdominal pain, fever, and pain during sexual intercourse. If PID is left untreated it can cause scarring of the fallopian tubes, leading to infertility.

Nongonococcal urethritis

Nongonococcal, also called non-specific, urethritis is inflammation of the urethra, usually caused by a sexually transmitted infection other than gonorrhea (p.418). In women, the main symptom is vaginal discharge. In men, there is often discharge from the penis and pain when urinating. The infection may spread to the epididymis (the coiled tube within the testis where sperm are stored), producing pain and swelling inside the scrotum.

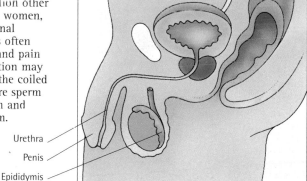

Urethra

Penis

Epididymis

Infertility

About one in six couples seeks treatment for the inability to conceive. In women, fertility depends on the occurrence of several processes, from ovulation to implantation of the fertilized egg in the uterus. Fertility in men depends on the production and transmission of healthy sperm. In both sexes, fertility starts to decline naturally after about the age of 30. There are also various problems in both men and women that can affect one or more of the stages of conception.

CAUSES OF FEMALE INFERTLITY

Egg release
Ovaries may fail to release mature eggs, or may release them at irregular intervals. The causes include hormone imbalance due to obesity or excessive weight loss, or to polycystic ovarian disease.

Blocked or damaged fallopian tubes
Narrowed or blocked tubes as a result of scarring from infection, endometriosis, or an ectopic pregnancy may sometimes prevent fertilization or implantation.

Abnormalities of the uterus
Structural abnormalities are a rare cause of infertility. The uterus may have been defective from birth, or may have been damaged by the formation of fibroids, by surgery, or by an infection.

Cervical problems
A hormone imbalance may cause thick cervical mucus, which blocks sperm as they travel along the woman's reproductive tract. Damage to the cervix may result in repeated miscarriages.

CAUSES OF MALE INFERTLITY

Abnormal sperm
The most common cause of male infertility is an inadequate number of sperm; sperm may also be misshapen or unable to move rapidly. These problems may be the result of hormone imbalance, drugs, or illness.

Difficult passage of sperm
Sperm must make their way through the tubal systems of the epididymis and the vas deferens before mixing with semen to be ejaculated. Blockage of these pathways may be a cause of male infertility.

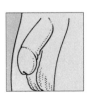

Ejaculation problems
Nerve reflexes may be damaged by spinal disorders or impaired by drugs. Following a prostate operation, ejaculate is sometimes propelled in reverse so that semen enters the bladder, resulting in infertility.

Blocked fallopian tubes

Blockage of the fallopian tubes is a common cause of infertility. Fertilization can take place only if a sperm and an egg meet in the tube. Blocked tubes may be due to scar tissue caused by pelvic inflammatory disease (p.419). Endometriosis (p.413), a disease in which fragments of uterine lining travel up the fallopian tubes, may also lead to scarring.

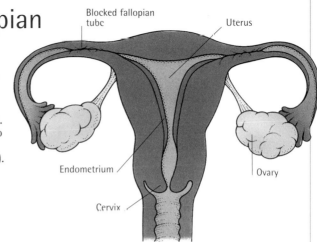

Antibodies to sperm

Infertility is sometimes due to the formation of antibodies to sperm by either partner. In a woman, antibodies to sperm may develop in the cervical mucus at the entrance to the uterus, damaging the sperm or preventing them from traveling further up the reproductive tract.

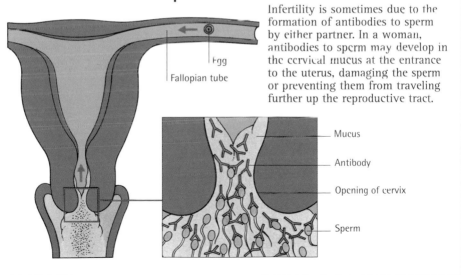

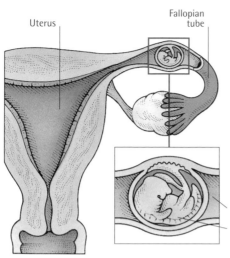

Uterus

Fallopian tube

Fallopian tube

Embryo

Ectopic pregnancy

If the fertilized egg implants outside the main cavity of the uterus, the pregnancy is known as ectopic. The cause is not always known, but ectopic pregnancies occur most often in women who have used an intrauterine contraceptive device, or who have had pelvic infections. If the pregnancy is not discovered at a very early stage, the fallopian tube may rupture; this is a life threatening condition, causing severe pain and bleeding from the vagina.

Miscarriage

A miscarriage is the loss of a fetus before week 20 of a pregnancy. Sometimes, this happens so early that the woman is unaware that she is even pregnant. The reasons for miscarriage are not always known, but common causes are fetal chromosomal abnormalities or developmental defects. The first sign of a threatened (possible) miscarriage is heavy vaginal bleeding; cramping pain in the back or lower abdomen may also occur. While the cervix remains closed, the fetus will be retained in the uterus and the pregnancy may continue. In almost two-thirds of all cases of threatened miscarriage, the pregnancy reaches full term without further problems.

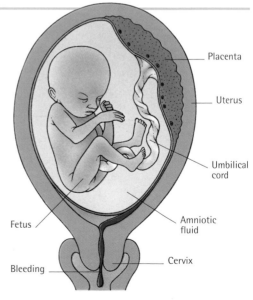

Placenta

Uterus

Umbilical cord

Fetus

Amniotic fluid

Cervix

Bleeding

Placental problems

A healthy placenta to nourish the fetus is essential for a normal pregnancy and a thriving baby. Very early in pregnancy, the placenta should develop in the upper wall of the uterus. Problems may occur if the placenta detaches from the uterus (a condition known as placental abruption) or if it is abnormally low (placenta praevia), which may lead to cervical obstruction, bleeding, or premature labor.

Placental abruption

In abruption, part of the placenta detaches from the uterine wall. This often causes sudden abdominal pain. Bleeding at the site occurs but there may not be visible blood loss from the vagina.

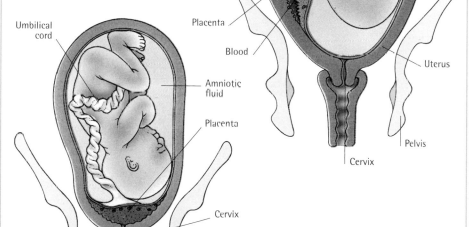

Umbilical cord

Placenta

Blood

Amniotic fluid

Placenta

Uterus

Cervix

Pelvis

Cervix

Placenta praevia

In this condition, the placenta may be positioned over or near the cervix rather than in the upper part of the uterus. If the placenta covers the entire cervix, it may detach at the start of labor.

Cancer

Cancers are a group of diseases characterized by abnormal cell growth and can occur anywhere in the body. In most forms of cancer, a tumor develops in particular tissues, such as the skin, or within an internal organ such as the lungs or stomach. Cancerous cells may spread from these growths to other sites in the body.

Tumor
A cancerous tumor is formed from a mass of irregularly shaped cells.

Cancerous tumors

A cancerous, or malignant, tumor is an abnormally growing tissue mass that invades surrounding tissues and can spread to more distant sites. Tumors most often form in internal organs but can also develop in skin, muscle, or bones. A cancerous tumor arising from epithelial tissue, such as skin, is called a carcinoma; one arising from connective tissue, such as muscle, is known as a sarcoma.

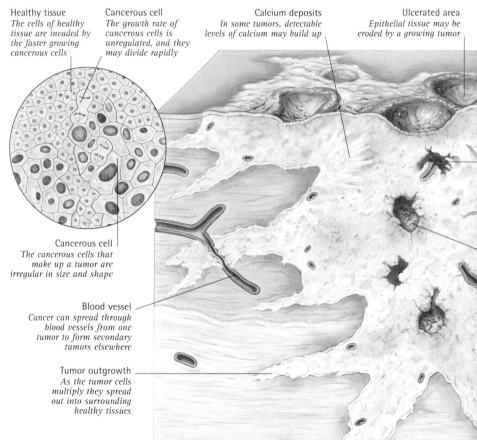

Healthy tissue
The cells of healthy tissue are invaded by the faster growing cancerous cells

Cancerous cell
The growth rate of cancerous cells is unregulated, and they may divide rapidly

Calcium deposits
In some tumors, detectable levels of calcium may build up

Ulcerated area
Epithelial tissue may be eroded by a growing tumor

Cancerous cell
The cancerous cells that make up a tumor are irregular in size and shape

Blood vessel
Cancer can spread through blood vessels from one tumor to form secondary tumors elsewhere

Tumor outgrowth
As the tumor cells multiply they spread out into surrounding healthy tissues

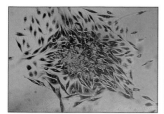

Tumor cells migrating
These cancerous cells have broken
away from a tumor and may
travel in blood or lymph vessels
to other sites in the body.

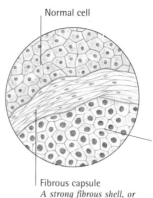

Normal cell

Noncancerous tumors
Noncancerous, or benign,
tumors do not contain
cancerous cells. Although they
can grow large enough to
cause pain, or even life-
threatening damage to nearby
organs, these tumors do not
invade surrounding tissues and
do not spread to other sites.

Tumor cell
*The cells of noncancerous
tumors are regular in
shape and size*

Fibrous capsule
*A strong fibrous shell, or
capsule, usually surrounds
a noncancerous tumor*

Epithelial layer
*Tumors often form in
epithelial tissue such as
skin and the linings of
internal organs*

Bleeding
*Multiplying cancerous
cells sometimes break
down the walls of blood
vessels, causing bleeding
within the tumor*

Nerve fiber
*A tumor may affect
nerve fibers,
causing pain*

Dead tissue
*If the tumor has
outgrown its blood
supply, tissues
within it die*

Lymph vessel
*Cells spreading from
a cancerous tumor
may be carried in the
lymph vessels to other
sites in the body*

Area
enlarged
above

Noncancerous tumor
*Noncancerous tumors
tend to grow slowly but
can become very large*

Surrounding tissue
*Tissue around a non-
cancerous tumor
is not invaded but
it may be damaged
or distorted*

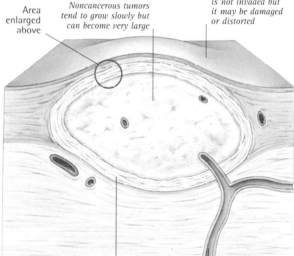

Fibrous capsule
*Contained within a capsule,
a noncancerous tumor can
often be removed easily*

Blood vessel
*Blood vessels engulfed by the
tumor supply the tissue mass
with oxygen and nutrients*

How cancer develops

Cancer starts when certain genes that control cell division and function are irreparably damaged by chemicals and other agents called carcinogens. Common carcinogens include tobacco smoke, asbestos fibers, and radiation. The damaged genes, known as oncogenes, are usually repaired; however, if they are repeatedly exposed to a carcinogen then, over time, the repair process is likely to fail. If this happens, the affected cell begins to function abnormally.

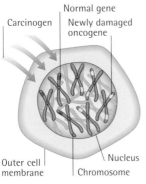

Carcinogen
Normal gene
Newly damaged oncogene
Outer cell membrane
Nucleus
Chromosome

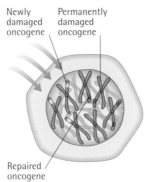

Newly damaged oncogene
Permanently damaged oncogene
Repaired oncogene

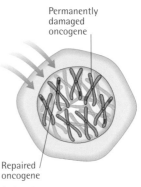

Permanently damaged oncogene
Repaired oncogene

1 Damage from carcinogens
Carcinogens enter the cell, damaging genes on the chromosomes. Newly damaged oncogenes are usually repaired.

2 Damage accumulates
The effects of repeated damage become cumulative. Over time, some of the oncogenes are no longer repaired.

3 Cell becomes cancerous
If oncogenes fail to be repaired, the cell can no longer function normally and eventually becomes cancerous.

Formation of a tumor

When a cancerous cell divides, it passes on its damaged oncogenes. The abnormal cells multiply rapidly to form a mass. A tumor's growth rate is measured by the time it takes for its cells to double in number. The formation of a detectable tumor takes 25–30 doublings.

Nucleus containing damaged oncogenes

CANCEROUS CELL

Damaged oncogenes passed on when cell divides

FIRST DOUBLING

Abnormal cells multiply, eventually forming solid tumor

SECOND DOUBLING

Spread of cancer in lymph

In a process called metastasis, cancerous cells can spread from their original (primary) site through lymph vessels. A growing tumor can damage nearby lymph vessels, allowing cancerous cells to enter a vessel and travel in the fluid. If a cell is trapped in a lymph node it may start to form a new (secondary) tumor.

1 Tumor breaches lymph vessel
As a tumor grows, it may invade lymph vessels. If cancerous cells split from the tumor they can be transported in the lymph.

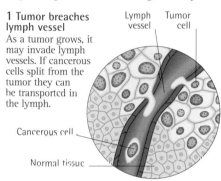

Lymph vessel · Tumor cell · Cancerous cell · Normal tissue

2 Cancerous cells in lymph node
Cancerous cells from the primary tumor may lodge in a lymph node and multiply. Immune cells in the node may slow the spread of the cancer.

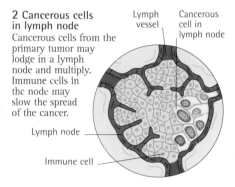

Lymph vessel · Cancerous cell in lymph node · Lymph node · Immune cell

Spread of cancer in blood

A cancer established in one part of the body can spread through the bloodstream to new sites, particularly to areas richly supplied with blood, such as the brain, liver, lungs, and bones. In this process, called metastatis, cancerous cells detach from the original (primary) tumor and form new (secondary) tumors elsewhere.

1 Tumor breaches blood vessel
If a tumor breaches the wall of a blood vessel, cancerous cells may become detached and pass into the bloodstream.

Blood vessel · Tumor cell · Cancerous cell

2 Cancerous cells in capillary
A cancerous cell carried in the bloodstream can lodge in a capillary (small blood vessel), where it may develop into another tumor.

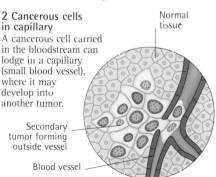

Normal tissue · Secondary tumor forming outside vessel · Blood vessel

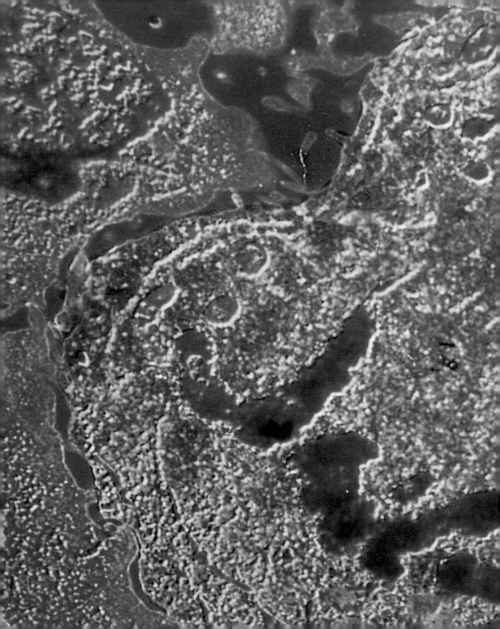

GLOSSARY
and
INDEX

Glossary 432
Index 438

Glossary

This glossary contains terms for which an easily located definition may be helpful. The entries are arranged alphabetically; terms in *italics* refer to other entries within the glossary.

A

Abscess
A walled cavity, containing pus, that is surrounded by inflamed or dying tissue.

Accommodation
The process by which the eyes adjust to focus on nearby or distant objects.

Acute
Medical conditions that begin abruptly and last for a short time. See also *Chronic*.

Adenoids
A collection of lymphatic tissue on each side of the back of the upper part of the throat; part of the body's immune system.

AIDS (acquired immune deficiency syndrome)
A syndrome that occurs after infection with *HIV* (the human immunodeficiency *virus*). AIDS is spread by sexual intercourse or infected blood and results in loss of resistance to infections and some *cancers*.

Allergen
Any substance causing an allergic reaction in a person previously exposed to it.

Alveoli
Tiny air sacs in the lungs through the walls of which gases diffuse into and out of blood during *respiration*.

Alzheimer's disease
A progressive *dementia*, due to loss of nerve cells in the brain, that affects more than 10 percent of people over 65.

Anemia
A disorder in which the amount of *hemoglobin* in the blood is reduced.

Aneurysm
A swelling of an *artery* caused by pressure of blood flow through a weakened section of the vessel wall.

Angina
Pain or tightness in the center of the chest that is usually brought on by exercise; caused by an inadequate supply of blood to the heart muscle.

Antibody
A soluble protein that attaches to harmful microorganisms in the body, such as *bacteria*, and helps to destroy them.

Aorta
The central and largest *artery* of the body, arising from the heart's left *ventricle* and supplying oxygenated blood to all other arteries except the *pulmonary artery*.

Aortic valve
A triple-cusped valve at the origin of the *aorta* that allows blood to leave the left *ventricle* of the heart but prevents backward flow.

Appendix
The wormlike structure attached to the initial part of the large intestine. It has no known function.

Aqueous humor
The fluid that fills the front chamber of the eye between the back of the *cornea* and the front of the iris and lens.

Arrhythmia
An irregular heartbeat that is caused by a defect in the electrical impulses or pathways that control heart contractions.

Arteriole
Small terminal branch of an *artery* leading to even smaller *capillaries*.

Artery
An elastic, muscular-walled tube that transports blood away from the heart to all other parts of the body.

Arthritis
Inflammation in one or more joints that causes pain and restriction of movement.

Articulation
A joint, or the way in which jointed parts are connected.

Asthma
A disease featuring variable narrowing of the air tubes so that breathing becomes intermittently difficult.

Atherosclerosis
A degenerative disease of *arteries* in which raised plaques of fatty material limit blood flow and cause *blood clots*.

Atrium (pl. atria)
The two thin-walled upper chambers of the heart.

Autoimmune disease
A disease caused by a defect in the immune system, which reacts against the body's own tissues and organs.

Autonomic nervous system
The part of the nervous system that controls unconscious functions such as heart rate. It has two divisions: *sympathetic* and *parasympathetic*.

Axon
The long, fiberlike process, or projection, of a nerve cell that conducts nerve impulses to or from the cell body. Bundles of many axons form *nerves*.

B

Bacteria
Types of microorganisms consisting of one cell. Only a few of the many different species cause disease.

Basal ganglia
Paired masses of nerve cell bodies, or nuclei, that lie deep in the brain and are concerned with the control of movement.

Benign
Usually mild, tending not to spread. See also *Malignant*.

Bile
A greenish-brown fluid produced by the *liver* and stored in the *gallbladder*. Bile helps the digestion of fats.

Biliary system
The network of bile vessels formed by the ducts from the *liver* and the *gallbladder*, and the gallbladder itself.

Blood clot
A meshwork of *fibrin*, *platelets*, and blood cells that forms when a blood vessel is damaged.

Boil
An inflamed, pus-filled area of skin. Boils usually form inside an infected hair follicle.

Bolus
A chewed-up quantity of food that can be swallowed and passed through to the stomach.

Bone marrow
The fatty tissue within bone cavities, which may be red or yellow. Red bone marrow produces *red blood cells*.

Bradycardia
A slow heart rate. This is normal in athletes but may signal disorders in others.

Brain stem
The lower part of the brain that houses the centers controlling vital functions such as breathing and heartbeat.

Bronchi
Air tubes that lead from the *trachea* to the lungs and divide into smaller bronchioles.

Bronchial tree
The *trachea* and branching system of air tubes in the lungs, consisting of progressively smaller *bronchi* and bronchioles.

Bronchitis
Inflammation of the lining of the *bronchi*, resulting in a cough that produces large amounts of sputum (phlegm).

C

Cancer
A localized growth from uncontrolled cell reproduction, which can spread (if untreated) to other parts of the body.

Capillary
One of the tiny blood vessels that link the smallest *arteries* and smallest *veins*.

Carcinoma
A *cancer* of a surface layer (epithelium). Carcinomas commonly occur in the skin, the breast, the large intestine, the linings of the air tubes, the *prostate gland*, and the *uterus*.

Cartilage
A tough, fibrous connective tissue, also known as gristle.

Central nervous system
The brain and spinal cord. The central nervous system receives and analyzes sensory data and then initiates a response.

Cerebellum
The region of the brain behind the *brain stem*. The cerebellum is concerned with balance and the control of fine movement.

Cerebrum
The largest part of the brain, which is made up of two hemispheres. The cerebrum contains the nerve centers for thought, personality, the senses, and voluntary movement.

Cholecystitis
Inflammation of the *gallbladder* that is commonly the result of obstructed outflow of *bile* due to a *gallstone*.

Chromosomes
Threadlike structures that are present in all nucleated body cells and carry the genetic code for the formation of the body. A normal human body cell carries 46 chromosomes arranged in 23 pairs.

Chronic
Persistent medical conditions that usually cause some long-term deterioration in the body. See also *Acute*.

Cirrhosis
Replacement of healthy *liver* tissue by bands of scar tissue, resulting in hardening and impairment of function. Cirrhosis may be caused by excessive alcohol consumption.

Cochlea
The coiled structure in the inner ear containing the organ of Corti, which converts sound vibrations into nerve impulses for transmission to the brain.

Collagen
An important structural protein that is present in bones, *tendons*, *ligaments*, and other connective tissues. Collagen fibers are twisted into bundles.

Colon
The major part of the large intestine that extends from the cecum to the rectum. Its main function is to conserve water in the body by absorbing it from the bowel contents.

Congenital
Present at birth. Congenital disorders may be hereditary or may have occurred as a result of disease or injury during fetal life or during birth.

Cornea
The transparent dome at the front of the eyeball that is the eye's main focusing lens.

Coronary
A term meaning "crown." Refers to the *arteries* that encircle and supply the heart with blood.

Corpus callosum
The wide curved band in the brain of about 20 million nerve fibers that connects the two hemispheres of the *cerebrum*.

Cranial nerves
The 12 pairs of nerves that emerge from the brain and *brain stem*. They include the nerves for smell, sight, head and eye movement, facial movement and sensation, hearing, and taste.

Crohn's disease
A *chronic* inflammatory disease of the gastrointestinal tract that can cause pain, fever, diarrhea, and loss of weight.

Cyst
A walled cavity, commonly spherical, filled with secreted fluid or semi-solid matter; some types occur in the skin. Cysts are usually *benign*.

Cystitis
Inflammation of the bladder, caused by infection. Produces frequent, painful urination and sometimes incontinence.

D

Dementia
The loss of mental powers and memory as a result of degenerative brain disease or narrowed *arteries* supplying blood to the brain. See also *Alzheimer's disease*.

Dermis
The inner layer of skin, made of connective tissue and containing various structures such as blood vessels, nerve fibers, hair follicles, and sweat glands.

Diaphragm
The dome-shaped muscular sheet separating the chest from the abdomen. On contraction, the dome flattens, increasing the volume of the chest.

Diastole
The period in the heartbeat cycle when the *ventricles* are relaxed; alternates with *systole*.

Dopamine
A *neurotransmitter* involved in the control of body movement. Dopamine is related to a substance used in the treatment of *Parkinson's disease*.

Duodenum
The first part of the small intestine, into which stomach contents empty. Ducts from the *gallbladder*, *liver*, and *pancreas* enter the duodenum.

E

Ectopic pregnancy
Implantation of the fertilized egg outside the uterine lining, commonly in the fallopian tube.

Embolus
Material such as *blood clots*, air bubbles, fat, or *tumor* cells carried in the blood and causing blockage of an *artery*.

Embryo
The developing baby from conception until the eighth week of pregnancy. See also *fetus*.

Endorphin
A substance produced within the body that relieves pain.

Enzyme
A protein acting as a catalyst to accelerate a chemical reaction.

Epidermis
The outer layer of the skin; its cells flatten and become scale-like towards the surface.

Epiglottis
A flap of *cartilage* located at the entrance of the *larynx*, which covers the opening during swallowing and prevents food or liquid from entering the airways.

Epilepsy
Episodes of unregulated electrical discharge in a specific area of, or throughout, the brain

Esophagus
The muscular tube down which food is carried, extending from the *pharynx* to the stomach.

Estrogen
A female *sex hormone*, produced mainly in the ovaries. Stimulates the development of secondary sexual characteristics in women and regulates menstrual cycle.

F

Fetus
The developing baby from about the eighth week after fertilization until the time of birth. See also *Embryo*.

Fever
A body temperature registering above 97.6°F (37°C), measured in the mouth, or 99.8°F (37.7°C), measured in the rectum.

Fibrin
An insoluble protein that is converted from the blood protein fibrinogen to form a fibrous network – a stage in the creation of a *blood clot*.

Fibroid
A *benign* tumor of fibrous and muscular tissue growing in the wall of the *uterus*, mostly occurring in women over 30. Fibroids are often multiple and can cause discomfort.

Fistula
An abnormal channel between any part of the interior of the body and the surface of the skin or between two internal organs.

G

Gallbladder
The small, fig-shaped bag, lying under the *liver*, in which *bile* secreted by the liver is stored.

Gallstones
Solid masses of cholesterol, calcium, and *bile* that form in the *gallbladder*. They vary in size and are more likely to occur in women than in men.

Gastric juice
A mixture containing digestive *enzymes* and hydrochloric acid, produced by the cells of the stomach lining.

Gastritis
Inflammation of the stomach lining from any cause, such as infection, alcohol, or irritating food substances.

Gastrointestinal tract
Part of the digestive system: a muscular tube consisting of the mouth, *pharynx*, *esophagus*, stomach, and intestines.

Gene
A section of a *chromosome* that is the basic unit of inheritance. Genes contain information for growth and development.

Glaucoma
A rise in pressure of the fluids within the eye that, if untreated, causes internal damage and can permanently affect vision.

Glial cell
A type of cell that provides support for *neurons*.

Gray matter
The regions of the brain and spinal cord that are composed primarily of *neuron* cell bodies, as opposed to their projecting fibers, which form the *white matter*.

H

Heart valves
The four structures of the heart that allow the passage of blood in one direction only through the heart chambers.

Hematoma
An accumulation of blood anywhere in the body due to a ruptured blood vessel.

Hemoglobin
The protein that fills the *red blood cells* and combines with oxygen, which it transports from the lungs to all parts of the body.

Hemophilia
An inherited bleeding disorder that is caused by shortage of a particular blood protein.

Hemorrhage
Escape of blood from a blood vessel, usually as the result of an injury.

Hemorrhoids
Ballooning of veins in the anus (external hemorrhoids) or lower rectum (internal hemorrhoids).

Hepatitis
Inflammation of the *liver*, usually due to a viral infection, alcohol, or toxins. Symptoms include *fever* and *jaundice*.

Hippocampus
Part of the brain concerned with learning and long-term memory.

HIV (human immunodeficiency virus)
This *virus*, which is the cause of *AIDS*, destroys certain cells of the immune system, thereby seriously undermining its effectiveness.

Homeostasis
Active processes by which an organism maintains a constant internal environment despite external changes.

Hormones
Chemicals, released into the blood from some glands and tissues, that act on receptor sites elsewhere in the body.

Hypothalamus
A small structure at the base of the brain where the nervous and hormonal body systems interact.

I–J

Ileum
The final part of the small intestine, in which absorption of nutrients is mainly completed.

Immune deficiency
Failure of the immune system's function from causes such as *AIDS*, *cancer* therapy, or aging.

Interferons
Protein substances produced by cells to protect against viral infections and some *cancers*.

Irritable bowel syndrome
Recurrent abdominal pain, bloating, and intermittent diarrhea or constipation.

Jaundice
Yellowing of the skin and whites of the eyes due to deposition of *bile* pigment in the blood, usually because of *liver* disease.

K

Kaposi's sarcoma
A slow-growing *tumor* of blood vessels that affects many people with *AIDS*. It is characterized by firm, bluish brown nodules on the skin and can affect internal organs.

Keratin
A hard protein found in hair, nails, and the skin's outer layers.

Kidneys
The two reddish-brown, bean-shaped structures at the back of the abdominal cavity that filter the blood and remove wastes.

Killer T cells
White blood cells that can destroy damaged, infected, or *malignant* body cells.

L

Larynx
The structure in the neck at the top of the *trachea*, also known as the voice box, that contains the vocal cords.

Leukemia
A group of blood disorders in which abnormal *white blood cells* proliferate in bone marrow, crowding out healthy cells.

Ligament
A band of tissue consisting of *collagen*, a fibrous, elastic protein. Ligaments support bones, mainly at joints.

Limbic system
A part of the brain that plays a role in automatic body functions, emotions, and the sense of smell.

Liver
The large organ in the upper right abdomen that performs vital chemical functions, including processing of nutrients from the intestines; synthesis of sugars, proteins, and fats; detoxification of poisons; and conversion of waste to *urea*.

Lymphatic system
A network of transparent lymph vessels and *lymph nodes*. The lymphatic system returns excess tissue fluid to the circulation and helps to combat infections and *tumor* cells.

Lymph nodes
Small, oval glands that occur in groups along lymph vessels and contain *white blood cells*, which have a role in fighting infection.

Lymphocyte
A type of small *white blood cell* that is a vital part of the body's immune system. Lymphocytes provide protection against viral infections and *cancer*.

M

Malignant
Tending to spread and to result in death unless effectively treated. See also *Benign*.

Medulla
The inner part of an organ, such as the *kidneys*, or adrenal glands. The term also refers to the part of the *brain stem* lying above the start of the spinal cord, in front of the *cerebellum*.

Meiosis
The stage in the formation of sperm and egg cells when chromosomal material is randomly redistributed and the number of *chromosomes* is reduced to 23 instead of the 46 found in other body cells.

Meninges
Three membranes (the pia mater, the arachnoid, and the dura mater) that surround and protect the brain and spinal cord.

Meningitis
Inflammation of the *meninges*, usually as a result of a viral or bacterial infection.

Menopause
The end of the reproductive period in a woman when egg development in the *ovaries* and menstruation have ceased.

Metabolism
The sum of all the physical and chemical processes that take place in the body.

Metastasis
The spread or transfer of disease, especially *cancer*, from its original site to another, where the disease process continues.

Migraine
The effects of narrowing and then widening of some of the *arteries* of the scalp and brain, usually on one side. Symptoms include visual disturbances, nausea, and severe headache.

Miscarriage
A spontaneous ending of a pregnancy before the *fetus* is mature enough to survive outside the *uterus*.

Mitochondria
A class of microscopic cell organ concerned with provision of energy for the various cell functions and containing genetic material.

Mitosis
The process of division of most body cells. In the process, one cell produces two daughter cells, each of which has the identical genetic makeup of the parent cell.

Mitral valve
The valve on the left side of the heart lying between the upper and lower chambers.

Mole
A flat, raised, and/or hairy *congenital* blemish, birthmark, growth, or pigmented spot on the skin; also called a nevus.

Motor cortex
The part of the surface layer of each hemisphere of the *cerebrum* in which voluntary movement is initiated. It can be mapped into various areas linked to particular parts of the body.

Motor neuron
A type of *neuron* (nerve cell) that carries impulses to muscles to produce movement.

Motor neuron disease
A group of rare disorders in which *motor neurons* suffer progressive destruction, resulting in loss of movement.

Mucous membrane
The soft, skinlike mucus-secreting layer lining the tubes and cavities, such as the respiratory tract, in the body.

Myocardium
The heart muscle, in which the fibers make up a network that will contract spontaneously.

Myofibrils
Cylindrical elements within muscle cells (fibers). Each myofibril consists of thinner filaments, which move to produce muscle contraction.

N

Nephron
A unit of the *kidney* that consists of a filtration capsule, and a series of tubules. Nephrons filter waste products from the blood and reabsorb salts and water as required. Each kidney contains about one million nephrons.

Nerves
Filamentous projections of individual *neurons* (nerve cells). Nerves carry electrical impulses to and from the brain and spinal cord and the rest of the body.

Neuron
A single nerve cell that transmits electrical impulses.

Neurotransmitter
A chemical that is released from a nerve fiber, creating an electrical "message" that passes from one nerve to another or to a muscle.

O

Olfactory nerves
The *nerves* of smell that run from the nose directly into the underside of the brain.

Optic nerves
The two nerves of vision that run from the *retina* of each eye and carry visual information to the brain. Each optic nerve is made up of about one million nerve fibers.

Ossicle
One of three tiny bones of the middle ear. These are called the malleus, incus, and stapes, and convey vibrations from the eardrum to the inner ear.

Osteoarthritis
A degenerative joint disease that is characterized by damage to the *cartilage*-covered, load-bearing surfaces in the joint.

Osteon
The rod-shaped unit, also called a Haversian system, that is the building block of hard bone.

Osteoporosis
Loss of bone substance due to bone being reabsorbed faster than it is formed. Bones become brittle and easily fractured.

Otitis media
Inflammation in the middle ear cavity, often caused by bacterial or viral infection that has spread from the nose or throat.

Ovaries
The two structures, on each side of the *uterus*, that produce eggs and female *sex hormones*.

Ovulation
The release of an egg each month from a follicle that has matured within the ovary. Ovulation usually occurs in the middle of the menstrual cycle.

Ovum
The egg cell, which develops in the ovary and, if fertilized, may become an *embryo*.

P

Pancreas
A gland situated behind the stomach that secretes digestive *enzymes* and *hormones* that regulate blood glucose levels.

Pancreatitis
Inflammation of the *pancreas* that causes severe pain in the upper abdominal area.

Paralysis
Loss of movement (and, in some cases, sensation) due to a *nerve* or a muscle disorder.

Parasympathetic nervous system
One of the two divisions of the *autonomic nervous system*. It maintains and restores energy, for example by slowing the heart rate during sleep.

Parkinson's disease
A progressive neurological disorder that features involuntary tremor, slowness of movements, muscle rigidity, shuffling walk, and a stooping posture. The intellect may also be affected in late stages of the disease.

Pelvic inflammatory disease
Persistent infection of the female reproductive organs. It may have no obvious cause but often occurs after a sexually transmitted disease.

Pelvis
The basinlike ring of bones to which the lower *spine* is attached and with which the thigh bones articulate.

Peptic ulcer
Erosion of the membrane lining the *esophagus*, stomach, or *duodenum* by excess stomach acid and digestive *enzymes*. It is most commonly the result of infection with a bacterium.

Pericardium
The tough, fibrous two-layered sac that encloses the heart and the roots of the major blood vessels that emerge from it.

Peripheral nervous system
All the *nerves* and their coverings that fan out from the brain and spinal cord, linking them with the rest of the body. The system consists of *cranial nerves* and spinal nerves.

Peristalsis
A coordinated succession of contractions and relaxations of the muscular wall of a tubular structure, such as the intestines, that moves the contents along.

Peritoneum
A double-layered membrane that lines the inner wall of the abdomen, covers the abdominal organs, and secretes fluid to lubricate their movement.

Peritonitis
Inflammation of the *peritoneum*, which may be caused by bacterial infection, pancreatic *enzymes*, *bile*, or chemicals.

Pharynx
The passage leading from the back of the nose and mouth to the *esophagus*. It consists of the nasopharynx, oropharynx, and laryngopharynx.

Pituitary gland
A pea-sized gland which hangs from the underside of the brain and secretes *hormones* that help to control many of the other glands in the body.

Placenta
The disk-shaped organ that forms in the *uterus* during pregnancy. It links the blood supplies of the mother and baby via the *umbilical cord*.

Plasma
The fluid part of blood from which all cells have been removed. Contains proteins, salts, and various nutrients that regulate the volume of blood.

Platelets
Fragments of large cells, called megakaryocytes, that are present in large numbers in blood and are needed for *blood clotting*.

Pleura
A double-layered membrane. The inner layer covers the lungs; the outer layer lines the chest cavity. Lubricating fluid enables movement between the two.

Pneumonia
Inflammation of the smaller air passages and *alveoli* of one or both lungs. Most commonly due to viral or bacterial infection; can also be caused by other microorganisms such as fungi or protozoa.

Pneumothorax
The presence of air in the space between the two layers of the *pleura*, causing the inward collapse of a lung.

Primary
A term for a disorder that has originated in the affected organ or tissue. See also *Secondary*.

Progesterone
A female *sex hormone* that is secreted by the *ovaries* and the *placenta*. Progesterone prepares the uterine lining to receive and retain a fertilized egg.

Prostaglandins
A group of fatty acids made naturally in the body that act much like *hormones*.

Prostate gland
The structure at the base of the male bladder that secretes some of the fluid in semen.

Psoriasis
A common skin disease of unknown cause that features thickened patches of red, scaly, and inflamed skin.

Pulmonary artery
The *artery* conveying blood from the heart to the lungs for reoxygenating.

Pulse
The rhythmic expansion and contraction of an *artery* as blood is forced through it.

Pus
A yellowish green fluid that forms at the site of a bacterial infection. Contains *bacteria*, dead *white blood cells*, and damaged tissue.

R

Red blood cells
Small, biconcave disks, also known as erythrocytes, that are filled with *hemoglobin*. Each cubic millimeter of blood contains about five million red blood cells.

Respiration
The process by which oxygen is conveyed to, and carbon dioxide is removed from, the body cells.

Reticular formation
A network of nerve cells scattered throughout the *brain stem* that are concerned with alertness and direction of attention to external events.

Retina
A light-sensitive layer lining the inside of the back of the eye. It converts optical images to nerve impulses, which travel to the brain via the *optic nerve*.

Rheumatoid arthritis
A disorder that causes joint deformity, thought to be an autoimmune disease.

S

Saliva
A watery fluid secreted into the mouth, via ducts from the salivary glands, to aid chewing, tasting, and digestion.

Sarcoma
A *cancer* that arises from connective tissue (such as cartilage), muscle, fibrous tissue, or blood vessels.

Secondary
A term for a disorder that follows or results from another disorder (the *primary* disorder).

Septal defect
An abnormal opening in the central heart that allows blood to flow from the right side to the left side or vice versa.

Sex hormones
Steroids that bring about the development of secondary sex characteristics. Sex hormones also regulate sperm and egg production and menstrual cycle.

Sinoatrial node
A cluster of specialized muscle cells located in the heart's right *atrium* that act as its natural pacemaker.

Sphincter
A muscle ring, or thickening of the muscle coat, surrounding an opening in the body, such as the outlet between the stomach and the duodenum.

Spine
The column of 33 ringlike bones, called vertebrae, that divides into seven cervical, 12 thoracic, five lumbar, and the four fused vertebrae of the sacrum and coccyx.

Sympathetic nervous system
One of the two divisions of the *autonomic nervous system*. In conjunction with the *parasympathetic nervous system*, it controls many involuntary activities of the glands, organs, and other parts of the body.

Systole
The period in the heartbeat cycle when first the *atria* and then the *ventricles* are contracted. It alternates with the relaxation period called *diastole*.

T

Tendinitis
Inflammation of a *tendon*, usually due to injury, that causes pain and tenderness.

Tendon
A strong band of *collagen* fibers that joins muscle to bone and transmits the pull caused by muscle contraction.

Tenosynovitis
Inflammation, usually from excessive friction, of the inner lining of a *tendon* sheath.

Testes
The pair of sperm-producing male sex glands that are suspended in the scrotum.

Testosterone
The principal *sex hormone* produced in the *testes* and in small amounts in the adrenal cortex and *ovaries*.

Thalamus
A mass of *gray matter* that lies deep within the brain. The thalamus receives and coordinates sensory information.

Thorax
The part of the trunk between the neck and the abdomen that contains the heart and lungs.

Thrombus
A *blood clot* that is usually the result of damage to the lining of a blood vessel.

Tonsils
Oval masses of lymphatic tissue, on the back of the throat, either side of the soft palate. The tonsils help to protect against childhood infections.

Trachea
The windpipe. A tube lined with a mucous membrane and reinforced with cartilage rings.

Transient ischemic attack
A "mini-stroke" that passes completely in 24 hours but can imply danger of a full stroke.

Tumor
A *benign* or *malignant* tissue swelling, especially an abnormal mass of cells resulting from uncontrolled multiplication.

U

Umbilical cord
The structure that connects the *placenta* and the *fetus*, thereby providing the immunological, nutritional, and hormonal link with the mother.

Urea
A waste product of protein breakdown, and the nitrogen-containing component of *urine*.

Ureter
One of the two tubes down which urine passes from the *kidneys* to the bladder.

Urethra
The tube that carries *urine* from the bladder to the exterior. The male urethra is much longer than that of the female.

Urinary tract
The waste system, consisting of the *kidneys*, *ureters*, bladder, and *urethra*, that forms *urine* and excretes it from the body.

Urine
The pale yellow fluid produced by the *kidneys* and excreted from the body via the *ureters*, bladder, and *urethra*.

Uterus
The hollow muscular organ of the female reproductive system in which the *fetus* develops.

V

Vagina
The highly elastic muscular passage from the *uterus* to the external genitals.

Vagus nerves
The tenth pair of *cranial nerves*. The vagus nerves help to control automatic functions such as heartbeat and digestion.

Vas deferens
One of two tubes that carry sperm from the *testes*. Each vas deferens joins a seminal vesicle (where sperm are mixed with fluid) to form an ejaculatory duct at the top of the *urethra*.

Vein
A thin-walled blood vessel that returns blood from body organs and tissues to the heart.

Vena cava
One of two large *veins* that empty into the right *atrium*.

Ventricles
The two lower chambers of the heart; also, the four cavities in the brain that are filled with cerebrospinal fluid.

Virus
A small infectious agent capable of invading and damaging cells and reproducing within them.

Vocal cord
One of two membrane sheets stretched across the *larynx* that vibrate to produce sounds when air passes between them.

W–Z

White blood cells
The colorless blood cells that play various roles in the body's immune system.

White matter
The regions of the brain and spinal cord made up of nerve fibers. See also *Gray matter*.

X chromosome
A sex *chromosome*. All the body cells of females have two X chromosomes.

Y chromosome
A sex *chromosome*, necessary for the development of male characteristics. Male cells have one Y and one X chromosome.

Zygote
The cell that is produced when an egg is fertilized by a sperm.

Index

Page numbers with the suffix g refer to the glossary.

A

abdomen 19, 24–25
abdominal cavity 25, 209
abdominal hernia 394
abdominal muscles 202, 205
abdominal wall 24
abducent nerve 125
abductor digiti minimi muscle 34, 35, 82, 84, 90
abductor hallucis muscle 34, 82
abductor pollicis brevis muscle 82, 90
abductor pollicis longus muscle 31
abnormal sperm 420
abrasion 310
abscess 432g
  brain 339
  breast 412
  liver 398
absence seizure 332
accessory nerve 124
accessory parotid gland 218
accessory saphenous vein 170
accommodation 155, 432g
Achilles tendon 34, 35, 84
acinar cells 163, 230
acne vulgaris 310
acoustic nerve 342
acquired immune deficiency syndrome (AIDS) 366, 367, 374, 375, 432g
acromion 19, 62, 328
acrosome 258
ACTH (adrenocorticotropic hormone) 160
active immunization 370
acute 432g
acute bronchitis 381
Adam's apple see laryngeal prominence
Addison's disease 373
adductor brevis muscle 82
adductor longus muscle 82
adductor magnus muscle 33, 84
adductor pollicis muscle 90
adenine 50
adenoid 189, 432g
adenosine diphosphate (ADP) 214
adenosine triphosphate (ATP) 46, 49, 214
adenoviruses 366
ADH (antidiuretic hormone) 161
adipose tissue 242
  see also fat
ADP (adenosine diphosphate) 214
adrenal gland 39, 130, 158, 160, 162, 239
adrenaline see epinephrine

adrenocorticotropic hormone (ACTH) 160
aging 296–299, 324, 350
agonist muscle 95
AIDS (acquired immune deficiency syndrome) 366, 367, 374, 375, 432g
air embolus 359
air passages 196–197
alcoholic hepatitis 396
alcoholic liver disease 396
aldosterone 162
allergen 372, 432g
allergic asthma 372, 389
allergic reaction 372, 389
alpha cells 163
alveolus 198, 199, 200, 382, 388, 432g
Alzheimer's disease 333, 432g
amino acids 46, 210, 211, 213
  in glomerular filtrate 245
  protein synthesis 50, 52
amniotic cavity 260, 261
amniotic fluid 262, 268
amniotic sac 269
amoebae 398
ampulla 150
ampulla of Vater 230, 231, 399
amygdala 136
amylase 211, 218
amyloid 333
anal canal 25, 27, 235
analgesic drugs 145
anal sphincter 27, 28, 29, 235, 240
anaphase 53
anastomosis, muscle 93
anconeus muscle 31, 85
anemia 432g
androgens 165
aneurysm 335, 360, 432g
angina 347, 432g
angular artery 13
angular vein 13, 171
ankle
  bone see talus
  injury 317, 322
anococcygeal ligament 28
Anopheles mosquito 368
ansa cervicalis nerve 13
antagonist muscle 95
anterior chamber of the eye 153
anterior fissure of the spinal cord 120
anterior jugular vein 17
anterior lateral malleolar artery 35
anterior lobe of the pituitary 160
anterior nucleus 137
anterior tibial artery 35, 170
anterior tibial vein 35, 170
antibiotics 365
antibody 52, 189–193, 370, 432g
antibody defenses 192
antidiuretic hormone (ADH) 161
antigens 192, 193, 366, 371
anus 28, 41, 208, 235, 405
aorta 171, 174–175, 178, 201, 346, 432g
  abdominal 25, 238
  arch of 172, 173, 177, 347

ascending 23
  blood circulation 180
  coarctation of 352
  descending 40, 171, 173
  fetal circulation 282
aortic ring 175
aortic valve 173, 178, 179, 350, 432g
Apgar score 283
aponeurosis 18
appendicitis 403
appendicular skeleton 60
appendix 24, 25, 208, 232, 403, 432g
aqueous humor 153, 343, 432g
arachnoid 115, 116, 120, 338, 340
arachnoid granulations 115, 116
arcuate artery 170, 243
arcuate nucleus 137
arcuate vein 243, 244
areola 255, 266
arm 30–31, 62
arrector pili muscle 54
arrhythmia 349, 356, 432g
arteriole 55, 182, 183, 190, 244, 432g
arteriovenous malformation 335
artery 101, 116, 117, 432g
  effect of age on 297
  structure 182–183
arthritis 323, 432g
articular cartilage 76, 78
  bone growth 74, 286
  degeneration 323, 324, 325
articular chondrocytes 324
articulation 432g
artificial heart valves 351
arytenoid cartilage 16, 204
ascending aorta 23
ascending colon 24, 25, 232
ascending nerve tracts 123
aspergillosis 369
aspirin 145
association areas of brain 142
asthma 372, 389, 432g
astrocyte 103, 117
asymptomatic carrier 375
atheroma 346
atherosclerosis 346, 347, 432g
atlas 69, 78
ATP (adenosine triphosphate) 46, 49, 214
atria 432g
atrial fibrillation 349
atrial septal defect 353
atrial systole 179
atriopeptin 159
atrioventricular node 176, 177, 349
atrium 172–173, 176, 179, 349
  fetal circulation 282–283
auditory association cortex 142
auditory impulses 138
auditory meatus 15
auricular nerve 12
auriculotemporal nerve 12, 99

autoimmune disorders 373, 432g
  fibrosing alveolitis (IPF) 386
  inflammatory bowel disease 401
  multiple sclerosis 332
  pericarditis 355
  rheumatoid arthritis 323
  vitiligo 313
autonomic nervous system (ANS) 100, 131, 133, 177, 432g
axial skeleton 60
axillary artery 16, 171
axillary nerve 30, 99
axillary nodes 189, 255
axillary vein 16, 171
axis 78
axon 48, 102, 103, 106, 161, 432g
  spinal cord 120
  white matter 111

B

baby
  delivery 276–278
  growth and development 286–289
  heart disorders 352–353
  newborn 280–281
bacteria 364, 398, 432g
  drug-resistant strains 365
bacterial meningitis 339
balance 150–151
ball-and-socket joint 79
basal cell carcinoma 312
basal cell layer 55, 147, 312, 391
basal ganglia 111, 136, 333, 432g
base triplet 50, 52
basilar artery 334
basilic vein 31, 171
basophil 184
B cell (B lymphocyte) 192–193
benign 432g
benign tumors 427
berry aneurysm 335
beta cell 163
bicarbonate 232, 245, 246
biceps brachii muscle 18, 19, 30, 38, 83
biceps femoris muscle 32, 33, 34, 84
bile 208, 229, 231, 432g
bile duct 228, 229, 231
bile pigment 400
bile salts 211, 215
biliary colic 400
biliary system 231, 432g
binocular vision 154
bipolar nerve cell 103
birth canal 274, 275
  see also vagina
blackhead 310
bladder 238, 247
  female 28, 42, 240, 252, 268
  male 26, 42, 241, 251
  nerve pathways 135
  see also incontinence
bladder sphincter 130, 132, 247
blastocyst 259, 260, 264

bleeding
  in brain 335
  in skull 340
blood
  cells 38, 185, 246
  circulation 40, 170–171, 180
  components 184
blood-brain barrier 117
blood clot 346, 358, 359, 432g
blood pressure 42, 335, 361
blood vessels 130, 180, 181, 182-183
body proportions, changes in childhood 287
boils 311, 432g
bolus 211, 216, 220, 221, 432g
bone 60–75
  aging 296, 317, 318–319
  cancellous 72, 91
  cortical 72, 73, 318
  foot 64
  growth 74, 160, 319
  hand 64
  lever systems 94
  reabsorption 319
  repair 75
  shapes 63
  spongy 72, 91, 296, 316, 318
  structure 72–73
  tendon links 91
  see also fractures
bone marrow 72, 188, 192, 286, 432g
  see also red bone marrow
bone tissue 61
Bowman's capsule 244, 245
brachial artery 30, 171
brachialis muscle 30, 83
brachial plexus 12, 16, 99, 127
brachial vein 171
brachiocephalic vein 17, 22, 23
brachioradialis muscle 30
bradycardia 432g
brain 39, 66, 98, 99
  abnormal electrical activity 332
  blood-brain barrier 117
  blood supply 117
  development 118–119
  hemispheres 12, 109
  information centers 112
  inner structure 110
  nerve tracts 112
  outer structure 108–109
  peripheral nervous system 100
  ventricles 115
brain abscess 339
brain infections 338
brain stem 143, 197, 205, 221, 432g
  functions 138
  information centers 113, 140, 197
  inner brain structures 110
  nerve pathways 112
  parasympathetic nerves 133
  sympathetic nerves 131
brain tumor 339
Braxton-Hicks contractions 271

breast 161, 255
  disorders 412
  newborn 280
  puberty 291
breast abscess 412
breathing 202, 219
  pressure changes 203
breech delivery 269
Broca's area 142
bronchial lining 381, 391
bronchial tree 197, 433g
bronchial tubes 130, 132
bronchiole 198–199, 200, 381, 382, 389
bronchitis 381, 388, 433g
bronchopneumonia 382
bronchus 22, 196, 197, 433g
  in acute bronchitis 381
  in chronic bronchitis 388
  inflammatory response 191
  smoke damage 391
buccinator muscle 38, 87
bundle-branch block 349

C

caged-ball valve 351
calcaneal tendon 34, 35
calcaneal tuberosity 34
calcaneus 60, 62, 64, 91
calcium 215, 319
  regulation 159, 164
calf 35
callus 75
calyx 242
canaliculi 229
cancellous bone 72, 91
cancer 424, 429, 433g
  breast 412
  cervix 414
  colon 405
  lung 390
  ovary 415
  pancreas 399
  prostate 417
  skin 312
  spread of 429
  testicular 416
cancerous tumors 426–427
candidiasis 369
canine teeth 216, 287
capillary 92, 117, 174, 201, 335, 433g
  network 183, 200
capitate 64
capsule
  Bowman's 244, 245
  fibrous 190, 427
  internal 112, 113
  joint 76
  lymph node 190
  renal 242
carbohydrates 210, 213
carbon dioxide 200, 201, 214, 233
carcinogens 390, 428
carcinoma 426, 433g
cardiac arrest 348
cardiac muscle 38, 93, 348
  disease 354

cardiac nerves 177
cardiomyopathy 354
cardioregulatory center 177
cardiovascular system 40, 168–185
  disorders 344–361
carotid artery 12, 16, 22, 23, 334
carotid canal 62
carpals 61, 64
carpal tunnel syndrome 327
carrier, recessive gene 305
cartilage 68, 69, 71, 74, 286, 433g
  articular 74, 76, 78, 286
  arytenoid 16, 204
  corniculate 16
  costal 61, 65
  cricoid, 15, 16, 164
  end of long bones 73
  meniscus 76, 322
  nasal 14
  thyroid 15, 16, 17, 164, 204
  torn 322
  tracheal 16, 204
cataract 298
cauda equina 121
caudate nucleus 111
cecum 24, 25, 224, 232, 403
celiac trunk 25
cell
  body 102, 103, 107
  division 53, 300–301
  nucleus 46, 48, 50, 53
  structure 46–47
  transport mechanisms 49
  types 48
cell membrane 47, 49, 50
  permeability 49
cellular defenses 193
cellulose 215
cementum 217
central canal 114, 120
central nervous system (CNS) 98, 125, 433g
  functions 140–155
central sulcus 108
centriole 46, 300
centromere 50
cephalic vein 16, 19, 31, 171
cerebellum 114, 121, 127, 140, 143, 433g
  brain development 118–119
  outer brain structure 109
cerebral aqueduct 115
cerebral artery 334, 336
cerebral cortex, 108, 111, 116
  brain development 119
  memory 143
  movement and touch 141
  nerve pathways 112
  parasympathetic nerves 133
  sympathetic nerves 131
cerebral hemisphere 12, 109, 117
  embryonic 118
cerebrospinal fluid 114–115, 116, 120
  spinal cord protection 122
cerebrum 108, 111, 116, 121, 433g
  brain development 118–119
  nerve pathways 113
cervical artery 12

cervical canal 29
cervical cancer 414
cervical dilation 273
cervical nerves 127, 128
cervical obstruction 423
cervical plexus 127
cervical vertebrae 21, 61, 68, 70
cervix 29, 253, 258, 259, 268, 277
  cancer of 414
  childbirth 271, 273, 285
  gonococcal infection of 418
  infertility 420
chambers of the heart 172–173
characteristics, genetic 302
cheek bones see zygomatic bones
cheek muscle 38
chest see thorax
chest pain 347, 348, 355, 385
chewing 216, 218
chickenpox 366
childbirth 268–279, 409
  gonococcal infection 418
child development 288–289
chloride 232
  ions 107
cholangitis 400
cholecystitis 400, 433g
cholesterol 215, 346, 359, 400
chondrocyte 74
chordae tendineae 175, 178
chorda tympani 146
chorionic villus 264
choroid 152, 153
choroid plexus 114
chromatin 46, 53
chromosomal abnormality 352
chromosome 47, 50, 53, 259, 303, 428, 433g
chronic 433g
chronic bronchitis 388
chyme 226, 227, 232, 233
chymotrypsin 211
cigarette smoke 390, 391
cilia 147, 391
ciliary body 152, 153
ciliary muscle 153, 155
cingulate gyrus 110, 136
circulation 40, 170–171, 180, 356
  effect of age on 297
  fetal 282
  lymph 188–189
  newborn 283
cirrhosis 396, 398, 433g
cisterna chyli 189
clavicle 16, 19, 61, 328
  fractured 317
clitoris 28, 29, 132, 252, 254
closed fracture 316
coarctation of the aorta 352
coccygeal nerves 126
coccygeal plexus 126
coccygeal vertebrae 21
coccyx 21, 60, 62, 70
  female 28, 29
  male 27, 241, 251
cochlea 146, 298, 342, 433g
cochlear nerve 149, 342
cold, common 366, 379
cold sores 366

colic vein 181
collagen 433g
  fibers 91, 319, 358
collapsed lung 385
collar bone see clavicle
collateral ganglion 134
collecting duct 243, 244
collecting tubule 244
Colle's fracture 317
colon 24, 25, 29, 181, 232,
  252, 433g
  cancer 405
  disorders 401–405
  see also large intestine
colonic movement 234
colostrum 285
color blindness 305
columnar cells 57, 391
comminuted fracture 316
commissure 111
common aneurysm 360
common bile duct 163, 228, 231,
  400
common carotid artery 12, 16, 22,
  23, 171
common cold 366, 379
common hepatic artery 25, 171
common hepatic duct 228, 231
common iliac artery 171
common iliac node 189
common iliac vein 171
common palmar digital nerve 98
common peroneal nerve 33, 98
communication 88–89
complement system 193
complete breech 269
compound fracture 316
conducting fibers 176, 349
conductive deafness 342
condyle 60
cones of the retina 155
congenital 433g
congenital arterial defects 335
congenital heart defects 352–353
conjugation 365
conjunctiva 152, 153
conjunctivitis 366
connective fibrous tissue 55, 92,
  178
  tendons 90–91
consciousness 138
consolidation of memory 143
constrictive pericarditis 355
coracobrachialis muscle 19
coracoid process 19, 61
cornea 14, 152, 343, 433g
corniculate cartilage 16
corona radiata 112
coronary 433g
coronary artery 174, 346, 347,
  348, 349
coronary sinus 175
coronary system 174
coronary vein 174, 175
coronoid process 12
corpus callosum 110, 111, 136
corpus cavernosum 26, 250
corpus luteum 165, 252, 291,
  292
corpus spongiosum 26, 250

corpus striatum 333
corrugator supercilii muscle 83,
  87, 89
cortex
  adrenal 162
  cerebral 110, 111, 116
  renal 243
cortical bone 72, 73, 318
Corti, organ of 149
cortisol 162
costal cartilage 61, 65
coughing reflex 205, 219
cranial nerves 100, 112, 124–125,
  433g
cranial vault 66, 67
cremaster muscle 26, 251
cricoid cartilage 15, 16, 164
cricothyroid ligament 16
cricothyroid muscle 17
crista ampullaris 150, 151
Crohn's disease 401, 433g
croup 366
crown, tooth 217
crush fracture 316
cryptococcosis 369
cubital nodes 189
cuboid 64
cuboidal epithelium 57
cuneiform bones 60, 64
cupula 151
cusp, heart valve 178
cuticle 56
cuts 310
cyst 311, 325, 412, 433g
cystic duct 400
cystitis 408, 433g
cytoplasm 46, 50, 52, 53, 364,
  369
cytosine 50
cytoskeleton 47

D

deafness see hearing, loss of
deep brachial artery 30
deep femoral artery 170
deep inguinal node 188
deep muscles 82, 84–85, 86–87
deep palmar arch 171
deep peroneal nerve 35, 98
defecation 235
delivery 276–278
delta cell 163
deltoid muscle 83, 85, 86, 87, 95
  anatomical relationships 12,
  16, 19, 21, 30
dementia 304, 409, 433g
dendrite 102, 103
dental development 287
dental nerve 217
dentine 217
deoxygenated blood 173, 175,
  196, 200
depressor anguli oris muscle 87,
  89
depressor labii inferioris muscle
  87, 89
dermatones 128
dermatophytosis 369

dermis 55, 310, 433g
descending aorta 40, 171, 173
  in fetus 282
descending colon 24, 25, 232
descending genicular artery 170
descending nerve tracts 123
detoxification 181, 229
detrusor muscle 238
developmental milestones
  288–289
diabetes mellitus 373
diabetic nephropathy 408
diaphragm 41, 65, 433g
  anatomical relationships 22,
  23, 25
  cough reflex 205
  function of 196, 198, 202, 203
diaphysis 73, 74, 286
diastole 179, 433g
diastolic pressure 361
dietary fiber 213, 215, 405
diffusion 49
digastric muscle 12, 17, 87
digestion 210–211
digestive enzymes 158, 209, 212,
  223, 230
  breakdown of food 210–211,
  214
digestive organs 208–209
digestive system 24, 26, 41,
  206–235
  disorders 392–405
  effects of AIDS 375
  food breakdown 210–211, 212
digestive tract 208–209
digital artery 170
digital sheath, fibrous 84, 90
digital tendons 31
digital vein 170
dilated cardiomyopathy 354
disaccharides 211, 213
disk
  embryonic 260
  intercalated 93
  intervertebral 68, 69, 71
  meniscus 76, 322
  Merkel's 261
  prolapsed 321
dissecting aneurysm 360
distal convoluted tubule 243,
  244, 245, 246
diverticula 404
diverticular disease 404
diverticulitis 404
diverticulosis 404
DNA (deoxyribonucleic acid) 46,
  50–51, 365, 366, 371
dominant gene 304
dopamine 333, 433g
dorsal carpal arch 171
dorsal digital artery 170
dorsal digital nerves 98
dorsal digital vein 170
dorsal horns of spinal cord 123
dorsal interossous 31, 35, 84
dorsal metatarsal artery 170
dorsal metatarsal vein 170
dorsal venous arch 170
dorsalis pedis artery 35
dorsomedial nucleus 137

double helix (DNA) 50
Down's syndrome 352, 353
droplet infection 379
ductus arteriosus 282, 283
ductus venosus 282
duodenum 25, 209, 226, 228,
  230, 231, 399, 433g
  digestive processes in 210–211,
  223, 224
  hepatic portal system 181
dura mater 127, 147
  bleeding in skull 340
  brain 114, 116
  infection 338
  spinal cord 120, 122, 123,
dust disease see silicosis

E

ear 148–149
  bud 118
  canal 148, 342
  disorders 342
  ossicles 67, 148
eardrum 148, 298, 342
ECG (electrocardiography)
  recording 176
ectoderm 260
ectopic pregnancy 422, 433g
eczema 313
edema 356
EEG (electroencephalogram) 332
egg 165, 258, 292–293, 299
egg cells 43, 48, 252
ejaculation problems 420
ejaculatory duct 251
elastic layer 183
elbow 30, 78
electrocardiography (ECG) 176
electroencephalogram (EEG) 332
ellipsoidal joint 78
embolus 334, 336, 346, 358, 359,
  433g
embryo 259, 433g
  brain development 118–119
  growth 261
embryonic disk 260
emphysema 388
empyema 400
encephalitis 338
enamel 217
endocardium 172, 178
endocrine glands 158–159
endocrine system 39, 43, 156–167
endoderm 261
endometriosis 413, 421
endometrium 259, 260, 262,
  292, 413, 414
  changes in 294
endoplasmic reticulum 46, 50
endorphins 145, 433g
endothelial cells 117
energy 214
engagement in pregnancy 272
enlarged prostate 241, 417
Entamoeba histolytica 368
enzymes 46, 52, 229, 433g
  cardiac 348
  drug-inactivating 365

eosinophil 184
epicondyle 31, 61
epicranial aponeurosis 87
epidermal cyst 311
epidermis 55, 144, 310, 311, 312, 433g
epididymis 26, 165, 241, 250, 251, 295
epidural space 122
epiglottis 15, 16, 164, 197, 204, 205, 219, 433g
epilepsy 332, 433g
epimysium 92
epinephrine 162
epineurium 101
epiphyseal growth plate 74, 286
epiphysis 73, 74, 286
epiploic vessels 24
epithelium 55, 57, 146, 147, 225
erector spinae 20, 27, 29
erythrocyte 184, 185
erythropoiesis 185
erythropoietin 158, 185
esophagus 41, 181, 197, 219, 220–221, 435g
    anatomical relationships 15, 25
    function of 209, 211
    herniation into 394
    peristalsis in 221
estrogen 158, 165, 252, 291, 293, 299, 435g
ethmoid bone 66, 67
ethmoid sinuses 380
Eustachian tube 148, 342
exhalation 202, 203
extensor carpi radialis brevis muscle 31
extensor carpi radialis longus muscle 30, 31
extensor carpi ulnaris muscle 31, 85
extensor digiti minimi muscle 31
extensor digitorum muscle 31, 85
extensor digitorum brevis muscle 35, 82, 84
extensor digitorum longus muscle 35, 82, 84
extensor digitorum longus tendon 82, 91
extensor hallucis brevis muscle 35, 82
extensor hallucis longus muscle 35, 82
extensor hallucis longus tendon 35, 82, 91
extensor muscles 31
extensor pollicis brevis muscle 31
extensor retinaculum muscle 31
external anal sphincter 29, 235
external auditory meatus 15
external carotid artery 12, 334
external genitalia, female 28, 254
external iliac artery 26, 29
external iliac nodes 188
external iliac vein 26, 29, 181
external intercostal muscle 18, 20, 22, 83, 85
external jugular vein 13, 17, 171

external oblique abdominal muscle 18, 20, 22, 24, 32, 83, 85
    female pelvis 28
    male pelvis 26
external respiration 201
extradural hemorrhage 340
eye 14, 57
    accessory structures 153
    cavities 153
    newborn 281
    structure 152
eye bud 118

F

facet joint 69, 71
facial artery 13, 171
facial bones 64
facial expressions 88–89, 333
facial hair 290
facial nerve 12, 99, 124, 146
facial vein 13, 17, 171
falciform ligament 24, 181, 228
fallopian tube 28, 253, 413
    blocked or damaged 420
    ectopic pregnancy 422
    site of fertilization 258, 259
Fallot's tetralogy 353
false rib 65
false vocal cord 16, 204, 205
farsightedness 343
fascicles 92, 101
fat 16, 55
    embolus 359
fats 213
    breakdown 210, 211, 231
    production 46
fat-storing cell 229
fatty acids 210, 211, 213, 214
fatty liver 396
fatty plaques 297, 334, 346, 358
fauces 146
feces 210, 215, 232, 233, 234, 235, 404
feedback mechanism 167
female genitals 254
female pelvic cavity 28–29, 254
female reproductive system 43, 252–253
    pelvic inflammatory disease 419
female sex hormones 165, 252
female urinary system 42
femoral artery 40, 170, 239
femoral hernia 402
femoral nerve 98
    branches 98
femoral vein 170, 239
femur 38, 60, 62, 76
    fractured 317
fertility 43, 165, 420
fertilization 258, 259
fetal circulation 282
fetal heart monitoring 278
fetus 261, 262–263, 268, 434g
    circulation 282
    heart defects 352–353
    position in uterus 269

fever 434g
fiber, dietary 213, 215, 405
fibrin 358, 434g
fibroadenoma 412
fibroblast 75, 387
fibrocartilage 325
fibrocystic disease 412
fibroid 414, 420, 434g
fibrosing alveolitis 386
fibrosis, idiopathic pulmonary (IPF) 386
fibrous
    capsule 190, 427
    connective tissue 55
    digital sheath 84, 90
fibula 35, 38, 60, 62, 76, 77
fight-or-flight response 136, 162
filamentous fungi 369
filaments, cytoskeleton 47
filtrate 245, 246
filtration in glomerulus 246
filum terminale 99
fimbriae 28, 253
fingers 31, 64
fissures 108
fistula 434g
    gallbladder 400
fixed joint 77
flagellum 258, 364, 368
flat bone 63
flexor carpi ulnaris muscle 31, 85
flexor digiti minimi muscle 34
flexor digitorum brevis muscle 34
flexor digitorum longus muscle 34, 35, 82, 84
flexor digitorum profundus muscle 83
flexor digitorum profundus tendon 90
flexor digitorum superficialis tendon 90
flexor hallucis brevis muscle 34
flexor hallucis longus muscle 34, 35, 84
flexor pollicis brevis muscle 90
flexor retinaculum 34
floating rib 65
follicle 292–293
follicle-stimulating hormone (FSH) 160, 291, 292, 299
fontanelle 281
food
    absorption 212, 224–225
    breakdown 210–211, 212, 224
    components 213
    energy provision 214
    transit times 233
foot 35
    bones 64
    movement 100
    tendons 91
foramen magnum 62
foramen ovale 282, 283
forceps delivery 277
forearm 31
forebrain 118
forehead 67
fornix, brain 110, 136
fossa navicularis 27
fourth ventricle 114–115

fractures
    repair 75
    sites 317
    spinal 320
    types 316
frank breech 269
frontal bone 12, 66, 67
frontal lobe 12, 108, 110, 117
frontal sinuses 380
frontalis muscle 15, 87, 88, 89
frowning 89
fructose 211
FSH (follicle-stimulating hormone) 160, 290–291, 299
fungi 369

G

galactose 211
galea aponeurotica 86
gallbladder 24, 181, 208, 211, 228, 231, 400, 434g
gallstones 400, 434g
gametocytes 368
ganglion 101, 130–131, 132–133, 134
gas exchange 201
gastric artery 25, 171
gastric gland 222
gastric juice 222, 226, 434g
gastric pit 222
gastric vein 181
gastric volvulus 394
gastritis 434g
gastrocnemius muscle 32, 33, 34, 82, 84
gastrocnemius tendon 34
gastroesophageal junction 394
gastrointestinal system see digestive system
gastrointestinal tract 434g
gemellus inferior muscle 84
gemellus superior muscle 84
gene 50, 434g
    damage 428
    dominant 304
    transmission of 302
    recessive 304, 305
    resistant 365
genetic engineering 371
genetic exchange 300–301
genioglossus muscle 14
geniohyoid muscle 14
genital herpes 366, 418
genitals
    newborn 280
    puberty 290
genital ulcers 418
genital warts 414
germinal center of lymph node 190
glans penis 250
glaucoma 343, 434g
glenoid cavity 19
glial cell 103, 108, 120
gliding joint 79
globus pallidus 111
glomerular filtration 245
glomerulonephritis 408

glomerulus 243, 244, 245, 246, 408
glossopharyngeal nerve 125, 146
glottis 219
glucagon 163
glucose 211, 214, 245, 246
glue ear *see* otitis media
gluteal artery 33
gluteal fascia 28, 33
gluteus maximus muscle 27, 28, 29, 32, 33, 84
gluteus medius muscle 26, 29, 33, 85
glycerol 210, 211, 213
glycogen 213, 229
goblet cell 57, 225, 388
Golgi complex 46
gonadocorticoids 162
gonadotrophin releasing hormone (GnRH) 290, 291
gonorrhea 418, 419
gooseflesh 54
gracilis muscle 28, 33, 34, 82, 84
grand mal seizure *see* tonic-clonic seizure
granulosa cells 258
Grave's disease 373
gray matter 434g
    brain 109, 111, 113, 131, 133, 140
    spinal cord 120, 123
greater auricular nerve 12
greater occipital nerve 12
greater omentum 24
great saphenous vein 35, 170
greenstick fracture 316
growth, childhood 286
growth hormone (GH) 160
guanine 50
gums 217
gyrus 108

H

hair 52, 144
    follicle 54, 55, 56, 310, 311
    growth 56
    nasal 197
    shaft 144
hamate 64
hand 31, 64
    movement, nerve pathways 140
    newborn 281
    venous network 170
hard bone 72
    see also cortical bone
haustral churning 234
haustrum 25
Haversian canal 73
Haversian system 72, 73
head 12–15
    injury 340
    newborn 281
headache 337, 338, 339
hearing 148–149
    loss of 298, 342

heart 40, 65, 170–171, 180, 196, 201
    blood collection 175
    blood supply 174
    chambers 23, 172–173
    disorders 344–361
    double pump 173
    fetal circulation 282, 352–353
    function 176
    heartbeat cycle 179, 349
    hormone production 159
    nervous control 130, 132, 177
    newborn 281, 283
    rate 100, 130, 132, 177, 349
    skeleton 175
    structure 172–173
heart attack 348, 361
heartburn 394
heart failure 356–357
heart murmurs 351
heart muscle 38, 93, 348
    disease 354
heart rhythm disorders 349, 356
heart valve 172–173, 178, 434g
    disorders 350–351
heel 91
Helicobacter pylori 395
hematoma 434g
heme 52, 185
hemiplegia 341
hemispheres of brain 109, 117
hemoglobin 52, 185, 434g
hemolytica anemia 373
hemophilia 305, 434g
hemorrhage 434g
hemorrhoids 405, 434g
Henle, loop of 243, 244, 246
hepatic artery 25, 228, 229, 297
hepatic flexure 25
hepatic portal system 181
hepatic vein 228
hepatitis 371, 398, 434g
hepatocyte 229
heredity 302, 305
hernia 394, 402
herpes, genital 418
herpes simplex virus 418
herpesviruses 366
heterozygous gene *see* dominant gene; recessive gene
hiatal hernia 394
hindbrain 118
hinge joint 78, 91
hip girdle 26
hip joint 79
hippocampus 110, 136, 143, 434g
histamine 145, 191, 372, 389
HIV (human immunodeficiency virus) 374, 434g
homeostasis 432g
homologous chromosomes 300
hormones 39, 42, 43, 52, 160–161, 223, 434g
    adrenal 162
    feedback mechanisms 167
    how they work 166
    production 158–159
    regulation of heart rate 177
    role in menopause 299
    sex hormones 43, 165, 437g

horns of spinal cord 123
host cell 367
hot flashes 299
human immunodeficiency virus (HIV) 374, 434g
human life cycle 256–305
human papillomavirus 311, 414
humeral circumflex artery 30
humerus 19, 30, 61, 62, 78, 79, 95
hydrocele 416
hydrochloric acid 211
hydrogen 233
hydrogen sulphide 233
hyoid bone 14, 16, 17, 87, 164, 204
hypermetropia 343
hypertension 335, 361
hypertrophic cardiomyopathy 354
hyphae 369
hypoglossal nerve 125
hypophyseal portal system 161
hypothalamus 110, 137, 159, 160, 434g
    feedback mechanisms 167
    heart rate regulation 177
    hormone production 159
    role in puberty 290–291

I

IBS (irritable bowel syndrome) 401, 434g
idiopathic pulmonary fibrosis (IPF) 386
IgE (immunoglobulin E) molecule 372
ileocecal valve 232
ileocolic vein 181
ileum 181, 209, 224, 233, 403, 434g
    anatomical relationships 24, 25
iliac artery 26, 27, 29
iliac crest 18, 20, 32
iliac vein 26, 27, 29
iliacus muscle 26, 28
iliocostalis erector spinae 20
iliocostalis thoracis muscle 85
iliohypogastric nerve 99
ilioinguinal nerve 99
iliopsoas muscle 82
iliotibial tract 32, 33, 84
ilium 27, 29, 61, 62, 66
immune deficiency 374–375, 434g
immune disorders 372–375
immune response, specific 192–193
immune system 24, 191–193, 372, 373
    see also lymphatic system
immunization 370
immunoglobulin 372
impetigo 313
impulses
    heart 176
    nerve 104, 106, 107, 112
incisor 216, 287
incompetence, heart valve 350

incontinence 409
incus 67, 148
infection 362–369
    urinary tract 408
inferior concha 67
inferior extensor retinaculum 35
inferior gemellus muscle 33
inferior gluteal artery 33
inferior mesenteric vein 181
inferior oblique muscle 14, 95
inferior phrenic artery 25
inferior rectus muscle 14
inferior thyroid vein 17, 171
inferior vena cava 171, 172, 173, 174, 175, 181, 238, 359
    anatomical relationships 25
    fetus 282
    in hepatic portal system 181
    role in blood circulation 180
infertility 419, 420–421
inflammatory bowel disease 401
inflammatory response 191
influenza 366, 367, 380
information processing in brain 140
information storage 143
infraspinatus muscle 21, 30, 85
inguinal ligament 83, 238, 252
inguinal lymph node 40
inhalation 202, 203
inner ear 148
insoluble fiber 215
insulin 163
intercalated disk 93
intercostal muscle 18, 20, 196, 202
intercostal nerve 99
interferons 434g
interlobular blood vessels 243, 244
internal capsule 112, 113
internal iliac artery 27, 29
internal iliac nodes 189
internal iliac vein 27, 29, 181
internal intercostal muscle 18, 22, 83
internal jugular vein 13, 16, 22, 23, 171
internal mammary lymph node 255
internal oblique abdominal muscle 20, 24, 26, 83, 85
internal obturator muscle 84
internal pudendal vein 33
internal respiration 201
internal spermatic fascia 26
interossei dorsales muscle 85
interossei palmares muscle 52
interosseous artery 171
interphase 53
interstitial cell 290, 295
interventricular foramen 115
intervertebral disk 68, 69, 71
    prolapse 321
intestinal lining 233
intestinal movement 227, 234
intestinal obstruction 402
intestinal villi 225
intestines 130, 132, 158, 224
    in newborn 280

intracerebral hemorrhage 335
intramural fibroid 414
intussusception 403
involuntary muscle 38, 54
IPF (idiopathic pulmonary
    fibrosis) 386
iris 152, 153, 343
iron 185, 215, 229
irregular bones 63
irritable bowel syndrome (IBS)
    401, 434g
ischial tuberosity 28
ischium 61, 66
islet of Langerhans 163, 230

J

jaundice 400, 434g
    in newborn 280
jaw bone 216, 217
jaws 21 6
jejunum 25, 209, 224
joint capsule 76
joints 76–79
    aging 324
    ball-and-socket 79
    disorders of 322, 323, 324–325
    ellipsoidal 78
    fixed 77
    gliding 79
    hinge 78, 91
    lever systems 94
    pivot 78
    saddle 79
    semi-movable 77
    structure 76
    synovial 76, 78–79
jugular vein 13, 16, 17, 22, 23

K

Kaposi's sarcoma 375, 434g
keratin 56, 311, 434g
kidneys 42, 130, 158, 239, 434g
    blood supply 239
    effect of age on 297
    red cell production 185
    structure 242–243
    tubules 161, 243
kidney stones 409
killer T cells 192–193, 434g
knee 34, 76–77, 78
    arterial network 170
    venous network 170
kneecap see patella
knee-jerk reflex 129
Krebs cycle 214
Kupffer cell 229

L

labia 28, 29, 252, 254
labor 268, 273–279
labyrinth 148
lacrimal bone 66, 67
lacrimal ducts 153
lacrimal gland 133, 153, 189

lactase 210, 211
lactation 285
lacteal 225
lactiferous duct 255
lactose 211, 213
lacunae 73
lamellae 73, 318
Langerhans, islet of 163, 230
lanugo 263, 280
large intestine 41, 208, 211, 215,
    232–233, 234–235
    see also colon
laryngeal prominence 16, 17
laryngitis 378
laryngopharynx 15, 197
larynx 15, 16, 164, 197, 204, 205,
    219, 378, 434g
lateral aortic nodes 189
lateral fissure 108
lateral horn 123
lateral malleolar network 35
lateral pectoral nerve 99
lateral plantar nerve 98
lateral preoptic nucleus 137
lateral rectus muscle 14, 153
lateral tibial nuclei 137
lateral ventricle 115
late telophase 53
latissimus dorsi muscle 18, 20,
    85
left-sided heart failure 356–357
leg 34–35
Legionella pneumophila 383
Legionnaire's disease 383
lens of the eye 152, 153, 155, 298
lentiform nucleus 111
lesser occipital nerve 12
leucocyte 48, 184
leukemia 366, 434g
leukotrienes 191
levator labii superioris alaeque
    nasi muscle 87
levator labii superioris muscle 87,
    89
levator scapulae muscle 15, 21,
    30, 86, 87, 95
lever systems 94
LH (luteinizing hormone) 160,
    290–291
life cycle, human 256–305
ligament 76, 77, 434g
    injury 322
lightening 272
limbic system 110, 136, 434g
linea alba 18, 26, 28, 83
linea arcuata 18
lingual tonsil 14, 146
lipase 210, 211
lipids 199
liver 24, 41, 65, 181, 208, 209,
    434g
    disease 396–398
    effect of age on 297
    function 229
    nervous control 130
    newborn 280
    structure 228
liver abscess 398
lobar bronchus 22
lobar pneumonia 382

lobe
    brain 108, 109, 117
    liver 228
    lungs 22, 23
    pituitary gland 160–161
lobule 229
lochia 284
long bones 63, 72–73
longissimus erector spinae 20
longissimus thoracis muscle 85
longitudinal fissure 109, 127, 143
long-term memory 143
loop of Henle 243, 244, 246
lower esophageal sphincter 220,
    223, 226
lumbar nerves 126, 128
lumbar plexus 126
lumbar vertebrae 21, 68, 70
lumbricals 34, 82
lumen 384
lunate 64
lung 22, 23, 41, 65, 196, 197, 202
    collapsed 385
    effects of AIDS on 375
    epithelial cells 57
    fetus 282
    nervous control 130
    newborn 280
    respiratory disorders 381–391
    structure 198–199
lung cancer 390
lunula 56
luteinizing hormone (LH) 160,
    290–291
lymph 188, 189
lymphatics 188, 229, 255, 429
lymphatic system 40, 186–190,
    435g
lymph capillary 188, 189
lymph node 188–189, 255, 429,
    435g
    structure 190
lymphocytes 184, 188, 189,
    190, 371, 379, 435g
    specific immune response
    192–193
lymphokine 192–193
lysosome 47

M

macrophage 188, 190, 200, 332,
    382, 387
macula 150, 151, 298
macular degeneration 298
magnesium 215
main bronchi 22, 381
malaria 368
male pelvic cavity 26–27
male reproductive system 43,
    250–251
male sex hormones 165
male urinary system 42
malignant 435g
malignant melanoma 312
malleolus 35, 60, 62
malleus 67, 148
maltase 210, 211
maltose 211, 213

mamillary body 136, 137
mammary glands 161, 255
mandible 61, 62, 66, 67, 87
    anatomical relationships 12,
    14, 16
    see also jaw bone
mandibular nerve 146
manubrium 61, 65
masseter muscle 87
mast cell 372, 389
mastoid process 12
maxilla 12, 14, 61, 66, 67
    see also jaw bone
maxillary artery 171
maxillary sinus 380
measles 366
meconium 280
medial plantar nerve 98
medial preoptic nucleus 137
medial rectus muscle 153
median nerve 99, 327
medulla 435g
medulla (oblongata) 113, 118–119,
    177
medullary canal 72, 318
meiosis 300 301, 435g
Meissner's corpuscle 145
melanin 160
melanocyte 160, 312
melanocyte-stimulating hormone
    (MSH) 160
melanoma, malignant 312
melatonin 159
membranous urethra 241
memory 140, 143
memory B cell 193
memory T cell 192
meninges 335, 338, 435g
    brain 114–115, 116
    spinal cord 120, 122
meningitis 339, 366, 369, 435g
meningococcal meningitis 339
menisci 76, 322
menopause 43, 299, 435g
menstrual cycle 292–293, 294
menstruation 43, 253, 291,
    292–293
mentalis muscle 14, 87, 89
Merkel's disk 144
merozoites 368
mesenteric infarction 402
mesenteric vessels 25
mesentery 25
mesoderm 260
messenger ribonucleic acid
    (mRNA) 51, 52
metabolism 435g
metacarpals 31, 61, 64, 79
metaphase 53
metastasis 390, 429, 435g
metatarsals 60, 62, 64, 79
methane 233
microglia 103
microorganism 364
microtubule 105
    cytoskeleton 47
microvilli 47
midbrain 113, 118, 136
middle ear 148
migraine 337, 435g

milk duct 285
milk-producing gland 285
minerals 213
miscarriage 422, 435g
mitochondria 435g
mitochondrion 46, 50, 102, 105, 214, 369
mitosis 53, 300, 435g
mitral ring 175
mitral valve 173, 178, 351, 435g
molar teeth 216, 287
moles 312, 435g
monocyte 184, 188
mononucleosis 366
monosaccharides 210, 211, 213
mons pubis 28
morphine 145
morula 259
motor cortex 140, 435g
motor nerve fiber 101, 129
motor nerve root 120, 122
motor nerves 100
motor neuron 435g
motor neuron disease 435g
mouth 41, 209, 211, 218
movement 141
mRNA (messenger ribonucleic acid) 51, 52
MS (multiple sclerosis) 332, 373
MSH (melanocyte-stimulating hormone) 160
mucocele 400
mucosa 222, 223, 224, 225, 395
mucous membrane 147, 197, 218, 222, 380, 435g
mucus
  digestive tract 223, 225
  nasal 147, 197
  respiratory tract 57, 205, 388, 389
multiple pregnancy 269, 278
multiple sclerosis 332, 373
multipolar nerve cell 103
mumps 366
muscle 52, 55, 80–95
  anastomosis 93
  contraction 93
  injuries 326
  lever systems 94
  stabilizing 95
  structure 92
  tendon links 90
  types 93
  working together 95
muscle cell 48
muscle fiber 92, 220, 224
  conducting 176
muscle injury 326
muscular system 38
musculocutaneous nerve 99
musculoskeletal disorders 314–329
myasthenia gravis 373
myelin sheath 101, 102, 107, 111, 332
mylohyoid muscle 14, 16, 87
myocardial infarction 348
myocarditis 354, 366
myocardium 173, 435g
myocyte 48

myofibrils 92, 93, 435g
myofilament 92
myopia 343

N

nail structure 56
narcotic drugs 145
nasal bone 12, 14, 66, 67, 87
nasal cartilage 14
nasal cavity 14, 147, 197
nasal glands 133
nasal hair 197
nasalis muscle 83, 87, 89
nasal mucus 147, 197
nasal passages 197
nasal septum 15
nasolacrimal duct 153
nasopharynx 15, 197
navel 18
navicular 60, 64
nearsightedness 343
neck 12–17
neck muscles 202
negative feedback 167
Neisseria gonorrhoeae 418
neonatal reflex 288
nephron 243, 244, 435g
nephropathy, diabetic 408
nerve cell *see* neurons
nerve fiber 98–99, 100, 102, 150
  myelinated 101, 102, 111, 112, 123
  regeneration 107
  in skin 54, 55
  in taste bud 146
  tracts 120, 123
  see also axon
nerve root sheath 123
nerves 435g
  brain-spinal cord tracts 112
  impulses 104, 106
  inhibition 107
  signal 129
  structure 101
  tracts 112, 120, 123
nervous system 39, 98–99
  effects of AIDS 375
  disorders 330–343
neural network 119
neural pathways 134–135, 341
neural tube 118, 261
neurofilament 105
neurons 48, 102, 108, 435g
  bodies 120
  behavior 104–105
  regeneration 107
  sending impulses 106
  support cells 103
  types 103
  see also nerves
neurosecretory cell 161
neurotransmitter 104–105, 106, 107, 333, 337, 435g
neutrophil 184, 191, 382
nevus 312
newborn baby 280–281
night sweats 299
nipple 255

Nissl body 102
node, lymph 188–189, 190, 255, 429, 435g
node of Ranvier 102
noncancerous tumors 427
nongonococcal urethritis 419
nonrapid eye movement (NREM) sleep 139
nonspecific urethritis 419
norepinephrine 162
nose, structure 14
NREM (nonrapid eye movement) sleep 139
nuchal ligament 86
nuclear membrane pore 51, 52
nucleic acids *see* DNA; RNA
nuclei, hypothalamic 137
nucleolus 47
nucleosome 50
nucleotide base 50, 52
nucleus
  cell 46, 48, 50, 53
  fungi 369
  muscle 93
  nerve cell 102
  protozoa 368
nutrients 181, 210–211
  see also food

O

obstruction
  cervical 423
  intestinal 402
obturator fascia 28
obturator internus 33
obturator nerve 98
occipital artery 12
occipital bone 62, 66, 67
occipital lobe 108, 109, 117
occipital nerve 12
occipital vein 12
occipitalis muscle 86, 87
occipitofrontalis muscle 83, 85
oculomotor nerve 125, 137
Oddi, sphincter of 231
olecranon 30, 31
olfactory bulb 136, 147
olfactory epithelium 147
olfactory nerves 124, 147, 435g
olfactory receptor cell 147
oligodendrocyte 103
omental appendices 25
omentum 24, 209
omohyoid muscle 12, 13, 16, 83, 87
oncogenes 428
open fracture 316
opponens pollicis muscle 90
optic chiasm 154
optic nerves 14, 99, 125, 154, 343, 435g
orbicularis oculi muscle 13, 14, 83, 85, 87, 88, 89
orbicularis oris muscle 13, 14, 87, 88, 89
orbital bones 153
organelle 46, 229
organ of Corti 149

oropharynx 15, 197
orthomyxoviruses 366
ossicles of the middle ear 67, 148, 435g
ossification 286
osteoarthritis 324–325, 435g
osteoblast 74
osteocyte 73, 319
osteon 72, 73, 318, 435g
osteophyte 325
osteoporosis 318–319, 435g
otitis media 342, 435g
outer ear 148
oval window 148
ovarian artery 28
ovarian cancer 415
ovarian cyst 415
ovarian ligament 253
ovarian vein 28
ovary 28, 43, 158, 160, 165, 259, 436g
  cancer of 415
  disorders of 415
  function of 252, 253
  puberty 291
  menopause 299
ovulation 252, 253, 291, 420, 436g
ovum 48, 252, 258, 292, 293, 299, 436g
oxygen
  exchange 200, 201
  fetal circulation 282, 283
  maternal-fetal transfer 262, 264
  transport 185
oxygenated blood 196, 200
oxyhemoglobin 185
oxytocin 161

P

pacemaker *see* sinoatrial node
Pacinian corpuscle 145
pain 145, 191, 324, 329
palatine bone 67
palantine tonsil 146
palm 31, 64
pancreas 158, 163, 211, 228, 436g
  cancer of 399
  function of 208, 230
  nervous control of 132
pancreatic amylase 211
pancreatic duct 163, 230, 399
papillae 146, 243
papovaviruses 366
parahippocampal gyrus 136
paralysis 341, 436g
paramyxoviruses 366
paranasal sinus 197
paraplegia 341
parasympathetic ganglia 133, 134, 35
parasympathetic nerves 132–133, 134, 135

parasympathetic nervous system 100, 436g
  coordination of response 135
  regulation of heart rate 177
parathyroid gland 159, 164
paraventricular nucleus 137
parietal bone 63, 66, 67
parietal lobe 108, 109, 117
parietal peritoneum 209, 238
Parkinson's disease 333, 436g
parotid duct 87, 218
parotid gland 16, 87, 218
passive immunization 370
patella 32, 60, 63
patellar ligament 32, 84, 129
patellar spinal reflex 129
pectin 215
pectineus 82
pectoral girdle 19
pectoralis major muscle 12, 16, 18, 19, 22, 83, 87
pectoralis minor muscle 22, 83
pedunculated fibroid 414
pelvic bones 66
pelvic cavity
  female 28–29, 413
  male 26–27
pelvic disproportion 270
pelvic floor muscles 28, 240, 241, 409
pelvic inflammatory disease (PID) 419, 421, 436g
pelvis 24, 25, 38, 66, 436g
  childbirth 270, 272
penis 27, 43, 132, 250
pepsin 211
peptic ulcer 395, 436g
peptidase 210, 211
peptide 210, 211
perforating artery 33, 170
perforating fibers 91
perforating vein 170
pericardial effusion 355
pericarditis 355
pericardium 23, 173, 355, 436g
perimysium 92
perineum 274
perineal muscle 28
perineurium 101
perineal ligament 217
periosteum 72, 91, 122, 123, 286, 318
peripheral nervous system (PNS) 98–99, 100, 125, 436g
  regeneration 107
peripheral spinal nerves 126–127
peristalsis 208, 221, 226, 227, 234, 235, 436g
peritoneum 26, 29, 209, 403, 436g
peritonitis 436g
permanent teeth 287
peroneal artery 35, 170
peroneal nerve 35
peroneal vein 170
peroneus brevis muscle 34, 35, 84
peroneus longus muscle 34, 35, 82, 84
peroneus tertius tendon 35, 82, 91
peroxisome 47

petit mal seizure 332
Peyer's patch 188
phagocytosis 191
phagosome 191
phalanges 60, 62, 64
pharyngeal tonsil 15
pharyngitis 378
pharynx 197, 209, 219, 220, 436g
phrenic artery 25
phrenic nerve 99
pia mater 116, 120, 338
picornaviruses 366
PID (pelvic inflammatory disease) 419, 421, 436g
piles 405
pilus 365
pineal gland 159
pinna 148, 342
piriformis muscle 33, 85
pisiform 64
pituitary gland 110, 137, 159, 160–161, 290, 291, 299, 436g
  feedback mechanisms 167
pivot joint 78
placenta 262, 264–265, 269, 275, 282, 436g
  delivery 276, 279
  formation of 260
  shared 278
placental abruption 423
placental hormones 264
placenta praevia 423
plantar aponeurosis 34
plantar arch 170
plantar interosseus 34
plantaris muscle 32, 84
plantar venous arch 170
plaque
  atherosclerotic 297, 346, 358
  psoriasis 313
plasma 182, 184, 436g
plasma cells 190, 192
plasmid 365
Plasmodium falciparum 368
platelets 182, 184, 350, 436g
platysma 13, 83, 85, 87, 89
pleurae (pleural membranes) 196, 199, 383, 385, 436g
  anatomical relationships 17, 22, 23
pleural effusion 383
plexus 127
Pneumocystis carinii 375
pneumonia 369, 382, 436g
pneumothorax 385, 436g
podocyte 245
polio (poliomyelitis) 366
polycystic ovary 415, 420
polyp 405
polypeptides 210, 211
polysaccharides 213
pons 113, 118, 119, 136
popliteal artery 33, 34, 170
popliteal nodes 188
popliteal vein 33, 34, 170
popliteus muscle 84
pore 54, 245
portal hypertension 397

portal system 181
portal vein 180, 181, 228, 229, 297, 397
positive feedback 167
posterior auricular muscle 86
posterior chamber of the eye 153
posterior femoral cutaneous nerve 33
posterior humeral circumflex artery 30
posterior lobe of the pituitary 161
posterior nucleus 137
posterior tibial artery 34, 40, 170
posterior tibial vein 34, 170
postganglionic axon 131, 133, 135
postganglionic neuron 134
postnatal period see puerperium
potassium 245, 246
potassium ions 107
prefrontal cortex 142
preganglionic axon 131, 133, 134
preganglionic neuron 134
pregnancy 266–267
  ectopic 422
premature baby 269
premolar teeth 216, 287
premotor cortex 142
prickle cell layer 55, 311
primary 436g
primary auditory cortex 142
primary motor cortex 142
primary olfactory cortex 142
primary ossification centers 206
primary somatic sensory cortex 142
primary teeth 216, 287
primary tumor 390
primary vesicle 118
primary visual cortex 142
progesterone 158, 165, 252, 291, 293, 436g
projection fibers 112
prolapsed uterus 413
pronator teres muscle 83
prophase 53
proprioceptors 125
prostaglandins 159, 191, 436g
prostate gland 27, 238, 251, 436g
  cancer 417
  enlarged 241, 417
prostatic urethra 241
protein 213, 246
  breakdown 210–211
  complement system 193
  manufacture 46, 50, 52, 160
protein hormone 166
protozoa 368
proximal convoluted tubule 243, 244, 245, 246
pseudostratified epithelium 57
psoas major muscle 26, 28
psoriasis 313, 436g
puberty 39, 43, 165
  boys 290
  girls 291
pubic hair 290, 291
pubic symphysis 60
  female 28, 240, 252, 254
  male 26, 66, 241

pubis 61, 66
pudendal nerve 98
pudendal vein 33
puerperium 284
pulmonary artery 22, 171, 173, 177, 178, 180, 436g
  gas exchange 196, 201
pulmonary circulation 180
pulmonary embolus 359
pulmonary hypertension 384
pulmonary ring 175
pulmonary trunk 23, 173
  fetus 282
pulmonary valve 172, 178, 351, 353
pulmonary vein 22, 171, 172, 173, 180
  gas exchange 196, 201
pulmonary ventilation 201
pulp 217
pulse 436g
puncture wound 310
pupil 153
  reaction 135
pus 436g
  see also empyema
putamen 111
pyelonephritis 408
pyloric sphincter 223, 224, 226, 231
pyramidalis 18

Q

quadratus femoris muscle 33, 84
quadriplegia 341

R

radial artery 171
radial nerve 30, 99
radial vein 171
radius 61, 62, 78
rapid eye movement (REM) sleep 139
RAS (reticular activating system) 138
rashes 313, 339
receptor
  pain 145
  sensory 144–145, 150
receptor site 104, 106, 147
recessive gene 304, 305
recto-uterine pouch 29
rectovesical pouch 27
rectum 181, 232, 240, 241, 251, 253, 268
  anatomical relationships 25, 26, 27, 28, 29
  defecation 234, 235
  function 208
rectus abdominis muscle 18, 24, 26, 28, 38, 83
rectus capitis posterior major muscle 95
rectus capitis posterior minor muscle 95
rectus femoris muscle 32, 38, 82

rectus sheath 18
red blood cells 38, 182, 184, 229, 368, 436g
production 185
red bone marrow 38, 185
reflex actions 113, 288
coughing 205
newborn 288–289
spinal 129
reflux, urine 408
regeneration of nerves 107
REM (rapid eye movement) sleep 139
renal artery 171, 239, 243
renal capsule 242
renal cortex 243
renal medulla 243
renal pelvis 242
renal pyramid 243
renal tubules 243
renal vein 239, 243
repetitive strain injury (RSI) 327
replacement heart valve 351
reproductive system 26, 28, 43, 248–255
disorders 410–423
respiration 201, 436g
respiratory centers 197
respiratory system 41, 194–205
disorders 376–391
epithelial lining 57
restrictive cardiomyopathy 354
reticular activating system (RAS) 138
reticular fibers 190
reticular formation 436g
retina 152, 153, 154, 155, 436g
retinaculum 31, 34, 35, 82, 84, 91
retromandibular vein 12
retroviruses 366
rheumatic fever 354, 355
rheumatoid arthritis 323, 386, 436g
rhomboid major muscle 20, 21, 30, 85, 86, 95
rhomboid minor muscle 21, 85, 86, 95
rib 18, 23, 24, 61, 62, 127, 197, 202
attachments 65
fractured 317
ribcage 18–19, 22, 23, 38, 62, 65, 255
ribonucleic acid (RNA) 46, 51, 366, 374
ribosome 46, 50, 52
right atrium 23
right-sided heart failure 356–357
right ventricle 23
risorius muscle 13, 88
RNA (ribonucleic acid) 46, 51, 366, 374
rods of the retina 155
root, tooth 217
round ligament 28
RSI (repetitive strain injury) 327
rubella 352
Ruffini's end-organ 145

S

saccharides 213
saccule 150
sacral canal 27, 29
sacral nerves 126, 128
sacral plexus 126
sacrum 21, 61, 62, 66, 68, 70, 121, 126
female pelvic cavity 29, 240
male pelvic cavity 26, 27, 66
saddle joint 79
saliva 16, 146, 211, 218, 436g
salivary gland 16, 131, 133, 189, 209, 218
see also parotid gland; submandibular gland
saphenous nerve 98
sarcolemma 93
sarcoma 426, 437g
sarcomere 92, 93
sartorius muscle 32, 33, 82
saturated fatty acids 213
scala tympani 149
scala vestibuli 149
scalenus muscle 15, 83, 87
scaphoid 64, 78
fractured 317
scapula 19, 30, 61, 62, 79, 95
spine of 21, 30
stabilizing muscles 95
sciatic nerve 32, 33, 98
sclera 14, 152, 153
scrotum 26, 27, 250, 251, 416
sebaceous gland 54, 255, 310
sebum 310
secondary 437g
secondary ossification centers 285
secondary sexual characteristics 290
secondary teeth 216
secondary tumors 426–427, 429
segmental bronchi 389
segmentation 227
seizures 332
semen 251
semicircular canal 149, 150, 298
semimembranosus muscle 32, 33, 34, 84
semi-movable joint 77
seminal vesicle 251
seminiferous tubules 26, 165, 250, 295
semispinalis capitis muscle 15, 85, 86, 87, 95
semitendinosus muscle 33, 34, 84
sensation 128
sensorineural deafness 342
sensory memory 143
sensory nerve fiber 101, 129
sensory nerve root 120, 123
sensory nervous system 100, 113
sensory receptors 144–145, 150
sensory root ganglion 120, 122
septum
heart 173
nasal 15
septum pellucidum 136
serosa 223, 224

serotonin 337
serous pericardium 355
serratus anterior muscle 18, 20, 22, 83
serratus posterior inferior muscle 20
serratus posterior superior muscle 21
Sertoli cells 290
serum 370
sesamoid bone 63
sex cells 300–301
sex chromosomes 303
sex determination 303
sex hormones 43, 165, 437g
sex-linked inheritance 305
sexually transmitted infection (STI) 418–419
Sharpey's fibers 91
shingles 366
short bone 63
short-term memory 143
shoulder 19, 30, 62, 79
shoulder blade 19, 30, 60, 62, 79
sigmoid colon 25, 27, 29, 209, 232
silicosis 387
simple fracture 316
sinoatrial node 176, 177, 179, 349, 437g
sinus
brain 114, 116, 117, 171
coronary 175
lymph node 190
paranasal 197, 380
sinusitis 378, 380
sinusoid 229
sinus tachycardia 349
skeletal muscle 82–85, 92–93, 100
skeletal system 38, 58–79
skeleton 60–62
skin 100, 130, 160
aging 296
disorders 308–313
effect of AIDS 375
newborn 280
structure 54–55
touch receptors 144–145
skin cancer 312
skin discoloration 313
skull 12, 38, 61, 66, 67, 114, 121
bleeding in 340
base 62
joints 77
sleep 139
see also brain stem; hypothalamus
sliding hiatal hernia 394
small intestine 208, 210, 220, 224–225, 227, 232
anatomical relationships 24, 25
small saphenous vein 33, 35, 40, 170
smell 147
smiling 88
smoking 390, 391, 428
smooth muscle 38, 93, 135, 223
sodium 232, 245, 246
sodium ions 106, 107
soft palate 14

soft-tissue inflammation 329
soleus muscle 34, 35, 82, 84
soluble fiber 215
somatic sensory association cortex 142
speech 142, 204
sperm 26, 43, 48, 250, 251
antibodies to 421
fertilization of ovum 258–259
infertility 420
production 165, 290, 295
spermatic cord 250, 416
spermatic fascia 26
spermatids 295
spermatocytes 295
spermatogonia 295
spermatozoon see sperm
sphenoid bone 12, 15, 63, 66, 67
sphenoid sinus 380
sphincter 437g
anal 27, 28, 29, 235, 240
bladder 130, 132
urethral 247
sphincter of Oddi 231
spinal cord 21, 39, 62, 68, 120–121
cerebrospinal fluid flow 114–115
damage 341
development 118–119
effect of intervertebral disk collapse 321
extent 121
fiber tracts 123
nerve pathways 98, 99, 112, 113, 131, 133, 134
peripheral nervous system 100
protection 122–123
spinal fractures 320
spinal joints 71
spinal nerves 68, 120, 121, 123, 126–127
spinal nerve root 121
spinal reflex 129
spinalis erector spinae 20
spinalis thoracis muscle 85
spindle thread 300, 301
spine 20–21, 38, 60–61, 62, 437g
curves 69
regions 70
structure 68–71
spiral fracture 316
spiral organ of Corti 149
spleen 24, 25, 40, 65, 181, 189
splenic artery 25
splenic flexure 25
splenic vein 181
splenius capitis muscle 15, 85, 86, 87, 95
splenius cervicis muscle 20, 85, 86
spongy bone 72, 91, 296, 316, 318
spongy urethra 26, 241
sporozoites 368
squamous cell 391
squamous cell carcinoma 312
squamous epithelial cell 55, 57, 311
stapes 67, 148

staphylococcal bacteria 311
starch 211, 213
stem cell 188, 192
stenosis, heart valve 350
sternocleidomastoid muscle 12, 13, 15, 17, 83, 85, 87
sternohyoid muscle 13, 17, 87
sternothyroid muscle 13, 83
sternum 19, 61, 65
steroid hormone 160, 166
STI (sexually transmitted infection) 418–419
stomach 41, 65, 181, 189, 220, 222–223, 228, 268
    anatomical relationships 24, 25
    functions 208, 209, 211, 226
    hormone production by 158
    muscles 223
    nervous control of 130, 132
    ulceration of 395
strain injury 326, 327
stratified squamous epithelium 55, 57
stress incontinence 409
stress response 130–131, 160, 162, 361
striation, muscle 93
stroke 333, 334, 358, 361
stylohyoid muscle 12, 87
styloid process 15
subarachnoid hemorrhage 335
subarachnoid space 116, 120, 122
subclavian artery 12, 16, 171
subclavian vein 16, 171, 189
subclavius muscle 19, 83
subcostal nerve 99
subcutaneous fat 55
subdural hemorrhage 340
sublingual gland 218
submandibular duct 218
submandibular gland 16, 87, 218
submental vein 17
submucosa 223, 223, 224, 395
submucosal fibroid 414
subserosal fibroid 414
subserous layer 223
substantia nigra 333
sucrase 210, 211
sucrose 211, 213
sugar–phosphate backbone, DNA 51
sulci 108
superficial muscle 82, 84, 86
superficial palmar arch 171
superficial peroneal nerve 98
superficial temporal artery 12, 171
superficial temporal vein 12, 171
superior gemellus muscle 33
superior gluteal artery 33
superior mesenteric artery 171
superior mesenteric vein 181
superior oblique muscle 14, 95
superior rectus muscle 14
superior sagittal sinus 117
superior temporal artery 13
superior temporal vein 13
superior thyroid artery 13
superior thyroid vein 13
superior vena cava 171, 172, 173, 174, 175, 178, 180

anatomical relationships 22
    pulmonary hypertension 384
suprachiasmic nucleus 137
supraclavicular lymph node 255
supraclavicular nerve 99
supraoptic nucleus 137
supraorbital nerve 13
supraspinatus muscle 30, 85, 328
sural nerve 35
surfactant 199
suspensory ligament 26
suture joint 66, 77
swallowing 218, 219, 221, 226
sweat gland 54, 55, 255
swelling 191, 329
Sylvian fissure 108
sympathetic ganglion chain 130, 134
sympathetic nerves 130–131
sympathetic nervous system 100, 437g
    regulation of heart rate 177
symphysis pubis see pubic symphysis
synapse 104, 131, 133, 134
synaptic cleft 104, 106
synaptic knob 103, 104, 105, 106
synaptic vesicle 105
synovial fluid 77, 324, 325
synovial joint 78–79
synovial membrane 77, 323, 324
systemic circulation 180
systemic lupus erythematosus 373
systole 437g
systolic pressure 361

T

talus 60, 63, 64
target cell 104, 106, 107, 166
target organs 144, 144
tarsals 60, 62, 64, 79
taste 146
taste buds 146
T cell (T lymphocyte) 192–193
tear gland see lacrimal gland
tear injury 326, 329
teeth 62, 209, 211, 216
    development 287
    structure 217
telophase 53
temporal artery 12, 13
temporal bone 66, 67
temporalis muscle 15, 83, 85, 86, 87
temporal lobe 108
temporal vein 12, 13
temporomandibular joint 216
tendinitis 328, 437g
tendon 90–91, 437g
    Achilles 34, 35, 84
    injury 328–329
tendon tears 329
tenia coli 223
tenosynovitis 328, 437g
tensor fasciae latae 32
teres major muscle 20, 30, 85, 95

teres minor muscle 30, 85
testes 437g
testicular artery 26, 239
testicular cancer 416
testicular disorders 416
testicular vein 26, 239
testis 39, 158, 160, 238, 241, 250, 295
    anatomical relationships 26, 27
    at puberty 290
    disorders of 416
    undescended 416
testosterone 158, 165, 437g
tetralogy of Fallot 353
thalamus 110, 113, 136, 437g
thigh 32–33
thigh muscle 129
third ventricle 115
thoracic cavity 22–23
thoracic duct 189
thoracic nerves 127, 128
thoracic vertebrae 21, 65, 68, 70
thoracolumbar fascia 20
thorax 18–19, 22–23, 437g
threatened miscarriage 422
throat see pharynx
thrombocytes 184
thrombosis 358
thrombus 334, 358, 359, 437g
thrush 369
thumb joint 79
thymine 50
thymus 40, 189
    newborn 281
thyrohyoid membrane 16
thyrohyoid muscle 16, 87
thyroid artery 13
thyroid cartilage 15, 16, 17, 164, 204
thyroid gland 39, 159, 164, 204
    anatomical relationships 14, 17
    control by pituitary gland 160
    feedback mechanisms 167
thyroid hormone 167
thyroid-stimulating hormone (TSH) 160, 167
thyroid vein 13, 17
thyrotropin-releasing hormone (TRH) 167
TIA (transient ischemic attack) 336
tibia 35, 60, 62, 77, 129
tibial artery 34, 35, 40
tibialis anterior muscle 35, 76, 82
tibialis anterior tendon 82
tibialis posterior muscle 84
tibial nerve 33, 34, 35, 98
tibial vein 34, 35
tilting disk valve 351
tinea 369
tobacco smoking see smoking
toes 64
tongue 146, 209, 218
tongue muscle 14
tonic-clonic (grand mal) seizure 332
tonsillitis 378

tonsils 14, 15, 40, 189, 437g
tooth structure 217
    see also teeth
touch 141
touch receptors 144–145
toxins 181, 364
trabeculae 61, 190
trachea 41, 134, 164, 197, 198, 204, 209, 437g
    anatomical relationships 14, 16, 17, 22, 23
    epithelial cells 57
tracheal cartilage 16, 204
transcription 51
transient ischemic attack (TIA) 336, 437g
transitional epithelium 57, 247
transit times 233
transport mechanisms 49
transport, oxygen 185
transverse cervical artery 12
transverse colon 24, 25, 209, 232
transverse fracture 316
transverse metacarpal ligament 90
transversus abdominis muscle 18, 24, 26
trapezium 64, 79
trapezius muscle 38, 83, 85, 87
    anatomical relationships 12, 15, 16, 21, 30
    stabilizing muscles 95
trapezoid 64
TRH (thyrotropin-releasing hormone) 167
triceps 30, 83, 85
triceps brachii muscle 19, 20, 30, 31
tricuspid ring 175
tricuspid valve 172, 170
trigeminal nerve 124
trigone 238
trimesters of pregnancy 266–267
triplet code, DNA 50, 51
triquetrum bone 64
trochlea 61
trochlear nerve 125
trophoblast 260, 264
true rib 65
true vocal cord 16
trunk 18–21, 62
Trypanosoma cruzi 368
trypsin 211
TSH (thyroid-stimulating hormone) 160, 167
tubule
    centriole 46
    renal 243
    seminiferous 26, 165, 250, 295
tumor 426–427, 437g
    formation 428
tunica adventitia 183
tunica albuginea 26
tunica intima 182, 183
tunica media 182, 183, 360
tunica vaginalis 26, 251
twin pregnancy 269, 278
tyrosine 166

U

ulcer
  genital 418
  peptic 395
ulcerative colitis 401
ulna 31, 61, 62, 78
ulnar artery 171
ulnar nerve 30, 31, 99
  deep branch 98
ulnar vein 171
umbilical artery 265, 282
umbilical cord 268, 269, 274,
  275, 437g
  cutting 276
  structure 265
  twins 278
umbilical vein 265, 282
umbilicus 18
unipolar nerve cell 103
unsaturated fatty acids 213
upper airway infections 378
urea 245, 246, 437g
ureter 42, 238, 239, 241, 242,
  247, 251, 437g
  anatomical relationships 26, 29
urethra 437g
  blockage of 408
  female 28, 42, 240, 252
  gonococcal infection 418
  male 26, 27, 42, 238, 241,
  250, 417
urethral orifice (outlet) 27, 29,
  240, 241
urethral sphincter 247
urge incontinence 409
uric acid 245
urinary bladder see bladder
urinary incontinence 409
urinary system 26, 42, 236–247
  disorders 406–409
  epithelial lining 57
urinary tract 437g
  disorders 408
urine 42, 239, 243, 247, 437g
  formation 246
  reflux 408
uterine contractions 268

uterus 161, 240, 253, 258, 259,
  268, 269, 437g
  abnormalities of 420
  anatomical relationship 28, 29
  blastocyst implantation
  260–261
  childbirth 284–285
  fibroids 414
  prolapsed 413
utricle 150
uvula 15, 146

V

vaccine 370, 371
vacuole 47
vacuum extraction 277
vagina 42, 43, 240, 253, 254,
  258, 259, 268, 437g
  after childbirth 284–285
  anatomical relationships 28,
  29
  at menopause 299
vagus nerves 99, 124, 146, 177,
  437g
valve
  cusp 178
  heart 172–173, 178, 350, 351
  lymph capillary 188, 189, 190
  vein 183
varicose veins 360, 405
vasa vasorum 183
vas deferens 26, 165, 251, 437g
vasopressin (ADH) 161
vastus lateralis muscle 32, 77,
  82, 84
vastus medialis muscle 77, 82
Vater, ampulla of 231, 399
veins 42, 55, 183, 190
veins 41, 170, 437g
  structure 183
venous sinus 114, 116, 171
ventral horns of spinal cord 123
ventricles 437g
  brain 115–116, 339
  heart 172–173, 175, 176, 179,
  349, 352, 353
ventricular failure 356–357
ventricular septal defect 353

ventricular systole 179
ventricular tachycardia 349
ventromedial nucleus 137
venule 55, 183, 190
vernix 263, 280
vertebrae 21, 68–69, 121, 127
  fractures 320
vertebral artery 334
vertebral body 122
vertebral column 21, 61, 62,
  68–71
  spinal cord protection 122
vesicles 47, 106
vestibular bulb 254
vestibular fold 16
vestibular gland 254
vestibular nerve 149
vestibule 148, 150
vestibulocochlear nerve 125, 149
villus
  chorionic 264
  intestinal 225
viral hepatitis 366, 398
viral meningitis 339
virus 437g
viruses 366–367
  genetically engineered 371
visceral peritoneum 209
vision 152
  effect of aging on 298
  loss 343
visual association cortex 142
visual impulses 138
visual pathways 154
visual problems 343
vitamin B 233
vitamin D 164
vitamin K 233
vitamins 213, 229
vitiligo 313, 373
vitreous humor 153
vocal cord 16, 204, 205, 219,
  437g
vocal fold 16
vocal ligament 204
voice box see larynx
voluntary muscle 38, 100
volvulus 394, 402
vomer 67

W

warts 311, 366
waste products 47, 245
water 211, 213, 245, 246
  absorption by colon 232, 233,
  234
Wernicke's area 142
wheezing 381, 389
whiplash injury 321
white blood cells 38, 48, 182,
  184, 229, 374, 437g
white matter 111, 113, 120, 131,
  133, 437g
windpipe see trachea
wisdom teeth 287
wounds 310
wrinkles 296
wrist 31, 64, 327

X

X chromosome 303, 305, 437g
xiphoid process 61, 65

Y

Y chromosome 303, 437g
yeast 369
yolk sac 260

Z

Z band 92
zinc 215
zygomatic arch 12, 61, 62, 87
zygomatic bones 14, 66, 67
zygomaticus major muscle 13,
  83, 87, 88
zygomaticus minor muscle 13,
  87, 88
zygote 259, 437g

Acknowledgments

Picture credits

Dr. D A Burns, Leicester Royal Infirmary: 313tl, 313tr. Mr. T C Hillard, Consultant Obstetrician & Gynecologist, Poole Hospital and courtesy of M Whitehead, The Atlas of the Menopause, Parthenon Publishing: 299tr, 299tcr. Science Photo Library: 360br; BSIP VEM 306-307; Dr. L Caro 365tl; CNRI 303tr, 336tr, 339bl, 380tr, 401br, 409br, 430-431; Barry Dowsett 383bl; Cecil H Fox 427tl; Astrid & Hanns Frieder Michler 416bl; Mehau Kuluk 414br; Moredun Animal Hospital 418br; Prof P Motta 61tr, 214tr, 231bl, 233c; Alfred Pasieka 112bl, 221tl; John Radcliffe Hospital 339br; J C Revy 8-9; Secchi-Lecaque/Roussell-UCLAF/CNRI 129tr; Andrew Syred 416bc; St John's Institute of Dermatology: 313bl, 313br.

Illustrations

Alison Brown Joanna Cameron, Simone End, Mick Gillah, Mike Good, Tony Graham, Andrew Green, Sandie Hill, Deborah Maizels, Janos Marffy, Patrick Mulrey, John Temperton, Richard Tibbitts, Halli Verrinder, Philip Wilson, DK Multimedia